Essentials of Aquaculture Practices

Norfazreena Mohd Faudzi •
Muhammad Dawood Shah •
Nurzafirah Mazlan •
Sitti Raehanah Muhamad Shaleh
Editors

Essentials of Aquaculture Practices

Springer

Editors
Norfazreena Mohd Faudzi
Higher Institution Centre of Excellence (HICoE)
Borneo Marine Research Institute, Universiti Malaysia Sabah
Kota Kinabalu, Sabah, Malaysia

Muhammad Dawood Shah
Higher Institution Centre of Excellence (HICoE)
Borneo Marine Research Institute, Universiti Malaysia Sabah
Kota Kinabalu, Sabah, Malaysia

Nurzafirah Mazlan
Higher Institution Centre of Excellence (HICoE)
Borneo Marine Research Institute, Universiti Malaysia Sabah
Kota Kinabalu, Sabah, Malaysia

Sitti Raehanah Muhamad Shaleh
Higher Institution Centre of Excellence (HICoE)
Borneo Marine Research Institute, Universiti of Malaysia Sabah
Kota Kinabalu, Sabah, Malaysia

ISBN 978-981-97-6698-7 ISBN 978-981-97-6699-4 (eBook)
https://doi.org/10.1007/978-981-97-6699-4

This Springer imprint is published by the registered company Springer Nature Singapore Pte Ltd.
The registered company address is: 152 Beach Road, #21-01/04 Gateway East, Singapore 189721, Singapore

Preface

Aquaculture is expanding in response to the increasing demand for animal protein, significantly contributing to food security and socioeconomic livelihoods. *Essentials of Aquaculture Practices* is a comprehensive book that delves into the fundamental principles and advanced techniques for successful aquaculture operations.

This book covers a broad range of topics, from the basics of aquaculture systems and nutrition to the latest advancements in disease management and genetic improvement. Expert authors provide in-depth insights into critical areas such as the importance of live feeds, the nutritional requirements of marine finfish, and the role of feeding stimulants in boosting aquaculture productivity.

The readers will also find valuable information on the challenges and solutions related to vibriosis, a common disease in aquaculture, and the vital importance of biosecurity in maintaining healthy and sustainable aquaculture communities. In addition, this book explores the social and economic dimensions of aquaculture, emphasizing the need for resilient communities and the potential of bioactive compounds from marine sources.

By addressing both technical and practical aspects, *Essentials of Aquaculture Practices* serves as an indispensable guide for cultivating a sustainable and prosperous future in aquatic food production. It is a valuable reference for students, researchers, policymakers, practitioners, and aquaculture operators.

We extend our heartfelt gratitude to the staff of the Borneo Marine Research Institute at Universiti Malaysia Sabah for their unwavering support and cooperation during the preparation of this book. We also want to thank the management team at Universiti Malaysia Sabah for their assistance in making this book a reality.

Kota Kinabalu, Sabah, Malaysia — Norfazreena Mohd Faudzi
Kota Kinabalu, Sabah, Malaysia — Muhammad Dawood Shah
Kota Kinabalu, Sabah, Malaysia — Nurzafirah Mazlan
Kota Kinabalu, Sabah, Malaysia — Sitti Raehanah Muhamad Shaleh

Contents

Editors and Contributors

About Editors

Norfazreena Mohd Faudzi is a senior lecturer at the Borneo Marine Research Institute, Universiti Malaysia Sabah (UMS) in Malaysia. Her academic journey includes the successful completion of an MSc and a PhD in Aquaculture from the Borneo Marine Research Institute, UMS, in 2013 and 2019, respectively. Recognized for her outstanding contributions to research and innovation, Dr. Norfazreena has been honored with numerous accolades. She has received one silver medal and two bronze medals from UMS, highlighting her exceptional work in these areas. She has been an active participant in both national and international conferences since 2009. Her passion for knowledge sharing has led her to be involved in various community works. By disseminating her research findings and expertise, Dr. Norfazreena contributes to the advancement of the scientific community and fosters meaningful collaborations across borders. Furthermore, Dr. Norfazreena's research work has been published in reputable journals.

Muhammad Dawood Shah is a senior lecturer at the Borneo Marine Research Institute, Universiti Malaysia Sabah (UMS), Malaysia. Dr. Dawood earned his MSc and PhD in Biotechnology from the Biotechnology Research Institute, UMS, in 2011 and 2016, respectively. In 2011, he received the excellence award for his MSc outstanding research work and best postgraduate student. From 2014 to 2024, he received 5 gold and 4 silver medals for his research work. In 2021, he was the winner of the Malaysian-Croatia Technology Exchange and Malaysia Technology Expo 2021 awards for his innovation. He has more than 45 peer-reviewed international publications and has delivered oral and poster presentations at numerous international conferences and symposia. Dr. Dawood has filed 2 patent applications, and he has established a research group network both nationally and internationally. Besides, his research work has also been published in the printing media.

Nurzafirah Mazlan is a senior lecturer at the Borneo Marine Research Institute, UMS. She obtained her PhD in Health Science from International Islamic University Malaysia (IIUM). She also holds a MSc in Industrial Biotechnology (UPM) and a BSc in Biomedical Science (Hons) from IIUM. Prior to joining UMS in 2021, she had worked at public and private academic institutions for 10 years and held various administrative positions. She performed research in a multidisciplinary area with core research in sea cucumber aquaculture, microbiology, and industrial biotechnology. She has won several accolades in local and international research competitions for her sea cucumber research. Her current interests are postharvest processing, sea cucumber product development, sea cucumber aquaculture, marine animal diseases, microplastic contamination, as well as lactic acid bacteria isolated from marine animals. She has supervised 12 postgraduate students and more than 30 undergraduates. She holds several local and industry grants and has established collaborations with local and international collaborators from Turkey, the UK, and Australia. She is also actively involved in marine conservation and passionate about building a sustainable future.

Sitti Raehanah Muhamad Shaleh is a researcher and lecturer at the Borneo Marine Research Institute (BMRI), Universiti Malaysia Sabah. Her journey in academia began with her pursuit of a Bachelor's degree in Fisheries Science, which she accomplished in 1994 from Universiti Pertanian Malaysia (UPM). Recognizing her potential as a budding scientist, she delved further into academia, embarking on a remarkable journey to attain a PhD in Aquatic Ecosystem Management from the same university. Her research interest is aquaculture with specific areas of live feed propagation (microalgae and zooplankton), sea cucumber breeding, and fish/

shrimp farming. She also conducted research on the impact of harmful algal blooms (HABs) on the aquaculture industry. Apart from doing research and supervising postgraduate students and teaching undergraduates, she also was entrusted to lead BMRI as the fourth director from 2019 to 2023.

Contributors

Roslianah Asdari Department of Aquatic Sciences, Faculty of Resource Science and Technology, Universiti Malaysia Sarawak, Kota Samarahan, Sarawak, Malaysia

Aiman Azseri Higher Institution Centre of Excellence (HICoE), Borneo Marine Research Institute, Universiti Malaysia Sabah, Kota Kinabalu, Sabah, Malaysia

Noora Barzkar Higher Institution Centre of Excellence (HICoE), Borneo Marine Research Institute, Universiti Malaysia Sabah, Kota Kinabalu, Sabah, Malaysia

Wei Sheng Chong Higher Institution Centre of Excellence (HICoE), Borneo Marine Research Institute, Universiti Malaysia Sabah, Kota Kinabalu, Sabah, Malaysia

Danielli Matias de Macêdo Dantas Department of Fisheries and Aquaculture, Federal Rural University of Pernambuco, Recife, Pernambuco, Brazil

Laenne Barbara Silva de Moraes Department of Fisheries and Aquaculture, Federal Rural University of Pernambuco, Recife, Pernambuco, Brazil

Isabella Ebi Higher Institution Centre of Excellence (HICoE), Borneo Marine Research Institute, Universiti Malaysia Sabah, Kota Kinabalu, Sabah, Malaysia

Hiam Elabd Aquatic Animals' Medicine Department, Faculty of Veterinary Medicine, Benha University, Moshtohor, Toukh, Egypt

Ching Fui Fui Higher Institution Centre of Excellence (HICoE), Borneo Marine Research Institute, Universiti Malaysia Sabah, Kota Kinabalu, Sabah, Malaysia

Alfredo Olivera Gálvez Department of Fisheries and Aquaculture, Federal Rural University of Pernambuco, Recife, Pernambuco, Brazil

Hariz Khairul Hisham WWF Malaysia, Lorong Centre Point, Pusat Bandar, Kota Kinabalu, Sabah, Malaysia

Rizal Ismail Higher Institution Centre of Excellence (HICoE), Borneo Marine Research Institute, Universiti Malaysia Sabah, Kota Kinabalu, Sabah, Malaysia

Saeid Tamadoni Jahromi Persian Gulf and Oman Sea Ecology Research Centre, Iranian Fisheries Sciences Research Institute, Agricultural Research Education and Extension Organization (AREEO), Bandar Abbas, Iran

Fikri Akmal Khodzori Higher Institution Centre of Excellence (HICoE), Borneo Marine Research Institute, Universiti Malaysia Sabah, Kota Kinabalu, Sabah, Malaysia

Rovina Kobun Faculty of Food Science and Nutrition, Universiti Malaysia Sabah, Kota Kinabalu, Sabah, Malaysia

Mohammad Tamrin Mohamad Lal Higher Institution Centre of Excellence (HICoE), Borneo Marine Research Institute, Universiti Malaysia Sabah, Kota Kinabalu, Sabah, Malaysia

Hon Jung Liew Higher Institution Centre of Excellence (HICoE), Institute of Tropical Aquaculture and Fisheries (AKUATROP), Universiti Malaysia Terengganu, Kuala Nerus, Terengganu, Malaysia

Leong-Seng Lim Higher Institution Centre of Excellence (HICoE), Borneo Marine Research Institute, Universiti Malaysia Sabah, Kota Kinabalu, Sabah, Malaysia

Heba Mahboub Department of Aquatic Animal Medicine, Faculty of Veterinary Medicine, Zagazig University, Zagazig, Sharkia, Egypt

Nurzafirah Mazlan Higher Institution Centre of Excellence (HICoE), Borneo Marine Research Institute, Universiti Malaysia Sabah, Kota Kinabalu, Sabah, Malaysia

Muhammad Addin Mazni WWF Malaysia, Lorong Centre Point, Pusat Bandar, Kota Kinabalu, Sabah, Malaysia

Lucyie Liana Misi Higher Institution Centre of Excellence (HICoE), Borneo Marine Research Institute, Universiti Malaysia Sabah, Kota Kinabalu, Sabah, Malaysia

Norfazreena Mohd Faudzi Higher Institution Centre of Excellence (HICoE), Borneo Marine Research Institute, Universiti Malaysia Sabah, Kota Kinabalu, Sabah, Malaysia

Géssica Cavalcanti Pereira Mota Department of Fisheries and Aquaculture, Federal Rural University of Pernambuco, Recife, Pernambuco, Brazil

Imelus Nius Higher Institution Centre of Excellence (HICoE), Borneo Marine Research Institute, Universiti Malaysia Sabah, Kota Kinabalu, Sabah, Malaysia

Aslam Zuhairi Nor'ashikin Higher Institution Centre of Excellence (HICoE), Borneo Marine Research Institute, Universiti Malaysia Sabah, Kota Kinabalu, Sabah, Malaysia

Rafidah Othman Higher Institution Centre of Excellence (HICoE), Borneo Marine Research Institute, Universiti Malaysia Sabah, Kota Kinabalu, Sabah, Malaysia

Fittrie Meyllianawaty Pratiwy Department of Fisheries, Faculty of Fisheries and Marine Science, Universitas Padjadjaran, Bandung, Indonesia

Sharifah Rahmah Higher Institution Centre of Excellence (HICoE), Institute of Tropical Aquaculture and Fisheries (AKUATROP), Universiti Malaysia Terengganu, Kuala Nerus, Terengganu, Malaysia

Faculty of Fisheries and Food Science, Universiti Malaysia Terengganu, Kuala Nerus, Terengganu, Malaysia

Nur Nashrah Abd Rahman Higher Institution Centre of Excellence (HICoE), Borneo Marine Research Institute, Universiti Malaysia Sabah, Kota Kinabalu, Sabah, Malaysia

Julian Ransangan Higher Institution Centre of Excellence (HICoE), Borneo Marine Research Institute, Universiti Malaysia Sabah, Kota Kinabalu, Sabah, Malaysia

Kenneth Francis Rodrigues Biotechnology Research Institute, Universiti Malaysia Sabah, Kota Kinabalu, Sabah, Malaysia

Nur Angelie Agnezara Roger Higher Institution Centre of Excellence (HICoE), Borneo Marine Research Institute, Universiti Malaysia Sabah, Kota Kinabalu, Sabah, Malaysia

Rita Rostika Department of Fisheries, Faculty of Fisheries and Marine Science, Universitas Padjadjaran, Bandung, Indonesia

Ejria Saleh Higher Institution Centre of Excellence (HICoE), Borneo Marine Research Institute, Universiti Malaysia Sabah, Kota Kinabalu, Sabah, Malaysia

Sari Asfarina Sariat Higher Institution Centre of Excellence (HICoE), Borneo Marine Research Institute, Universiti Malaysia Sabah, Kota Kinabalu, Sabah, Malaysia

Shigeharu Senoo Higher Institution Centre of Excellence (HICoE), Borneo Marine Research Institute, Universiti Malaysia Sabah, Kota Kinabalu, Sabah, Malaysia

Nur Syafiqah Shafiee Faculty of Health and Life Sciences, Management and Science University, Shah Alam, Selangor, Malaysia

Muhammad Dawood Shah Higher Institution Centre of Excellence (HICoE), Borneo Marine Research Institute, Universiti Malaysia Sabah, Kota Kinabalu, Sabah, Malaysia

Sitti Raehanah Muhamad Shaleh Higher Institution Centre of Excellence (HICoE), Borneo Marine Research Institute, Universiti Malaysia Sabah, Kota Kinabalu, Sabah, Malaysia

Rossita Shapawi Higher Institution Centre of Excellence (HICoE), Borneo Marine Research Institute, Universiti Malaysia Sabah, Kota Kinabalu, Sabah, Malaysia

Annita Yong Seok Kian Higher Institution Centre of Excellence (HICoE), Borneo Marine Research Institute, Universiti Malaysia Sabah, Kota Kinabalu, Sabah, Malaysia

Asep Agus Handaka Suryana Department of Fisheries, Faculty of Fisheries and Marine Science, Universitas Padjadjaran, Bandung, Indonesia

Kianann Tan Guangxi Key Laboratory of Beibu Gulf Marine Biodiversity Conservation, College of Marine Sciences, Beibu Gulf University, Qinzhou, Guangx, China

Wahidatul Husna Zuldin Higher Institution Centre of Excellence (HICoE), Borneo Marine Research Institute, Universiti Malaysia Sabah, Kota Kinabalu, Sabah, Malaysia

Introduction to Aquaculture 1

Norfazreena Mohd Faudzi, Sari Asfarina Sariat, Wahidatul Husna Zuldin, Julian Ransangan, Shigeharu Senoo, and Ching Fui Fui

Abstract

Aquaculture, the farming of aquatic organisms such as fish, shellfish, and aquatic plants, has become essential for meeting global demands for seafood while alleviating pressure on wild fish stocks. The chapter provides a comprehensive overview of the field of aquaculture, covering its history, importance, and environmental considerations. The chapter reviewed the importance of aquaculture for food security, economic development, and poverty alleviation, especially in developing regions. The environmental impacts of aquaculture, such as water pollution from fish waste and excess nutrients, are also examined in detail. Strategies for mitigating these impacts through sustainable practices and regulatory measures are explored, emphasizing the importance of responsible aquaculture management. The chapter also highlighted the future of aquaculture and the emerging trends and technologies that promise further to enhance the sustainability and efficiency of this vital industry.

Keywords

Food security · Poverty alleviation · Environmental impacts · Sustainable practices

N. Mohd Faudzi (✉) · S. A. Sariat · W. H. Zuldin · J. Ransangan · S. Senoo · C. F. Fui
Higher Institution Centre of Excellence, Borneo Marine Research Institute, Universiti Malaysia Sabah, Kota Kinabalu, Sabah, Malaysia
e-mail: fazreenafaudzi@ums.edu.my

N. M. Faudzi et al. (eds.), *Essentials of Aquaculture Practices*,
https://doi.org/10.1007/978-981-97-6699-4_1

1.1 Definition and Status of Aquaculture

Aquaculture, usually referred to as aquafarming, is the practice of raising aquatic animals in regulated conditions, including fish, mollusks, crabs, and aquatic plants (Young et al. 2019). These aquatic animals are bred, raised, and harvested to satisfy various human requirements, including food production and aquatic environment restoration and conservation. The Latin words *aqua* (water) and *cultura* (cultivation) are the source of the English word "aquaculture," which emphasizes the cultivation of aquatic species in water-based systems.

Aquaculture dates back to the earliest civilizations, who participated in various fish and seafood cultivation practices. According to archaeological findings, the Chinese engaged in primitive pond fish farming about 4000 years ago, during the Shang era (Rabanal 1988). Around 2500 BCE, the Egyptians participated in tilapia aquaculture in historic ponds (Bardach et al. 1972). It is known that the Romans, the Greeks, and other ancient nations built fishponds and oyster beds for trade. Techniques for aquaculture have changed over time and in various places (Fig. 1.1). In Asia, a rice-fish culture emerged in which fish were raised alongside rice crops in rice fields, benefiting both types of agriculture. Fishponds were a typical sight in monasteries in mediaeval Europe because they provided a steady supply of fish during holy days of fasting (Bond 2016).

Aquaculture had notable breakthroughs in the nineteenth and twentieth centuries and the produced relatively both affordable and high-priced species frequently consumed as a fresh product (von Brandt and Amundson 2024). Fish hatcheries were developed to generate vast quantities of fish fry to replenish depleted fisheries stock. Wild populations were intended to be improved by introducing aquaculture species such as trout, salmon, groupers, sea bream, catfishes, prawns, and mollusks (von Brandt and Amundson 2024). To increase production, recirculating aquaculture systems (RAS) and open net-pen cages are two examples of innovative technology used for intensive fish farming practices.

As the human population increases annually, aquaculture has become an important global sector to provide protein sources (FAO 2020). It has been reported that aquaculture has made a significant contribution to global fish supply from 1986 (14.9 million tons, 14.6% of total fisheries and aquaculture) to 2018 (82.1 million tons, 46% of total fisheries and aquaculture) (Fig. 1.2) (FAO 2022). As opposed to wild fisheries, which are frequently impacted by overfishing and environmental changes, aquaculture offers a controlled and sustainable method of producing seafood sources.

According to Verdegem et al. (2023), the global aquaculture production contributed by finfish, crustaceans, and other aquatic animals increased from 43.0 to 120.1 million metric tons, an increase of close to 180% from the year 2000 to 2019 (Table 1.1). In 2000, 90% of global production came from Asia, and the percentage had risen to 92% in 2019. Meanwhile, the production rates of other producing countries such as America, Europe, Africa, and Oceania are 3.5%, 2.7%, 2.0%, and 0.2%, respectively. Therefore, the statistics indicate that Asia has great potential for

Fig. 1.1 Culture systems have changed over time in various places. (**a**) Pen culture, (**b**) recirculating aquaculture system (RAS) for grouper culture, (**c**) cage culture in pond, and (**d**) the use of Biofloc system for tilapia culture

Fig. 1.1 (continued)

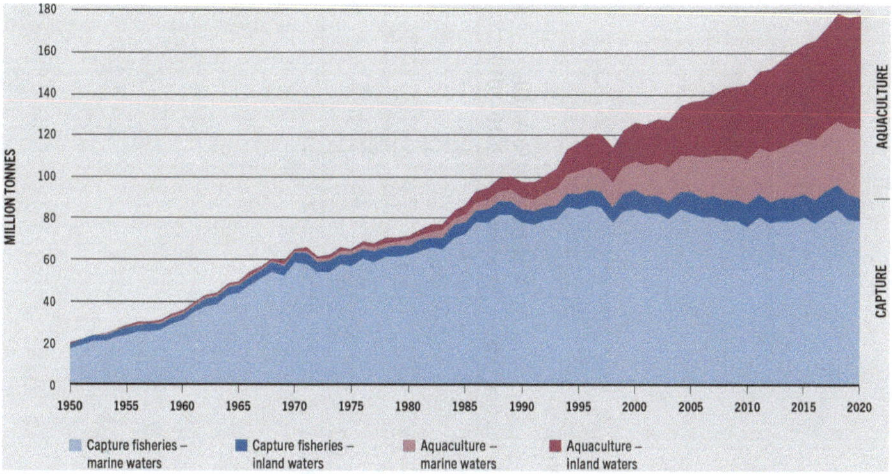

Fig. 1.2 World capture fisheries and aquaculture production (Source: FAO 2022)

the development of aquaculture, as seen by the large difference in productivity with other countries.

1.2 The Importance of Aquaculture

Aquaculture, the controlled culturing and harvesting of aquatic organisms, plays a crucial role in various aspects of human life and the global economy. The outlined importance of aquaculture can be categorized into several key areas.

1.2.1 Food Security and Nutrition

More than 850 million people are chronically hungry, with nearly one in seven people worldwide lacking sufficient food to live a healthy and active life (Kousar et al. 2021). Additionally, the percentage of undernourished individuals has been persistently increasing since 2014 (World Health Organization 2018). According to the Food and Agriculture Organization (FAO), the number of malnourished people reached approximately 834 million in 2017 (Unicef and World Health Organization 2017). Consequently, food insecurity has become a crucial issue that must be addressed, as it poses a significant risk to the achievement of the Sustainable Development Goals (SDGs).

Aquaculture significantly contributes to global food security by providing a sustainable and efficient source of protein. With the world's population continuing to grow, the demand for seafood has increased, and aquaculture helps meet this demand. By culturing various fish, shellfish, and other aquatic organisms, aquaculture diversifies the protein sources available for human consumption, reducing

Table 1.1 Global aquaculture production in 2000, 2010, and 2019 (Verdegem et al. 2023)

Country	2000		2010			2019		
	Production (Mt)	% global production	Production (Mt)	% global production	% AAGR 2000–2010	Production (Mt)	% global production	% AAGR 2010–2019
Asia	38.9	90	71.3	91.4	6.2	110.0	91.6	4.9
Americas	1.5	3.4	2.5	3.2	5.7	4.2	3.5	5.8
Europe	2.1	4.8	2.5	3.2	2.1	3.2	2.7	2.8
Africa	0.5	1.0	1.4	1.8	12.2	2.4	2.0	5.9
Oceania	0.1	0.3	0.2	0.3	3.9	0.2	0.2	1.1
World	43.0	100	78.0	100	6.1	120.1	100	4.9

dependence on wild fisheries and contributing to a more stable and secure food supply. Aquaculture can effectively use more natural resources than terrestrial agriculture, often requiring less freshwater and land. Moreover, it can also be integrated with other agricultural practices to create better food systems. The rapid expansion of the aquaculture industry can help overcome food insecurity and fulfill the nutritional needs of the growing global population.

Fish is a good source of nutrients for humans, such as proteins, fats, vitamins, and minerals, which are essential for good health as they provide energy in the repair and regulation process (Fig. 1.3) (Mishra 2020). Fish provides essential nutrients, especially proteins and fats, and is often referred to as a rich-nutrient food for poor people (Sujatha et al. 2013). The presence of major nutrients in fish makes it better than other animal foods. According to FAO (2022), more than 7.8 billion people in developed and developing countries consume at least 17% of their daily average animal protein from fish. Since fish are typically low in calories, they become a good option for maintaining a healthy weight (Pawar and Sonawane 2013).

Due to their excellent nutritional value, fish not only function as a food source but also prevent the occurrence of various human diseases in the world (Mishra 2020). The omega-3 fatty acids (eicosapentaenoic acid (EPA) and docosahexaenoic acid (DHA)) that are commonly found in fish are important for the human brain, heart, blood vessels, immune system, and endocrine organs (Mishra 2020). Other than that, they help in the prevention of cardiovascular disease, lupus, depression, and other mental illnesses (Pradeepkiran 2019). A variety of vitamins that are essential for human health are also found in fish; however, the amounts may vary according to the fish species. Fish are also important sources of vitamins A and D, as well

Fig. 1.3 Fresh fish at Kota Kinabalu Fish Market, Sabah, Malaysia, provides a rich source of high-quality protein and essential nutrients for a healthy diet

as vitamin B. Large amounts of vitamins A and D are commonly stored in the fish liver. Vitamin A helps with normal growth, the formation of bones and teeth, and the prevention of eye diseases (Pal et al. 2018). Meanwhile, adult osteoporosis and fractures are prevented by vitamin D, which can be found in fish and fish oils (Mishra 2020). Fish are also rich in minerals such as iodine, calcium, zinc, iron, and others that help in the synthesis of hemoglobin, oxygen transportation, and the reduction of antioxidant enzymes.

1.2.2 Economic Contributions

In both developed and developing regions, aquaculture is a major economic driver, generating income and employment opportunities (Parrao et al. 2021). The industry creates jobs throughout the value chain, from farming and processing to distribution and retail. Additionally, aquaculture fosters economic growth by providing export opportunities, supporting trade, and contributing to the overall economic well-being of communities involved in the industry. One of the most significant economic benefits of aquaculture is its capacity to create a large number of employment opportunities. The industry promotes direct employment on fish farms, where workers are required to feed, breed, and harvest aquatic organisms and maintain equipment and facilities (Fig. 1.4). It has been reported that over 20 million people in the world are engaged in the aquaculture industry on a full-time, part-time, or seasonal basis (FAO 2020).

These positions are essential for the daily operations of aquaculture facilities and can offer stable employment in regions where jobs may be challenging. Moreover, the aquaculture industry also promotes indirect employment in related industries such as feed production, equipment manufacturing, and transportation services that show increased demand due to the growth of aquaculture. Furthermore, processing facilities that prepare aquaculture products for the market create employment in quality control, fish cleaning, and packaging. As fish and seafood products are distributed from producers to customers, employment opportunities are further increased by retail and wholesale outlets. Seasonal employment especially during peak times such as the harvesting period also allows local populations to generate extra income, thus supporting the stability of the regional economy.

Other than that, by increasing export prospects and diversifying income sources, aquaculture makes a substantial contribution to economic growth. Regions can access international markets and boost their foreign exchange revenues and trade balances by producing high-value aquatic products (FAO 2022). Due to the demand these export activities create for products and services made locally, further economic activity stimulates and consequently supports local economies.

Fig. 1.4 Employment opportunities in hatchery daily operations

1.2.3 Environmental Benefits

Traditional wild fish catch methods have significant negative impacts on marine ecosystems, resulting in a range of environmental problems. One major consequence is the reduction of specific fish populations, which are frequently harvested at a rate that exceeds their capacity for natural reproduction, leading to the collapse of wild fish stocks (Pham et al. 2023). This overfishing has an impact on species that depend on these fish for food and affects the balance of marine food webs. Destructive fishing methods, such as bottom trawling, can make the problems worse. In addition to capturing the target species, dragging heavy nets around the seafloor, known as bottom trawling, also destroys fragile ecosystems, including seagrass beds and coral reefs (Wang et al. 2021; Pham et al. 2023). As habitats provide breeding and feeding grounds for a variety of species, they are very essential to the biodiversity of marine life. These environments become less able to sustain marine life over time as a result of the physical damage that bottom trawling causes (Pham et al. 2023). Traditional wild fish catch also leads to high bycatch rates, which are caused by unintentionally catching nontarget species such as juveniles and endangered species, which are frequently discarded dead or dying. This bycatch further disrupts marine ecosystems, causing declines in nontarget species populations and altering the ecological balance. As a result, these practices contribute to the deterioration of the biodiversity and health of the ocean.

Aquaculture is therefore introduced since it is essential in the negative effects of traditional wild fish catch methods and in promoting sustainable fisheries. Aquaculture reduces the pressure on wild fish populations by culturing fish and other aquatic organisms in controlled environments, thus allowing wild stock populations to recover and replenish. This practice guarantees a consistent supply of seafood while preserving the balance of marine ecosystems and preventing further depletion of natural resources. Through the implementation and support of sustainable aquaculture practices, we can mitigate the negative consequences of overfishing, protect marine biodiversity, and sustain the well-being of our oceans.

1.3 Challenges and Limitations of Aquaculture

Although aquaculture holds great promise for meeting the world's increasing demand for seafood, it faces challenges and limitations, such as environmental impacts, disease and health management, social and ethical considerations, and technological and knowledge gaps.

1.3.1 Environmental Concern

Despite the benefits, aquaculture can lead to habitat destruction. Large-scale fish production requires the modification of land to be used for aquaculture sites (Polpanich et al. 2009; Kurniawan et al. 2021; Rogers 2023). It has been reported that thousands of hectares of mangroves and coastal wetlands have been destroyed as a result of the construction of ponds for the aquaculture of marine finfish or shrimp (Boyd 2002). Mangroves have suffered significant losses in several countries such as the Philippines, Vietnam, Thailand, Bangladesh, and Ecuador (Boyd 2002). Mangroves are important as they provide habitat for a wide variety of marine and terrestrial animals, protect coastal areas from storms, and are essential in the subsistence of many coastal communities. Additionally, it serves as a nursery ground for a variety of aquatic animals including fish that are significant to the commercial fisheries (Rönnbäck 1999; Kathiresan and Rajendran 2002).

Aquaculture can lead to habitat deterioration through sedimentation and nutrient pollution, in addition to mangrove destruction. Feed and fertilizers are frequently used in the intensive farming of fish and shrimp which can cause an accumulation of nutrients and organic waste in the water (Zhang et al. 2014; Dauda et al. 2019; Kurniawan et al. 2021). The accumulation of nutrients leads to eutrophication, a process that lowers oxygen levels in the water and creates hypoxic conditions or "dead zones" where aquatic life cannot survive (Devlin and Brodie 2023). Moreover, the construction of aquaculture ponds and releasing untreated wastewater can speed up sedimentation, suffocating seagrass beds, coral reefs, and other sensitive habitats. The alteration of an area for aquaculture sites can disrupt the ecological balance, degrade biodiversity, and impair the functioning of ecosystems that are crucial for the well-being of coastal and marine environments. Effective management practices and legislative actions are essential to lessen these effects and sustainably encourage the development of aquaculture.

The use of fish meal and fish oil from wild fish capture in aquaculture feeds leads to overfishing and further depletes marine resources (Zlaugotne et al. 2022). This approach not only affects the target but also has cascading effects across the food chain. The primary food supply for larger predatory fish, marine mammals, seabirds, and small pelagic fish are essential to marine ecosystems and are often exploited for fish meal and fish oil. According to Pham et al. (2023), overfishing of these species can cause major disturbances in marine food chains which can lead to the decline of predator populations and changes in dynamics of marine ecosystems. Furthermore, the loss of these forage fish populations has the potential to weaken marine habitat resilience and threaten the ecological balance that sustains a variety of marine species. Currently, the aquaculture industry is investigating and utilizing alternate feed sources such as plant-based proteins, insect meal, and algal-based products to lessen these effects (Albrektsen et al. 2022; Zlaugotne et al. 2022). Aquaculture can become less dependent on wild fish and promote a more balanced and eco-friendly approach by using sustainable substitutes.

Other than that, the introduction of non-native species is one of the methods to increase fish production. Non-native species are often chosen due to their superior growth and high market value (Ju et al. 2019). It has been reported that a total of 5612 records of non-native species were introduced in aquaculture (FAO 2019). In many developing countries such as Brazil, Cuba, India, and the Philippines, the non-native species contributes to about 60–95% of aquatic food production (Shelton and Rothbard 2006). However, the escape of non-native species into the wild poses a threat to the environment. Even though it is difficult to assess, non-native species escaping from aquaculture operations have grown to be a worldwide concern (FAO 2019; Ju et al. 2019). This is because of the non-native species competing with native species, which causes local population declines and disrupts the biodiversity and ecological balance (Ju et al. 2019). Moreover, hybrid offspring produced by the interbreeding between non-native and native species may have a lower level of fitness and adaptability toward the environment. The escape of non-native species also can alter the habitat or ecosystem which has an indirect impact on other aquatic organisms. While the introduction of non-native species in aquaculture can significantly boost aquaculture production and economic returns, it is crucial to manage and mitigate the associated environmental risks to preserve biodiversity and ecological health.

1.3.2 Disease and Health Management

Disease and health management are other issues that limit the development of aquaculture. Disease outbreaks in cultured fish are generally the result of intensified cultural practices that lack of fundamental understanding of the intricate balance between host, pathogen, and environment. These outbreaks involved problems such as disease spread, the overuse and misuse of antibiotics, vaccination challenges, and a shortage of effective treatments. Furthermore, environmental factors, including water quality, temperature fluctuations, and pollution, hurt fish health, consequently making the effort of disease and health management more difficult.

The main challenge in managing the health of aquaculture is the quick spread of disease, especially in small areas with high stocking density. Good biosecurity measures such as monitoring stocking density, feeding activity, and water quality are essential to be applied to prevent the occurrence of disease and its spreading in the system quickly. Aside from that, stress caused by poor handling and a lack of genetic diversity in cultured fish further increases susceptibility to diseases.

Antibiotics have been widely used for controlling diseases in aquaculture, especially in developing countries (Cabello et al., 2006; Vignesh et al. 2011). Even though the toxicity level is low, an environmental concern is highlighted as the chemical compounds are not broken down but remain active long after being excreted (Vignesh et al. 2011). Thus, the continuous use of antibiotics led to various hazards not only for aquatic organisms but also for humans (Sneeringer et al. 2019). Other than that, it is also promoting the development of antibiotic-resistant strains of bacteria, which diminishes the efficacy of antibiotics in aquaculture. It has been highlighted that the use of antibiotics has been discouraged, especially in the food production industry (Ridzuan et al. 2022). Strict guidelines on the use of antibiotics must be implemented strictly, ensuring that the antibiotics are only given under the supervision of professional guidance. Adopting responsible guidelines, such as treatment duration and appropriate dosage, lessens resistance development.

Vaccination is the process of stimulating protective immune responses in fish against pathogenic microorganisms by exposing them to nonpathogenic forms or components of microorganisms. Vaccination for fish started in the 1940s as disease prevention, especially for bacterial and viral diseases (Snieszko and Friddle 1949). The introduction of vaccination has dramatically reduced the use of antibiotics in Norwegian salmon culture as it is cost-effective and sustainable for controlling diseases (Horzinek et al. 1997; Ma et al. 2019). Even though vaccination has the potential for disease prevention, the implementation of the vaccine in large-scale fish farming is difficult (Ma et al. 2019). Previously, injectable or immersion vaccine methods were widely used, but they were expanded to include the development of oral vaccination, especially for mass vaccination that is suitable for all sizes and ages of fish (Ma et al. 2019; Ridzuan et al. 2022). Despite offering flexibility in application, oral administration has its drawbacks, including a shorter duration of protection and an inaccurate dose of antigen that reflects inconsistent responses.

Aquaculture health and disease management are further complicated by environmental considerations. Variations in water quality, including temperature fluctuations and broader ecological changes, have a substantial effect on the susceptibility of fish to disease. The problems are made worse by climate change, whose changes in the water temperature make some infections more common and exacerbate these challenges by altering water temperatures and increasing the prevalence of certain pathogens. Understanding these environmental factors and their effects on the dynamics of disease is important for implementing adaptive management strategies.

1.3.3 Social and Ethical Considerations

Despite the many social benefits of aquaculture, social effects of aquaculture have occurred. Coastal communities serve a variety of user groups, such as aquaculture farmers, leisure fishermen, and tourists. The utilization of common areas for aquaculture purposes might lead to a conflict between these different user groups. It has been reported that local communities in Norway and Scotland initially supported salmon farming, but the expansion of the farm has led to an increase in conflicts (Krause et al. 2015). Even though the government attracts larger and export-oriented farms to increase tax income and foreign exchange, it has caused conflicts with the local communities (Engle and van Senten 2022). According to Lebel et al. (2002), large-scale aquaculture farms have disrupted access to public lands and water in Vietnam and Thailand. Therefore, aquaculture growth has been shown to discriminate against poverty in communities in coastal areas of developing countries (Toufique and Gregory 2008).

Ethical concerns about aquaculture have attracted a lot of interest, reflecting larger food-related issues such as controversy over genetically modified (GM) food, food safety, and the acceptability of multifunctional food in society (Grigorakis 2009). The ethical concern is further complicated by the responsibility of food producers, who contribute to public health issues such as obesity, animal welfare concerns, and food insecurity. Other than that, the expansion of the aquaculture industry often leads to habitat alteration, pollution, and genetic interactions with wild populations (Ciliberti et al. 2023). Ensuring responsible practices that minimize environmental degradation and protect biodiversity is critical to ethical and sustainable aquaculture. It is necessary to take into account a balance between production needs and fish welfare. Evaluating fish welfare is a difficult challenge as it is different for each fish species. The guidelines have to ensure appropriate space for aquatic organisms, responsible resource usage, minimal environmental effects, and proper training for workers to assess and manage fish welfare (Ciliberti et al. 2023). The issues highlight the importance of ethical considerations in aquaculture practices, as the implications extend beyond the health and safety of consumers, cultured fish, and the environment as well.

1.4 Future Prospects in Aquaculture

Aquaculture has bright prospects for the future, including ways to promote economic growth and global food security (Bryke 2023). As a reliable and sustainable source of protein, which is vital for human nutrition and health worldwide, it helps ensure food security. With 87 million tons of fish produced in 2020 and 155 million tons needed by 2050, the aquatic food supply must increase to satisfy the future demand (Demavivas 2023). Managing wild fisheries, growing aquaculture sustainably, and reducing loss and waste are all necessary to meet this demand.

1.4.1 Alternative Species and Diversification

The introduction of alternative species and diversification are the two main ideas in wild fisheries management that also lead to a sustainable aquaculture industry (Thomas et al. 2021). It involves species selection through identifying and farming species that are underutilized or have a lower environmental impact, which can contribute to sustainable aquaculture, for instance, culturing seaweed and bivalves, which have minimal ecological impacts and aid in improving the water quality. Besides, alternative species may be more resilient to changing ocean conditions, providing a buffer against the impacts of climate change on traditional fisheries (Mason et al. 2022).

Meanwhile, the diversification of aquaculture species offers a broader range of seafood options that helps stabilize market prices and supplies, protecting the industry from the volatility associated with overreliance on a few species (Longo et al. 2019). Species diversification works by expanding the variety of species that are farmed or harvested to lessen reliance on a small number of well-known species, which may result in overfishing and ecological imbalance (Pham et al. 2023). By farming a wider variety of species, aquaculture can alleviate the intense fishing pressure on overexploited wild stocks such as Atlantic cod or bluefin tuna (Froehlich et al. 2023).

Apart from that, genetic improvements such as gene selection, editing, clustered regularly interspaced short palindromic repeats, and CRISPR technology in aquaculture are powerful tools that complement the diversification of species by enhancing the traits of existing species (Gutási et al. 2023). This approach not only helps in producing more resilient and efficient aquaculture species but also supports biodiversity in several ways (Nguyen 2023). The approach involves mainly selective breeding and the utilization of genetic markers. Through selective breeding, aquaculture species can be developed with increased resistance to common diseases. This reduces the need for antibiotics and other treatments, promoting healthier ecosystems. Moreover, the use of genetic markers helps identify and select individuals with desirable traits, speeding up the breeding process for disease-resistant aquaculture species (Griot et al. 2021).

Genetic improvement in aquaculture not only boosts the efficiency and resilience of farmed species but also plays a crucial role in maintaining and promoting biodiversity. By developing species that are better suited to the challenges of disease and environmental changes and by reducing pressure on wild populations, genetic improvement supports the long-term sustainability of aquaculture (Momin et al. 2024). Integrating genetic advancements with practices such as species diversification and sustainable management helps build a robust and adaptive aquaculture industry that is capable of meeting future food security demands while preserving the ecological balance.

1.4.2 Technological Advancement

Technological advancements in aquaculture have revolutionized the industry, making it more efficient, sustainable, and capable of meeting the growing global demand for seafood (Tom et al. 2021). Technologies such as automation, artificial intelligence, and data analytics are improving operational efficiency and reducing costs in aquaculture. These innovations span a wide range of areas, from farming techniques and genetic improvements to monitoring systems and environmental management.

Advanced farming systems such as recirculating aquaculture systems (RAS) and offshore or deep-sea aquaculture are key components in the technological advancement of farming techniques (Yue and Shen 2022). The RAS technology allows for continuous recycling and treatment of water within the system, which significantly reduces water usage and environmental impact. These systems also provide a controlled environment that can optimize conditions for specific species, improving growth rates and health (Zhang et al. 2023). Furthermore, the technology can minimize the risk of disease transmission from external sources, enhance biosecurity, and reduce the need for antibiotics. Meanwhile, offshore farming involves the cultivation of fish and other marine organisms in deep water, away from coastal ecosystems, which can reduce competition for space and minimize environmental impacts. The advanced engineering approach has led to the development of robust, submersible cages and automated feeding systems that withstand harsh ocean conditions.

In terms of monitoring and environmental management, their technological advancement has been progressing well, which involves automated feeding systems that apply precision feeding, in which the automated feeders deliver the right amount of food at the right time, reducing waste and optimizing growth (Chinaeke-Ogbuka et al. 2021). It also involves real-time adjustments, whereby the systems can adjust feeding schedules based on real-time data on fish behavior and environmental conditions. In terms of environmental monitoring and control, sensors and IoT (Internet of Things) devices are curated, such as advanced sensors that can monitor water quality parameters such as temperature, pH, oxygen levels, and salinity, providing real-time data to farmers. Furthermore, the application of IoT technology can also enable remote monitoring and management of aquaculture operations, enhancing efficiency and response times (Jais et al. 2024). The application of underwater drones and remotely operated vehicles (ROVs) is also beneficial for inspecting and maintaining aquaculture facilities, particularly in offshore environments, which can reduce the need for divers. The devices are also useful for surveillance in monitoring fish health and behavior, as well as detecting any structural issues in farming equipment (Ubina and Cheng 2021).

1.4.3 Future Market and Demands

The future of aquaculture is assured of significant growth, driven by rising global demand for seafood, advances in technology, and a push for sustainable food production. As the world population grows and wild fish stocks become increasingly depleted, aquaculture is expected to play a crucial role in meeting the global need for protein (Troell et al. 2023). The global population is projected to reach around 9.7 billion by 2050. This growth will drive an increased demand for nutritious and sustainable food sources, including seafood. In fact, seafood consumption per capita is also on the rise, particularly in developing countries where economic growth and urbanization increase access to diverse diets.

There is also a shift toward a healthy and sustainable diet among the public. Seafood is recognized for its health benefits, including high-quality protein, essential fatty acids, and other nutrients. This recognition is boosting its popularity among health-conscious consumers (Herrero et al. 2023). Other than that, consumers are increasingly concerned about the environmental impact of their food choices. Aquaculture, especially when sustainably managed, offers a more controlled and eco-friendlier alternative to wild-caught fish. Recently, there has been a growing market for a wider variety of aquaculture species beyond traditional staples such as salmon and tilapia. Species such as shrimp, oysters, seaweed, and even less common fish are gaining popularity. The market for processed and value-added seafood products, such as pre-packaged meals and ready-to-eat options, is expanding, driven by consumer demand for convenience. Therefore, future advancements in sustainable aquaculture are necessary to fulfill the demands of the advanced market.

1.5 Conclusion

The chapter provides a comprehensive overview of the diverse field of aquaculture. It has covered key topics that include the status and importance of aquaculture as well as potential future developments in aquaculture. The book has highlighted the importance of aquaculture in meeting global food demand, its effects on the environment, and its potential for sustainable development. It is encouraged to further explore this dynamic and evolving field to contribute to the responsible management and growth of aquaculture practices.

Acknowledgement The study was funded by Universiti Malaysia Sabah (Skim Penyelidikan Lantikan Baharu, Grant Code: SLB2218) and Higher Institution Centre of Excellence (HICoE), Borneo Marine Research Institute, Universiti Malaysia Sabah (Grant Code: HIC2401).

References

Albrektsen S, Kortet R, Skov PV, Ytteborg E, Gitlesen S, Kleinegris D, Mydland LT, Hansen JØ, Lock EJ, Møkøre T, James P, Wang X, Whitaker RD, Vang B, Hatlen B, Daneshvar E, Bhatnagar A, Jensen LB, Øverland M (2022) Future feed resources in sustainable salmonid production: a review. Rev Aquac 2022:1–23

Bardach JE, Ryther JH, McLarney WO (1972) Aquaculture—the farming and husbandry of freshwater and marine organisms. John Wiley & Sons, Inc, p 351

Bond J (2016) Introduction. In: Bonow M, Olsen H, Svanberg I (eds) Historical aquaculture in Northern Europe. Elanders, Stockholm, pp 1–16

Boyd CE (2002) Mangroves and coastal aquaculture. In: Stickney RR, McVey JP (eds) Responsible Marine Aquaculture. CABI Publishing, New York, NY, pp 145–158. Ch. 9

Bryke K (2023) Aquaculture: a sustainable solution for fish cultivation and global food security. J Agric 6(4):91–93

Cabello FC (2006) Heavy use of prophylactic antibiotics in aquacul- ture: a growing problem for human and animal health and for the environment. Environmental Microbiology 8:1137–1144

Chinaeke-Ogbuka I, Anoliefo E, Ajibo C, Ogbuka C (2021) Design and implementation of an automated feeding system for poultry farms. In 4th European International Conference on Industrial Engineering and Operations Management

Ciliberti R, Alfano L, Petralia P (2023) Ethics in aquaculture: animal welfare and environmental sustainability. J Prev Med Hyg 64(4):E443–E447

Dauda AB, Ajadi A, Tola-Fabunmi AS, Akinwole AO (2019) Waste production in aquaculture: sources, components and managements in different culture systems. Aquac Fish 4(3):81–88

Demavivas K (2023) 4 Ways to ensure the future of aquaculture is sustainable. World Economic Forum. https://www.weforum.org/agenda/2023/03/4-ways-to-ensure-the-future-of-aquaculture-is-sustainable/

Devlin M, Brodie J (2023) Nutrients and eutrophication. In: Reichelt-Brushett A (ed) Marine pollution—monitoring, management and mitigation. Springer textbook in earth science, geography and environment

Engle CR, van Senten J (2022) Resilience of communities and sustainable aquaculture: governance and regulatory effects. Aust Fish 7(5):268

FAO (2019) Database on introductions of aquatic species. Food and Agriculture Organization of the United Nations

FAO (2020) The state of world fisheries and aquaculture: sustainability in action. Food and Agriculture Organization of the United Nations

FAO (2022) The state of world fisheries and aquaculture: towards blue transformation. Food and Agriculture Organization of the United Nations

Froehlich HE, Montgomery JC, Williams DR, Kuempel CD, Halpern BS (2023) Biological life-history and farming scenarios of marine aquaculture to help reduce wild marine fishing pressure. Fish Fish 24(6):1034–1047

Grigorakis K (2009) Ethical issues in aquaculture production. J Agric Environ Ethics 23(4):345–370

Griot R, Allal F, Phocas F, Brard-Fudulea S, Morvezen R, Haffray P, François Y, Morin T, Bestin A, Bruant S, Cariou S, Peyrou B, Brunier J, Vandeputte M (2021) Optimization of genomic selection to improve disease resistance in two marine fishes, the European sea bass (*Dicentrarchus labrax*) and the gilthead sea bream (*Sparus aurata*). Front Genet 12:665920

Gutási A, Hammer SE, El-Matbouli M, Saleh M (2023) Review: recent applications of gene editing in fish species and aquatic medicine. Animals 13:7

Herrero M, Hugas M, Lele U, Wirakartakusumah A, Torero M (2023) A shift to healthy and sustainable consumption patterns. In: von Braun J, Afsana K, Fresco LO, Hassan MHA (eds) Science and innovations for food systems transformation. Springer, Cham
Horzinek MC, Schijns VECJ, Denis M, Desmettre P, Babiuk LA (1997) General description of vaccines. In: Pastoret PP, Blancou J, Vannier P, Verschueren C (eds) Veterinary vaccinology. Elsevier Press, Amsterdam, pp 132–152
Jais MNA, Abdullah AF, Mohd Kassim MS, Abd Karim MM, Muhadi A (2024) Improved accuracy in IoT-based water quality monitoring for aquaculture tanks using low-cost sensors: Asian seabass fish farming. Heliyon 10:8
Ju RT, Li X, Jiang JJ, Wu J, Liu J, Strong DR, Li B (2019) Emerging risks of non-native species escapes from aquaculture: call for policy improvements in China and other developing countries. J Appl Ecol 57(1):85–90
Kathiresan K, Rajendran N (2002) Fishery resources and economic gain in three mangrove areas on the south-east coast of India. Fish Manag Ecol 9:277–283
Kousar S, Ahmed F, Pervaiz A, Bojnec S (2021) Food insecurity, population growth, urbanization and water availability: the role of government stability. Sustain For 13:12336
Krause G, Brugere C, Diedrich A, Ebeling MW, Ferse SCA, Mikkelsen E, Agundez JAP, Stead SM, Stybel N, Troell M (2015) A revolution without people? Closing the people-policy gap in aquaculture development. Aquaculture 447:44–55
Kurniawan SB, Ahmad A, Rahim NFM, Said NSM, Alnawajha MM, Imron MF, Abdullah SRS, Othman AR, Ismail NI, Hasan HA (2021) Aquaculture in Malaysia: water-related environmental challenges and opportunities for cleaner production. Environ Technol Innov 24:101913
Lebel LN, Tri H, Saengnoree A, Pasong S, Buatama U, Thoa LK (2002) Industrial transformation and shrimp aquaculture in Thailand and Vietnam: pathways to ecological, social, and economic sustainability? AMBIO J Hum Environ 31:311–323
Longo SB, Clark B, York R, Jorgenson AK (2019) Aquaculture and the displacement of fisheries captures. Conserv Biol 33(4):832–841
Ma J, Bruce TJ, Jones EM, Cain KD (2019) A review of fish vaccine development strategies: conventional methods and modern biotechnological approaches. Microorganisms 7(11):569
Mason JG, Eurich JG, Lau JD, Battista W, Free CM, Mills KE, Tokunaga K, Zhao LZ, Dickey-Collas M, Valle M, Pecl GT, Cinner JE, McClanahan TR, Allison EH, Friedman WR, Silva C, Yáñez E, Barbieri MÁ, Kleisner KM (2022) Attributes of climate resilience in fisheries: from theory to practice. Fish Fish 23(3):522–544
Mishra SP (2020) Significance of fish nutrients for human health. Int J Fish Aquat Res 5:47–49
Momin CM, Baidya S, Debbarma S, Yadav NK, Chandravashi S, Debnath A, Deb S, Lal J, Lavkush VA (2024) Genetic improvement initiatives in aquaculture. Int J Adv Biochem Res 8(4S):441–447
Nguyen NH (2023) Genetics and genomics of infectious diseases in key aquaculture species. Biology 13:1
Pal J, Shukla BN, Maurya AK, Verma HO (2018) A review on role of fish in human nutrition with special emphasis fatty acid. Int J Fish Aquat Stud 6(2):427–430
Parrao CG, Shisler S, Moratti M, Yavuz C, Acharya A, Eyers J, Snilstveit B (2021) Aquaculture for improving productivity, income, nutrition and women's empowerment in low- and middle-income countries: a systematic review and meta-analysis.Campbell. Syst Rev 17(4):e1195
Pawar SM, Sonawane SR (2013) Fish muscle protein highest source of energy. Int J Biodivers Conserv 5(7):433–435
Pham CV, Wang HC, Chen SH, Lee JM (2023) The threshold effect of overfishing on global fishery outputs: international evidence from a sustainable fishery perspective. Aust Fish 8(2):71
Polpanich O., Shing WC., Collin JA., Talib MA. and Samarakoon L. 2009. Mangrove conversion and aquaculture development in Sabah, Malaysia: Remote sensing- and GIS-based approach for assessing land use change in mangroves. Proceeding of the 30th Asian Conference on Remote Sensing, Beijing, China. pp: 1–6
Pradeepkiran JA (2019) Aquaculture role in global food security with nutritional value: a review. Transl Anim Sci 3(2):903–910

Rabanal HR (1988) History of aquaculture. ASEAN/UNDP/FAO Regional Small-Scale Coastal Fisheries Development Project. pp. 13
Ridzuan MSM, Abdullah A, Ramly R, Mansor NN, Ramli N, Nawi MF (2022) Current status and advances of fish vaccines in Malaysia. Veterinary World 15(2):465–482
Rogers AJ (2023) Aquaculture in the ancient world: ecosystem engineering, domesticated landscapes and the first blue revolution. J Archaeol Res:1–65
Rönnbäck P (1999) The ecological basis for economic value of seafood production supported by mangrove ecosystems. Ecol Econ 29:235–252
Shelton WL, Rothbard S (2006) Exotic species in global aquaculture—a review. Israeli J Aquac-Bamidgeh 58:3–28
Sneeringer S, Bowman M, Clancy M. The US and EU Animal pharmaceutical industries in the age of antibiotic resistance. USDA; Washington, DC, USA: May, 2019. USDA Economic Research Service Report Number 264
Snieszko SF, Friddle SB (1949) Prophylaxis of furunculosis in brook trout (*Salvelinus fontinalis*) by oral immunization and sulfamerazine. Prog Fish Cult 11:161–168
Sujatha K, Joice AA, Senthilkumaar P (2013) Total protein and lipid content in edible tissues of fishes from Kasimodu fish landing centre, Chennai, Tamil Nadu. European Journal of Experimental Biology. 3(5):252–257
Thomas M, Pasquet A, Aubin J, Nahon S, Lecocq T (2021) When more is more: taking advantage of species diversity to move towards sustainable aquaculture. Biol Rev Camb Philos Soc 96:767–784
Tom AP, Jayakumar JS, Biju M, Somarajan J, Ibrahim MA (2021) Aquaculture wastewater treatment technologies and their sustainability: a review. Energy Nexus 4:100022
Toufique KA, Gregory R (2008) Common waters and private lands: distributional impacts of floodplain aquaculture in Bangladesh. Food Policy 33:587–594
Troell M, Costa-Pierce B, Stead S, Cottrell RS, Brugere C, Farmery AK, Little DC, Strand Å, Pullin R, Soto D, Beveridge M, Salie K, Dresdner J, Moraes-Valenti P, Blanchard J, James P, Yossa R, Allison E, Devaney C, Barg U (2023) Perspectives on aquaculture's contribution to the sustainable development goals for improved human and planetary health. J World Aquacult Soc 54(2):251–342
Ubina NA, Cheng S (2021) A review of unmanned system technologies with its application to aquaculture farm monitoring and management. Drones 6(1):12
Unicef and World Health Organization (2017) The state of food security and nutrition in the world 2017: building resilience for peace and food security. Food and Agriculture Organization, Rome
Verdegem M, Buschmann AH, Latt UW, Dalsgaard AJT, Lovatelli A (2023) The contribution of aquaculture systems to global aquaculture production. J World Aquacult Soc 54(2):206–250
Vignesh R, Karthikeyan B, Periyasamy N, Devanathan K (2011) Antibiotics in aquaculture: an overview. South Asian J Exp Biol 1(3):1–7
von Brandt ARFT, Amundson CH (2024) Aquaculture. *Encyclopedia Britannica*, 25 Apr. 2024, https://www.britannica.com/topic/aquaculture. Accessed 12 June 2024
Wang Z, Leung KMY, Sung YH, Dudgeon D, Qiu JW (2021) Recovery of tropical marine benthos after a trawl ban demonstrates linkage between abiotic and biotic changes. Commun Biol 4:212
World Health Organization (2018) The state of food security and nutrition in the world 2018: building climate resilience for food security and nutrition. Rome, Italy, Food and Agriculture Organization
Young N, Brattland C, Digiovanni C, Hersoug B, Johnsen JP, Karlsen KM, Kvalvik I, Olofsson E, Simonsen K, Solas AM, Thorarensen H (2019) Limitations to growth: social-ecological challenges to aquaculture development in five wealthy nations. Mar Policy 104:216–224
Yue K, Shen Y (2022) An overview of disruptive technologies for aquaculture. Aquaculture and Fisheries 7(2):111–120
Zhang R, Chen T, Wang Y, Short M (2023) Systems approaches for sustainable fisheries: A comprehensive review and future perspectives. Sustainable Production and Consumption 41:242–252

Zhang Q, Achal V, Xu Y, Xiang WN (2014) Aquaculture wastewater quality improvement by water spinach (*Ipomoea aquatica* Forsskal) floating bed and ecological benefit assessment in ecological agriculture district. Aquacult Eng 60:48–55

Zlaugotne B, Pubule J, Blumberga D (2022) Advantages and disadvantages of using more sustainable ingredients in fish feed. Heliyon 8(9):e10527

Aquaculture Diets: Live Feed and Sustainable Aquafeed

2

Alfredo Olivera Gálvez, Danielli Matias de Macêdo Dantas, Laenne Barbara Silva de Moraes, and Géssica Cavalcanti Pereira Mota

Abstract

Aquaculture is fundamental to global food security, challenged by population growth and dwindling natural resources. Despite the environmental problems still linked to aquaculture, this activity can minimize the impacts of traditional fishing, such as overfishing, habitat degradation and pollution. The production of live feed and aquafeed are fundamental to the aquaculture industry, as they enable the production of fish and shellfish in captivity on a large scale. Live feed production is an important technique in aquaculture, as it provides a natural and nutritious feed source, which includes phytoplankton and zooplankton organisms, for farmed fish and shellfish. Aquafeed formulation is another important technique in aquaculture, providing nutrients and a balanced diet based on the nutritional needs of the cultivated organisms. The choice of aquafeed ingredients depends, in addition to nutritional requirements, on the availability and cost of the ingredient, being optimized with the use of alternative ingredients, such as algae, insects and agroindustrial by-products. These ingredients are sources of protein, energy, fatty acids and other bioactive compounds, which contribute to reducing dependence on ingredients of animal origin, such as fishmeal and oil, making aquaculture more sustainable and resilient. Aquaculture is a constantly evolving activity, and the adoption of sustainable and innovative practices is essential to ensure global food security and the conservation of natural resources. This chapter explores the aspects of live feed and aquafeed production, highlighting their nutritional, environmental and economic importance in the aquaculture industry and their impact on global food security and sustainable development.

A. O. Gálvez (✉) · D. M. de Macêdo Dantas · L. B. S. de Moraes · G. C. P. Mota
Department of Fisheries and Aquaculture, Federal Rural University of Pernambuco, Recife, Pernambuco, Brazil
e-mail: alfredo.galvez@ufrpe.br

N. M. Faudzi et al. (eds.), *Essentials of Aquaculture Practices*,
https://doi.org/10.1007/978-981-97-6699-4_2

Keywords

Aquafeed · Live feed · Agroindustrial by-products · Sustainable feed

2.1 Introduction

Aquaculture plays a fundamental role in global food security, given the population growth, high demand for food and decreasing natural resources. The production of fish, molluscs and crustaceans in controlled environments allows the production of high-quality proteins efficiently and sustainably (Garlock et al. 2022). With the ability to provide high-quality food, aquaculture makes it possible to complement traditional fishing, which faces challenges of overexploitation and negative environmental impacts (Sumaila and Tai 2019). Aquaculture appears as a promising solution that allows the creation of a controlled environment for the farming of fish, molluscs and crustaceans, minimizing negative impacts on aquatic ecosystems and contributing to the preservation of marine and freshwater biodiversity (Boyd et al. 2022). Furthermore, the development of aquaculture has enabled the production of different species, providing several sources of dietary protein.

Accordingly, the production of live feed plays a fundamental role in the creation of aquatic organisms. Live feed, such as phytoplankton and zooplankton, provide essential nutrients to growing organisms and are a fundamental part of the early stages of cultivation in many aquatic species (Haoujar et al. 2022). The use of live feed is crucial and often indispensable in obtaining high survival and growth rates, directly influencing aquaculture productivity.

The formulation of aquafeeds is another fundamental pillar in the aquaculture industry, as it involves the preparation of balanced and nutritious diets according to the needs of the farmed species (Tacon 2020). Constant research and innovation in aquafeed formulation, such as the search for alternative ingredients, have been fundamental to maximizing the growth efficiency of aquatic organisms, minimizing feed waste and reducing environmental impacts associated with feed production.

Aquaculture, especially from the production of live feed and the formulation of aquafeeds, represents one of the solutions to the feed security and sustainability challenges currently faced. This chapter will provide a comprehensive analysis of live feed and aquafeed in the aquaculture industry and its importance in global nutrition, highlighting the advances and innovations that have shaped this constantly evolving field.

2.2 History and Challenge of Aquatic Feeding

The evolution of food using aquatic resources is the result of the human need to obtain food. Since thousands of years ago, ancient communities realized that aquatic environments could be exploited to ensure their survival. This vision not only shaped existing cultures but was also a catalyst for the development of humanity.

Over time, human populations spread across different geographic regions, adapting their diets to local aquatic food sources. This has led to a great diversity of water-related dietary practices (Sicuro 2021).

Fishing, one of the oldest ways of obtaining food, involves capturing fish, crustaceans, molluscs and other aquatic organisms directly from rivers, lakes and oceans. The need to obtain food from the water led to the development of fishing techniques over the centuries, whether with the creation of devices to help capture animals or the manufacture of vessels suitable for fishing (Naylor et al. 2021).

Ancient cultures, such as Mesopotamian and Egyptian civilizations, depended on fishing and aquaculture for their populations' survival. The Tigris and Euphrates rivers in Mesopotamia served as crucial sources of fish, and the Egyptians developed rudimentary systems for farming fish (Costa-Pierce 2022). These ancient practices demonstrate the importance of aquatic resources in human survival. During the Middle Ages, European monasteries innovated in aquaculture, raising carp and trout in their lakes and ponds. In Japan, carp farming evolved as a cultural and gardening tradition, also helping to develop ornamental fish farming (Calixto et al. 2020).

Over time, society gave rise to more advanced ways of exploiting natural resources, through breeding methods. As a response to the demand for food, aquaculture emerged as a practice characterized by the controlled creation of aquatic organisms (Gephart et al. 2020). This activity grew significantly over the centuries, as breeding techniques were improved, especially through the development of feeding practices and the diversification of cultivated organisms. Until the nineteenth century, the feeding of captive aquatic organisms was predominantly based on natural food. With the evolution of aquaculture came the development of commercial feeds, aimed at meeting the nutritional needs of organisms and enhancing the economic viability of aquaculture. This has culminated in the present day, where research and innovation are focused on producing more sustainable aquafeeds (Fig. 2.1). The development of controlled breeding methods has allowed the large-scale production of fish, crustaceans and molluscs, whether in tanks, nurseries and recirculation systems, among others (Garlock et al. 2020).

Since the twentieth century, growing urban populations and increasing demand for protein have driven the development of intensive aquaculture practices and the industrialization of fishing. Technology has advanced, resulting in the formulation of specific feeds for aquaculture and quality monitoring systems. Currently, aquaculture plays a fundamental role in global nutrition, providing proteins to billions of people around the world (Hua et al. 2019).

Therefore, the history of aquatic nutrition is a reflection of the need for survival and the human ability to adapt. Over time, the practice of fishing and aquaculture was recorded in ancient civilizations, whose subsistence depended on these aquatic resources, thus shaping current dietary practices.

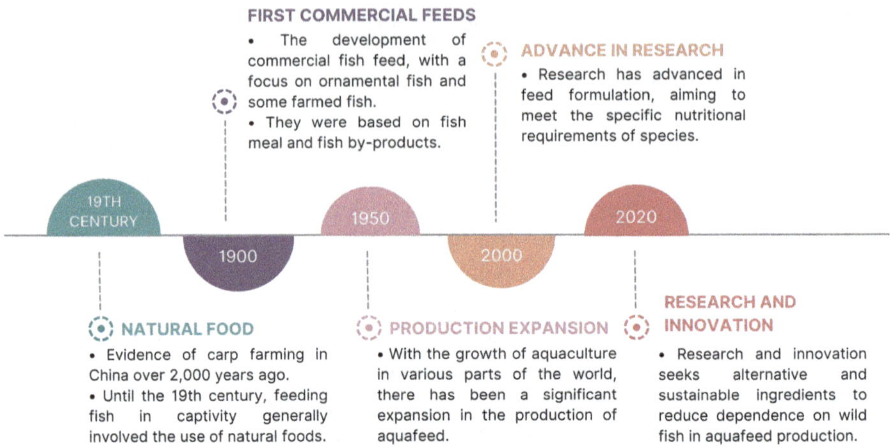

Fig. 2.1 Evolution of aquaculture based on the development of feed, from natural feed to aquafeed and its improvements

2.3 Live Feed

2.3.1 Importance of Live Feed in Aquaculture

Live feed organisms in aquatic ecosystems include both plants (i.e. phytoplankton) and animals (i.e. zooplankton), which are crucial as a feed source for economically important fish and shellfish. Phytoplankton and zooplankton correspond to photosynthetic microorganisms and phagotrophic consumer animals, respectively, which are adapted to spend part or all their lives in suspension in the water column, drifting in water currents (Reynolds 2006). Phytoplankton constitutes the base of the aquatic food chain, acting as primary producers and generally being consumed by zooplankton, which in turn become prey for larger organisms. These organisms are available in the natural environment and are important live feed sources for fish and shellfish larvae. Live feed is important for feeding larvae due to its small size, suitable for the size of their mouths, and the movement of the organisms, which stimulates the larval feeding response and increases attractability, in addition to exogenous enzymes that promote digestion. Furthermore, live feed is nutritionally important, as it contains essential compounds for the development of larvae, such as fatty acids, amino acids, proteins, lipids, carbohydrates, vitamins and minerals (Moraes et al. 2022).

In the natural environment, many aquatic species depend on phytoplankton and zooplankton during the early stages of their life cycle. Therefore, providing these live feeds in aquaculture, mimicking the natural diet, can result in better growth and health of farmed animals (Das et al. 2012). Live feed for aquaculture is typically grown in controlled environments, such as laboratories or hatcheries, under specific conditions to optimize growth and nutritional quality (e.g. proteins and fatty acids). Growing live feed can be challenging, as it is crucial to maintain water quality and nutrition for both the live feed and the target species. Challenges related to the use of live feed in aquaculture include the potential for disease outbreaks and the risk of vectorization of pathogens, the difficulty of maintaining a consistent and high-quality supply, the need for specialized facilities and equipment and the costs involved in maintaining these cultures (Robinson et al. 2005). Additionally, reliance on live feed can limit the scalability of certain aquaculture operations.

In aquaculture, live feed is often used in the early stages of a species' development when they are still too small to consume conventional commercial feeds. As organisms grow, there is a gradual transition to inert diets, starting with a combination of live and inert feed. However, partial or complete replacement of live feed is increasingly sought from incipient larval stages (Ljubobratovic et al. 2020). The latest research is focused on improving the efficiency and sustainability of aquaculture. This includes developing new methods for producing live feed, such as the use of effluents as a culture medium, optimizing productivity and nutritional profiles and exploring alternative feed sources to reduce dependence on live feed, such as using aquafeed.

2.3.2 Live Feed Used in Aquaculture

Microalgae possess a nutrient-rich composition, making them suitable for formulating a wide range of products across various fields, including animal nutrition. These microorganisms have the advantage of rapid growth rates, the capacity to accumulate substantial biomass and the capability to generate biologically active compounds. Numerous species serve as valuable sources of pigments and biomolecules known for their various beneficial properties. These include antibacterial, antiviral, anti-oxidant and immunomodulatory agents that enhance the animal immune system (Tayag et al. 2010; Rizwan et al. 2018; Sandeep et al. 2023). These effects, in turn, have a direct impact on the health, survival and growth performance of aquaculture animals. Furthermore, microalgae play a crucial role in aquaculture by removing nitrogenous compounds, increasing oxygen levels through photosynthesis and stabilizing pH, ultimately enhancing water quality, maintaining system equilibrium and promoting the overall health of aquatic organisms (Karthik et al. 2016; Tang et al. 2020).

Microalgae play a vital role in larval and fry nutrition, either being directly consumed by molluscs and penaeid shrimp or indirectly as a feed source for the live prey provided to small fish larvae (Hemaiswarya et al. 2011). Microalgae species vary according to the production of metabolites, such as proteins, fatty acids, vitamins and pigments, and cell characteristics, such as size, biovolume and tendency to flocculate. The species used as live feed varies according to the habitat, behaviour, eating habits, size and life stage of the farmed animal. Among the genera frequently used are *Chlorella*, *Scenedesmus*, *Ankistrodesmus*, *Arthrospira*, *Pavlova*, *Phaeodactylum*, *Skeletonema*, *Tetraselmis*, *Isochrysis*, *Chaetoceros*, *Nannochloropsis*, *Dunaliella* and *Conticribra* (*Thalassiosira*) (Fig. 2.2). Additionally, the combination of various algal species offers a more balanced

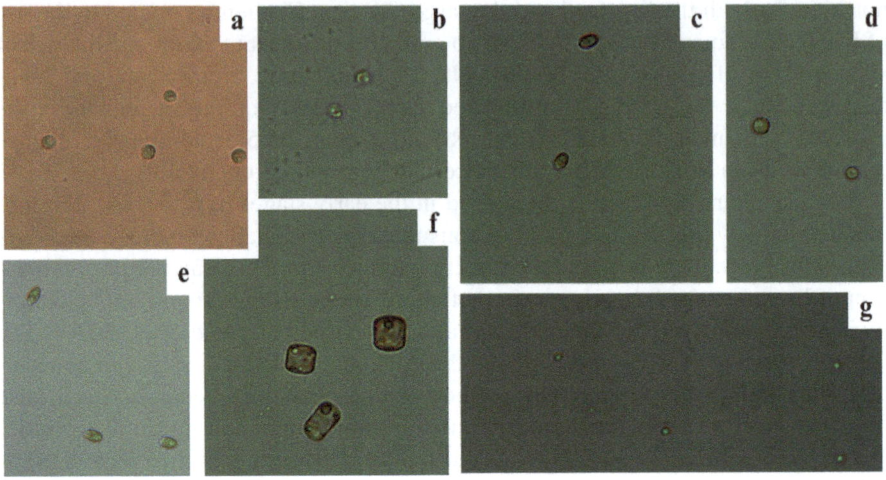

Fig. 2.2 Main genera of microalgae used as live feed in aquaculture: *Chlorella* (**a**), *Chaetoceros* (**b**), *Dunaliella* (**c**), *Isochrysis* (**d**), *Tetraselmis* (**e**), *Conticribra* (*Thalassiosira*) (**f**) and *Nannochloropsis* (**g**)

nutritional profile and enhances animal growth compared to a diet consisting of only a single algal species (Moraes et al. 2022). The main species of microalgae used as live feed for fish, crustaceans and molluscs of economic importance are described in Table 2.1.

The cultivation of microalgae requires specific facilities and equipment, formulated culture media, specialized labour and control of environmental parameters and possible sources of contamination, which results in high production costs (Borowitzka 1992). The scalability of cultivation must take into account, in addition to operating costs, the demand for the cultivated animals, the productive capacity of microalgae (i.e. growth speed, doubling time and maximum cell density), light efficiency, ability to control temperature, carbon transfer, resistance to hydrodynamic stress and ability to maintain axenic culture (Hemaiswarya et al. 2011). Moreover, the fluctuating demand for microalgae, contingent upon broodstock productivity and larval growth, can increase costs, as the microalgae already produced cannot be stored for future use (Moraes et al. 2022). Despite the high cost, live microalgae are essential for the good development of farmed organisms, especially in the initial larval stages, as well as for feeding and enriching zooplanktonic organisms used in the following stages.

Zooplankton have been utilized for the rearing of fry and larvae, especially for species that typically do not readily accept artificial feeds. Zooplanktonic organisms can swim freely in the water column and have high levels of digestive enzymes, which provide constant accessibility, feeding responses and appetite stimulation to fry and larvae. The main groups of zooplankton used as live feed for fish and shellfish are branchiopods (*Artemia* spp.), rotifers (*Brachionus* spp.), cladocerans (*Moina* spp. and *Daphnia* spp.) and copepods (e.g. *Tisbe* spp. and *Oithona* spp.) (Fig. 2.3).

Table 2.1 Species of microalgae and zooplankton used to feed larvae and fry

Aquaculture species	Microalgae species	Zooplankton species	References
Penaeus spp. and *Metapenaeus* spp.	*Chaetoceros* spp., *Tetraselmis* spp., *Isochrysis* spp., *Pavlova* spp., *Phaeodactylum* spp., *Dunaliella* spp., *Skeletonema* spp., *Conticribra* spp.	*Artemia* spp., *Brachionus* spp., *Pseudodiaptomus annandalei*, *Calanopia elliptica*	Jamali et al. (2015), Abbaszadeh et al. (2022), Moraes et al. (2022), Sundarraj et al. (2023)
Macrobrachium rosenbergii	*Chlorella* spp.	*Artemia* spp. *Brachionus* spp., *Daphnia* spp., *Moina* spp., *Oithona* sp., *Apocyclops dengizicus*	Manickam et al. (2020), Rasdi et al. (2021), Tan and Wang (2022), Hosain et al. (2023)
Sparus aurata	*Nannochloropsis gaditana*, *Nannochloropsis oculata*, *Isochrysis galbana*, *Chlorella* sp.	*Artemia* sp., *Brachionus plicatilis*, *Brachionus rotundiformis*	Cañavate and Fernández-Díaz (2001), Suzer et al. (2008), Rizkallah et al. (2023)
Lates calcarifer	*Nannochloropsis oculata*	*Brachionus plicatilis*, *Oithona rígida*, *Cyclopina kasignete*, *Artemia* sp.	Nankervis and Southgate (2006), Santhanam and Perumal (2012), Thépot et al. (2016), Rasdi and Qin (2018), Pham et al. (2023)
Bivalve molluscs (e.g. oysters, mussels, clams and scallops)	*Pavlova* spp., *Isochrysis* spp., *Chaetoceros* spp., *Skeletonema* spp., *Thalassiosira* spp., *Tetraselmis* spp., *Nannochloropsis* spp., *chlorella* spp.	–	Rico-Villa et al. (2006), Cheng et al. (2020)

Microalgae are used to enrich zooplankton, as these organisms may be deficient in nutrients such as vitamins, amino acids and fatty acids (e.g. docosahexaenoic acid (DHA), 22:6n-3, and eicosapentaenoic acid (EPA), 20:5n-3), essential for the development and growth of fish and shellfish. Therefore, the enrichment process is necessary for the mass production of zooplankton under hatchery conditions. This process involves the use of microalgae or fishmeal, canola oil and various oil immersion methods to elevate the levels of polyunsaturated fatty acids (PUFAs) and supply essential amino acids (EAAs) to the cultures (Samat et al. 2020). Enhancing the nutritional quality of rotifers, branchiopods, copepods and cladocerans via algal enrichment techniques is a common practice aimed at improving the quality of these feeds, which are typically deficient in nutrients. The high cost and challenges associated with producing, concentrating and storing live microalgae have spurred significant research efforts towards developing alternative forms of microalgae. Hence,

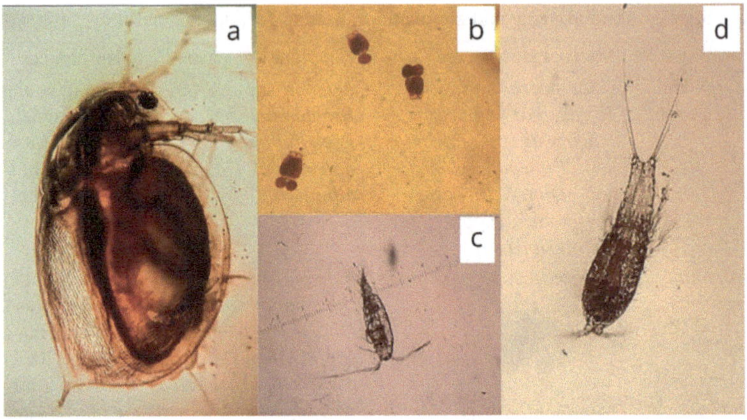

Fig. 2.3 Main groups of zooplankton used as live feed in aquaculture: Cladocerans, *Daphnia* sp. (**a**); rotifers, *Brachionus* sp. (**b**); copepods, *Calanus* sp. (**c**); and *Tisbe* sp. (**d**)

there has been an increasing trend in employing processed microalgae, such as paste or powder, as a more cost-effective substitute for live microalgae (Raja et al. 2018).

Zooplanktonic organisms enriched with microalgae are valuable sources of proteins, essential amino acids, lipids, essential fatty acids, minerals and enzymes for fry and larvae. The use of these organisms can lead to improved survival, enhanced reproduction of broodstock, accelerated growth, immune stimulation and enhanced colouration in prawns and fish (Manickam et al. 2020). Copepods are regarded as the nutritional standard for a broad spectrum of marine fish larvae. Nevertheless, *Artemia* and rotifers are frequently preferred as initial feeds due to their ease of cultivation, even though they offer lower nutritional value than copepods. Consequently, *Artemia*, rotifers and other live feeds are often enriched with microalgae or nutrients such as fatty acids, vitamins, minerals and probiotics to mimic the copepod's essential nutrient levels, thereby enhancing the rearing success of fish and crustacean larvae (Samat et al. 2020).

In addition to phytoplankton and zooplankton organisms, oligochaete worms (e.g. *Tubifex tubifex*, *Limnodrilus* species, *Branchiura sowerbyi* and *Lumbriculus variegatus*) and polychaetes (e.g. *Glycera dibranchiata*, *Diopatra aciculata*, *Nereis* (*Alitta*) *virens*, *Marphysa sanguinea* and *Arenicola defodiens*) are used as live food in aquaculture. These worms serve as supplementary sources of amino acids and fatty acids, particularly PUFA, for fish and shellfish, potentially enhancing growth rates and survival (Das et al. 2012). Polychaetes are considered indispensable for shrimp maturation due to their fatty acid profile and the presence of hormonally active substances (Kannappan et al. 2019). Besides being used as live food, worms are employed as partial substitutes for fishmeal in aquafeeds.

2.4 Aquafeed

2.4.1 Importance of Aquafeed in the Development of Aquaculture

Aquafeed refers to feeds specially formulated and produced to feed aquatic organisms, such as fish, shellfish and other animals raised in aquatic farming systems. Its main function is to meet the nutritional needs of these organisms, in a specific and personalized way for each species, ensuring not only healthy growth but also the quality and safety of aquatic products intended for human consumption, thus contributing to the sustainability and development of the food industry. Unlike natural feeding and the provision of live feed, this technology has a controlled and optimized composition in the animals' diet, with selected ingredients, including proteins, lipids, carbohydrates, vitamins and minerals, formulated to meet the specific needs of the organisms in cultivation. Furthermore, it is produced under controlled conditions, guaranteeing quality, periodicity and food safety (Salin et al. 2018).

As the population and global demand for food grow, aquaculture is becoming a crucial source of food and human nutrition. Aquafeed plays a fundamental role in this scenario, as it allows the efficient creation of aquatic organisms in intensive cultivation systems through a balanced feed that helps or solves problems linked to these systems. Proper use of aquafeed has a significant impact on the quality and sustainability of the production of aquatic organisms, as it helps to maximize growth, improve animal health, minimize waste generation and reduce pressure on natural stocks (Ghamkhar and Hicks 2020). The quality of ingredients and balanced nutrition offered by aquafeeds also have a direct impact on the quality of products, such as fish and seafood, ensuring they are safe and healthy for human consumption.

Enhancing the environmental and economic sustainability of aquafeeds can be achieved through two primary methods: (a) maximizing the efficient utilization of finite resources such as fishmeal and fish oil and (b) reducing waste production and emissions. Numerous ingredients have been previously suggested, examined and incorporated to implement these strategies while maintaining satisfactory feed production efficiencies (Salin et al. 2018). Therefore, the appropriate use of ingredients in aquafeed is essential for promoting sustainability and meeting the growing demand for seafood around the world.

2.4.2 Nutritional Components in Aquafeed

In the cultivation of different aquatic species, customizing the dietary offerings to align with the specific nutritional requirements of each species is of paramount significance, going beyond merely ensuring the diet's quality. A comprehensive aquatic organism diet should encompass essential constituents: proteins, amino acids, carbohydrates, lipids, fatty acids, vitamins and minerals. Each of these elements assumes unique and pivotal roles in aquatic organism nutrition, with direct implications for their growth, health and reproductive capabilities.

Considered fundamental in aquatic diets, proteins are essential for the growth and health of animals, as they provide the amino acids necessary for the synthesis of muscle tissues, organs and enzymes, in addition to playing a crucial role in strengthening the immune system. Adequate intake of these macromolecules helps in the repair and ongoing maintenance of body tissues; they are involved in the production of antibodies and the regulation of the immune system. A diet rich in proteins helps strengthen the immune response of aquatic organisms, making them more resistant to infections and diseases (Masagounder et al. 2016).

Protein stands out as the costliest high-quality nutrient among various nutrients, primarily determined by the amino acid composition of the ingredients. It is imperative for the diet to precisely meet the animal's amino acid requirements to prevent any decline in performance. Equally essential is the avoidance of amino acid over-formulation, as it would only result in escalated feed costs. Consequently, the formulation of diets must be precise regarding the amino acid profile of the raw materials used, ensuring an optimal balance of essential amino acids and avoiding both excess and deficiency (Masagounder et al. 2016).

The quantity and quality of proteins in the fish diet directly influence the texture, flavour and nutritional value of the meat. A diet rich in high-quality protein results in healthier fish, which tends to result in better-quality meat. Therefore, aquaculture systems must be designed to provide a balanced diet that meets the protein needs of the animals. Furthermore, protein balance can also affect the feed efficiency of farmed organisms, which not only maximizes meat production but also reduces costs and feed waste in aquaculture.

The protein sources used in the aquafeed formulation may vary depending on several factors, such as regional availability, species cultivated and cost of obtaining, among others. One of the most used ingredients is still fishmeal, but its use raises concerns about overfishing and the sustainability of marine resources, which is why there is a lot of research aimed at replacing this ingredient. Choosing sustainable protein sources in aquafeed can reduce the environmental impact of aquaculture. Therefore, aquaculture feeding has evolved significantly in the search for alternative sources of protein, such as insect meat, algae meal, grains and vegetable proteins. Replacing fishmeal contributes to the conservation of marine resources and improves production sustainability (Annamalai et al. 2021; Zarei et al. 2022).

Insects, such as *Hermetia illucens* and *Tenebrio molitor*, are rich in proteins and can be cultivated using organic waste, reducing environmental impact and making it a sustainable resource. Likewise, macroalgae and microalgae flours can be produced following the principles of sustainability, especially when cultivated in effluents. These organisms act as bio-remediators in effluents with high loads of nitrogenous and phosphate compounds, enabling the reuse of water and the production of biomass rich in proteins and other nutrients, being applied, especially, to feed herbivorous species (Haoujar et al. 2022). This process is illustrated in Fig. 2.4, where based on the concept of circular economy, fish farming effluents are used for the cultivation of microalgae. Subsequently, the processed biomass will be used in

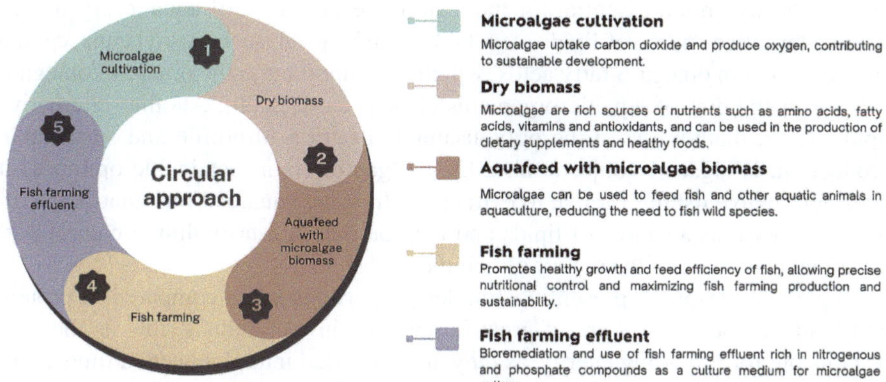

Fig. 2.4 Circular approach to the use of aquafeed with the addition of microalgae biomass grown in aquaculture effluent

the formulation of aquafeed, which will be offered to the farmed fish. In addition, several vegetable proteins, such as soy, corn and wheat, are used in the aquafeed formulation. However, its use in carnivorous fish diets has been limited due to challenges with digestibility and amino acid profile, which may require supplementation to ensure balance and meet adequate nutritional needs for the target species (Zarei et al. 2022).

Not the least, lipids are essential components of cell membranes and a long-term energy source for aquatic organisms, providing more calories compared to proteins and carbohydrates (Hossain et al. 2023). Organisms use this energy for activities such as swimming, digestion and maintaining vital functions, making lipids a fundamental component in aquafeed. Lipids are also precursors of hormones, including sex hormones and hormones involved in regulating growth and metabolism, making them important in the reproduction and sexual development of aquatic organisms. Furthermore, stored lipids can be mobilized to provide energy during periods without a feed supply (Torsabo et al. 2022). In addition to their nutritional importance, lipids can make feeds more palatable to aquatic organisms, which encourages adequate nutrition. In parallel, the adequate inclusion of lipids in diets can improve feed digestibility, ensuring that nutrients are absorbed efficiently.

Furthermore, the fatty acid profile is of great importance, and it is essential to balance the relationship between omega-3 and omega-6 fatty acids to promote the health and adequate growth of aquatic organisms. Consequently, adequate inclusion of fatty acids in aquaculture feed diets is necessary to promote healthy growth, successful reproduction and resistance to adverse conditions such as low temperatures. Therefore, when formulating aquatic feed, the quantity and quality of lipids must be considered to meet the specific needs of the species and growth phases (Hodar et al. 2020).

The lipid source commonly used in aquafeed is fish oil; however, like fishmeal, it can represent an unsustainable source for the industry. Fish oil is rich in omega-3 fatty acids, which are essential for the healthy development of aquatic organisms. As an alternative source of lipids, vegetables can be used, such as soybeans, canola and corn, rich in omega-6 fatty acids, which are important in the omega-3/omega-6 balance in the diet of aquatic organisms (Hodar et al. 2020). Another alternative lipid source, mainly due to the polyunsaturated fatty acid profile and sustainable production, is algae (Haoujar et al. 2022). Algae offers a sustainable option as it does not involve fishing for fish to extract oil. In some aquafeed formulations, krill oil is also used as a source of lipids and may be rich in astaxanthin, a natural pigment that can improve the colour of fish meat.

In parallel, despite representing smaller proportions when compared to protein and lipids, carbohydrates are of great importance in the animals' diet as they are responsible for providing energy, as they are converted into glycogen, a fundamental form of energy reserve for stressful situations. Furthermore, adequate inclusion of carbohydrates in the diet allows proteins to be spared for essential functions, such as the growth and maintenance of body tissues. When a sufficient supply of carbohydrates is present in the diet, organisms can forego utilizing proteins as an energy source, resulting in greater nutritional efficiency. On the other hand, the high digestibility of carbohydrates helps in the absorption of energy, contributing to feed efficiency and the good performance of organisms. In addition, to avoid health problems in animals, adequate proportions of carbohydrates can help control the variation in blood glucose in organisms (Zhang et al. 2022).

As for micronutrients, vitamins and minerals are vital for the general health of aquatic organisms. Vitamins perform specific functions in regulating metabolic processes, synthesizing hormones and maintaining the immune system. On the other hand, minerals are essential for bone development, nerve function and the maintenance of the osmotic balance of organisms. Vitamins and minerals impact the reproduction of organisms, vitamins work to improve fertility and egg health, while minerals aid in the production of healthy eggs and sperm (Liu et al. 2022a). Furthermore, both help aquatic organisms adapt to changes in environmental conditions, such as fluctuations in temperature or water quality. Vitamin and mineral deficiencies can lead to health problems such as deformities, bone fragility and susceptibility to infections, as well as inadequate growth (Masagounder et al. 2016).

Considering that the aquafeed formulation seeks to meet the basic nutritional requirements of organisms, it is necessary to consider, in addition to the species, the growth phases. Fish and shrimp in the early stages, for example, need feed with a higher protein content to promote rapid growth, while adults may need feed with different proportions of nutrients to maintain health and reproduction (Lemos and Weissman 2021). In addition, customizing feed based on the individual needs of aquatic organisms is important to optimize performance and minimize feed waste. This approach considers factors such as size, age, health status and growth rate. Adaptive feeding is a strategy that adjusts the amount and type of feed provided based on environmental conditions and the needs of organisms, reducing feed waste and improving feed efficiency (Prabu et al. 2017).

Another important strategy in modern aquaculture to optimize production and reduce waste generation is automated feeding. Through the use of automated systems, the amount and timing of feeding can be precisely controlled, ensuring that cultured organisms receive the right amount of feed, reducing overfeeding and minimizing water pollution (Wu et al. 2022).

Therefore, aquafeed formulation must involve choosing ingredients to meet nutritional requirements while considering issues of economic viability and sustainability. Therefore, the use of alternative and functional ingredients that offer macro- and micronutrients suitable for the species and growth phases, promotes, in addition to health, growth, resistance and productive capacity, thereby leading to a more sustainable aquaculture (Chen et al. 2021). Hence, the search for more sustainable alternatives is essential to guarantee a sustainable future for aquaculture.

Aquafeed processing technologies enable feed customization, allowing strategic adaptation to the preferences of the cultivated species and the specificities of the cultivation systems. Two of the most common methods of aquafeed production are extrusion and pelletization, each with its own characteristics and impacts on feed quality and digestibility (Musa et al. 2023).

Extrusion is a process that involves passing a mixture of ingredients through a heated cylinder and a die with a specific hole. The pressure, heat and moisture generated during the process turn the mixture into a semi-cooked mass that is pushed through the hole. This technology can inactivate antinutrients, which block or interfere with the way the body absorbs nutrients, improve the gelatinization of starches and proteins and create a uniform texture in the final granules. Therefore, extruded feed is more easily digested by animals, improving the use of nutrients and growth (Musa et al. 2023).

In contrast, pelletizing involves compressing the ingredient mixture into a matrix with holes. The combination of pressure and heat causes the formation of pellets that are cut into uniform sizes. Pelleting is known to increase the durability of feed, making it easier to handle and store (Zettl et al. 2019). However, this process is not as effective in inactivating antinutrients as extrusion and, thus, may require the addition of enzymes or additional treatments to improve the digestibility of pelleted feed (Riaz 2023).

The choice between extrusion and pelletization depends on the needs of the species and growing conditions. Extrusion is generally preferred for carnivorous species, as it results in more digestible feeds, where nutrients are better released and assimilated by carnivorous fish (Musa et al. 2023; Riaz 2023). On the other hand, pelleting may be the most appropriate choice for herbivorous species or when feed durability is a priority. Pellets produced by pelletizing are known for their physical resistance, making them ideal for organisms that prefer to graze or grind feed. Furthermore, pelletization facilitates the handling and storage of feed in different conditions (Zettl et al. 2019).

2.4.3 Active Feed

Active feed is an advanced concept in aquaculture that aims to improve various aspects of the cultivation of aquatic organisms, from their immunity to fillet quality, through bioactive compounds. These compounds may have anti-oxidant, anti-inflammatory, antibacterial, antiviral and immunological properties, among others. The active feed includes bioactive ingredients that strengthen the immune system, making it more resistant to diseases (Banerjee and Ray 2019; Romanova et al. 2022). In addition to immunity, they can influence the growth of animals, through healthy growth in a shorter period, increasing the productivity and sustainability of aquaculture operations.

Furthermore, this selected diet can also influence the appearance and colour of fish and shellfish. Active feeds formulated with ingredients rich in carotenoids can improve the colour of fish raised in aquaculture, especially applied to commercially valuable species, where colour directly influences market value. Many species of fish, such as salmon, obtain their characteristic colour from carotenoids in their diet. This technology is also widely used in the ornamental fish market, where colour is one of the value-adding factors (Jiang et al. 2019).

The use of natural bioactive, such as those from microalgae and bacteria, is preferred over synthetic compounds, as in addition to being more efficient, they contribute to sustainable development through the application of renewable resources. Among the bioactives derived from microalgae, carotenoids such as β-carotene, lutein and astaxanthin stand out, possessing anti-oxidant, anti-inflammatory and immunostimulant properties. These compounds enhance the health and growth of farmed organisms while contributing to skin or muscle pigmentation (Harith et al. 2024). Alternatively, there is the utilization of probiotics, which involve live microorganisms or components of bacteria that act through various mechanisms to confer health benefits to the host or its environment (Wang et al. 2019). The possible modes of action include the production of inhibitory compounds, competition for available chemicals or energy, competition for adhesion sites, enhancement of the immune response, improvement of water quality, interaction with phytoplankton; a source of macro- and micronutrients and enzymatic contribution to digestion (Verschuere et al. 2000). While probiotic research in aquaculture has traditionally focused on fish juveniles, more attention has recently been directed towards larvae of fish and shellfish (Verschuere et al. 2000).

On the other hand, bioactive compounds can influence the flavour, texture and nutritional profile of consumer fish and shellfish (Banerjee and Ray 2019). Fillet quality is an important aspect of aquaculture, especially in meeting consumer demands for healthy, high-quality products. Using active feed can improve the quality of the fillet, making it tastier and with a better texture. Furthermore, it results in healthier and more nutritious feed for consumers, through the transfer of beneficial compounds incorporated into the meat. Some studies report positive results linked to the addition of natural bioactive compounds to the feed of different species of fish and shrimp, as shown in Table 2.2.

Table 2.2 Responses of natural bioactive compounds in active feed for fish and shrimp

Species of fish and shrimp	Bioactive compounds	Responses	References
Litopenaeus vannamei	*Arthrospira platensis* proteins and pigments	Replacing up to 75% of fishmeal with *A. platensis* does not affect survival or growth, in addition to improving immunological parameters	Macias-Sancho et al. (2014)
Pelteobagrus fulvidraco	Lutein from *Arthrospira platensis*	Activates the anti-oxidant response, as well as alleviates oxidative stress and pigmentation disorder induced by exposure to air	Liu et al. (2022b)
Oreochromis sp.	Astaxanthin from *Haematococcus pluvialis*	Increase in weight gain and protein efficiency index, in addition to better skin pigmentation	Harith et al. (2024)
Pseudochromis fridmani	Astaxanthin from *Haematococcus pluvialis*	Changes the colour of fish away from a light, dull, violet colour to a dark, intense, magenta/raspberry colour	Jiang et al. (2019)
Salmo salar	*Bacillus velezensis* and *Rhodotorula mucilaginosa* probiotic compound	Beneficial effects as a replacement for antibiotics in terms of growth performance, non-specific immune responses and resistance to diseases caused by *Aeromonas salmonicida*	Wang et al. (2019)
Oreochromis niloticus	*Enterococcus faecium* probiotic compound	Improves growth and stimulation of the immune system, protecting against *Aeromonas hydrophila* infection	Tachibana et al. (2020)

2.5 Conclusion

The aquaculture industry relies on essential live feed production and aquafeed manufacturing procedures that allow for the mass cultivation of fish and shellfish in controlled environments. Although several methods are being explored to replace the use of live feed in aquaculture, it is crucial to highlight its indispensable role for larvae and fry. This importance arises from the specific nutritional needs and feeding behaviours of larvae and fry. However, the use of live feed entails considerable maintenance costs, significant investment in specialized facilities, the need for skilled labour, limited practicality and the potential risk of pathogen transmission. Therefore, the most viable approach lies in dietary supplementation with aquafeed. Aquafeed plays a fundamental role in promoting sustainable aquaculture; as the demand for food increases, aquaculture stands out in the supply of seafood. The future of aquafeed is focused on promoting food sustainability in multiple ways, aligned with global conservation and sustainability goals. Traceability is also an important innovation, allowing consumers to have greater confidence in the origin and quality of aquaculture products. Furthermore, increased research and development of alternative and lower-impact ingredients, such as algae and insects, is contributing to more sustainable aquafeeds, reducing pressure on marine resources and decreasing the ecological footprint of aquaculture. Therefore, to promote sustainable aquaculture, aquafeed needs to be developed based on environmentally responsible practices.

Acknowledgement This work was supported by the National Council for Scientific and Technological Development (CNPq) (Grant Code: PQ310898/2023-4).

References

Abbaszadeh A, Mozanzadeh MT, Qasemi A, Oujifard A, Nafisi Bahabadi M (2022) Effects of the addition of *Calanopia elliptica*, *Artemia franciscana*, and *Brachionus rotundiformis* in a nursery biofloc system on water quality, growth, gut morphology, health indices, and transcriptional response of immune and antioxidant-related genes in *Penaeus vannamei*. Aquac Int 30(2):653–676. https://doi.org/10.1007/s10499-021-00823-1

Annamalai SN, Das P, Thaher MI, Abdul Quadir M, Khan S, Mahata C, Al Jabri H (2021) Nutrients and energy digestibility of microalgal biomass for fish feed applications. Sustain For 13(23):13211. https://doi.org/10.3390/su132313211

Banerjee P, Ray DP (2019) Functional food: a brief overview. Int J Bioresour Sci 6(2):57–60. https://doi.org/10.30954/2347-9655.02.2019.2

Borowitzka MA (1992) Algal biotechnology products and processes—matching science and economics. J Appl Phycol 4:267–279. https://doi.org/10.1007/BF02161212

Boyd CE, McNevin AA, Davis RP (2022) The contribution of fisheries and aquaculture to the global protein supply. Food Secur 14(3):805–827. https://doi.org/10.1007/s12571-021-01246-9

Calixto ES, Santos DFB, Lange D, Galdiano MS, Rahman IU (2020) Aquaculture in Brazil and worldwide: overview and perspectives. J Environ Anal Prog 5(1):098–107. https://doi.org/10.24221/jeap.5.1.2020.2753.098-107

Cañavate JP, Fernández-Díaz C (2001) Pilot evaluation of freeze-dried microalgae in the mass rearing of gilthead seabream (*Sparus aurata*) larvae. Aquaculture 193(3–4):257–269. https://doi.org/10.1016/S0044-8486(00)00492-0

Chen X, Yi H, Liu S, Zhang Y, Su Y, Liu X, Li G (2021) Promotion of pellet-feed feeding in mandarin fish (*Siniperca chuatsi*) by *Bdellovibrio bacteriovorus* is influenced by immune and intestinal flora. Aquaculture 542:736864. https://doi.org/10.1016/j.aquaculture.2021.736864

Cheng P, Zhou C, Chu R, Chang T, Xu J, Ruan R, Chen P, Yan X (2020) Effect of microalgae diet and culture system on the rearing of bivalve mollusks: nutritional properties and potential cost improvements. Algal Res 51:102076. https://doi.org/10.1016/j.algal.2020.102076

Costa-Pierce BA (2022) The anthropology of aquaculture. Front Sustain Food Syst 6:843743. https://doi.org/10.3389/fsufs.2022.843743

Das P, Mandal SC, Bhagabati SK, Akhtar MS, Singh SK (2012) Important live food organisms and their role in aquaculture. In: Frontiers in aquaculture. Narendra Publishing House, Rohini, DL, pp 69–86

Garlock T, Asche F, Anderson J, Bjørndal T, Kumar G, Lorenzen K, Tveterås R (2020) A global blue revolution: aquaculture growth across regions, species, and countries. Rev Fish Sci Aquac 28(1):107–116. https://doi.org/10.1080/23308249.2019.1678111

Garlock, T, Asche, F, Anderson, J, Ceballos-Concha, A, Love, D. C, Osmundsen, TC, Pincinato, RBM. (2022). Aquaculture: the missing contributor in the food security agenda. Glob Food Sec 32:100620. https://doi.org/10.1080/23308249.2019.1678111

Gephart JA, Golden CD, Asche F, Belton B, Brugere C, Froehlich HE, Allison EH (2020) Scenarios for global aquaculture and its role in human nutrition. Rev Fish Sci Aquac 29(1):122–138. https://doi.org/10.1080/23308249.2020.1782342

Ghamkhar R, Hicks A (2020) Comparative environmental impact assessment of aquafeed production: sustainability implications of forage fish meal and oil free diets. Resour Conserv Recycl 161:104849. https://doi.org/10.1016/j.resconrec.2020.104849

Haoujar I, Haoujar M, Altemimi AB, Essafi A, Cacciola F (2022) Nutritional, sustainable source of aqua feed and food from microalgae: a mini review. Int Aquat Res 14:3. https://doi.org/10.22034/IAR.2022.1958713.1278

Harith ZT, Sukri SM, Remlee NF, Sabir FN, Zakaria NN (2024) Effects of dietary astaxanthin enrichment on enhancing the colour and growth of red tilapia, *Oreochromis* sp. Aquac Fish 9(1):52–56. https://doi.org/10.1016/j.aaf.2022.06.001

Hemaiswarya S, Raja R, Ravi Kumar R, Ganesan V, Anbazhagan C (2011) Microalgae: a sustainable feed source for aquaculture. World J Microbiol Biotechnol 27:1737–1746. https://doi.org/10.1007/s11274-010-0632-z

Hodar AR, Vasava RJ, Mahavadiya DR, Joshi NH (2020) Fish meal and fish oil replacement for aqua feed formulation by using alternative sources: a review. J Exp Zool India 23(1):13–21

Hosain ME, Amin SN, Arshad A, Kamarudin MS, Karim M, Shohaimi S, Romano N (2023) Effect of copepod (*Apocyclops dengizicus*) addition on growth, survival and proximate composition of *Macrobrachium rosenbergii* postlarvae cultured in brackish water biofloc system. N Am J Aquac. https://doi.org/10.1002/naaq.10308

Hossain MS, Small BC, Hardy R (2023) Insect lipid in fish nutrition: recent knowledge and future application in aquaculture. Rev Aquac 15(4):1664–1685. https://doi.org/10.1111/raq.12810

Hua K, Cobcroft JM, Cole A, Condon K, Jerry DR, Mangott A, Strugnell JM (2019) The future of aquatic protein: implications for protein sources in aquaculture diets. One Earth 1(3):316–329. https://doi.org/10.1016/j.oneear.2019.10.018

Jamali H, Imani A, Abdollahi D, Roozbehfar R, Isari A (2015) Use of probiotic Bacillus spp. in rotifer (*Brachionus plicatilis*) and Artemia (*Artemia urmiana*) enrichment: effects on growth and survival of Pacific white shrimp, *Litopenaeus vannamei*, Larvae. Probiotics Antimicrob. Proteins 7(2):118–125. https://doi.org/10.1007/s12602-015-9189-3

Jiang J, Nuez-Ortin W, Angell A, Zeng C, de Nys R, Vucko MJ (2019) Enhancing the colouration of the marine ornamental fish *Pseudochromis fridmani* using natural and synthetic sources of astaxanthin. Algal Res 42:101596. https://doi.org/10.1016/j.algal.2019.101596

Kannappan S, Sudheer NS, Aravind R, Balasubramanian CP (2019) Polychaete worms as maturation feed for shrimps. In: Balasubramanian CP, Jayanthi M, Vinay TN, Biju IF (eds) Hands-on training program on shrimp and mud-crab aquaculture, training manual series, n 18, pp 91–97

Karthik R, Thamizharasan K, Sankari D, Ram Cherma Kadiravan R, Ashwitha A (2016) Biochemical profile of shrimp larvae fed with five different micro algae and enriched Artemia salina under laboratory conditions. Int J Fish Aquat 4(4):376–379

Lemos D, Weissman D (2021) Moulting in the grow-out of farmed shrimp: a review. Rev Aquac 13(1):5–17. https://doi.org/10.1111/raq.12461

Liu A, Santigosa E, Dumas A, Hernandez JM (2022a) Vitamin nutrition in salmonid aquaculture: from avoiding deficiencies to enhancing functionalities. Aquaculture 561:738654. https://doi.org/10.1016/j.aquaculture.2022.738654

Liu C, Liu H, Zhu X, Han D, Jin J, Yang Y, Xie S (2022b) The effects of dietary *Arthrospira platensis* on oxidative stress response and pigmentation in yellow catfish *Pelteobagrus fulvidraco*. Antioxidants 11(6):1100. https://doi.org/10.3390/antiox11061100

Ljubobratovic U, Kosanovic D, Demeny FZ, Krajcsovics A, Vukotic G, Stanisavljevic N, Lukic J (2020) The effect of live and inert feed treatment with lactobacilli on weaning success in intensively reared pike-perch larvae. Aquaculture 516:734608. https://doi.org/10.1016/j.aquaculture.2019.734608

Macias-Sancho J, Poersch LH, Bauer W, Romano LA, Wasielesky W, Tesser MB (2014) Fishmeal substitution with Arthrospira (*Spirulina platensis*) in a practical diet for Litopenaeus vannamei: effects on growth and immunological parameters. Aquaculture 426:120–125. https://doi.org/10.1016/j.aquaculture.2014.01.028

Manickam N, Bhavan PS, Santhanam P, Muralisankar T (2020) Influence of wild mixed zooplankton on growth and muscle biochemical composition of the freshwater prawn

Macrobrachium rosenbergii post larvae. Aquaculture 522:735110. https://doi.org/10.1016/j.aquaculture.2020.735110

Masagounder K, Ramos S, Reimann I, Channarayapatna G (2016) Optimizing nutritional quality of aquafeeds. In: Aquafeed formulation. Academic Press, pp 239–264. https://doi.org/10.1016/B978-0-12-800873-7.00006-3

Moraes LBS, Santos RFB, Gonçalves Junior GF, Mota GCP, Dantas DMDM, Souza Bezerra R, Olivera Gálvez A (2022) Microalgae for feeding of penaeid shrimp larvae: an overview. Aquac Int 30(3):1295–1313. https://doi.org/10.1007/s10499-022-00857-z

Musa S, Aura CM, Tomasson T, Sigurgeirsson Ó, Thorarensen H (2023) A comparative study of the effects of pelleted and extruded feed on growth, financial revenue and nutrient loading of Nile tilapia (*Oreochromis niloticus L.*) cage culture in a lacustrine environment. J Appl Aquac 35(3):633–655. https://doi.org/10.1080/10454438.2021.2011528

Nankervis L, Southgate PC (2006) An integrated assessment of gross marine protein sources used in formulated microbound diets for barramundi (*Lates calcarifer*) larvae. Aquaculture 257(1–4):453–464. https://doi.org/10.1016/j.aquaculture.2006.03.006

Naylor RL, Kishore A, Sumaila UR, Issifu I, Hunter BP, Belton B, Crona B (2021) Blue food demand across geographic and temporal scales. Nat Commun 12(1):5413. https://doi.org/10.1038/s41467-021-25516-4

Pham HD, Siddik MA, Rahman MA, Huynh LT, Nahar A, Vatsos IN (2023) Effects of n-3 HUFA-enriched *Artemia* on growth, biochemical response, skeletal morphology and stress resistance of Asian sea bass (*Lates calcarifer*) larvae reared at high temperature. Aquaculture 574:739732. https://doi.org/10.1016/j.aquaculture.2023.739732

Prabu E, Felix S, Felix N, Ahilan B, Ruby P (2017) An overview on significance of fish nutrition in aquaculture industry. Int J Fish Aquat Stud 5(6):349–355

Raja R, Coelho A, Hemaiswarya S, Kumar P, Carvalho IS, Alagarsamy A (2018) Applications of microalgal paste and powder as food and feed: An update using text mining tool. Beni-Suef Univ J Basic Appl Sci 7(4):740–747. https://doi.org/10.1016/j.bjbas.2018.10.004

Rasdi NW, Qin JG (2018) Copepod supplementation as a live food improved growth and survival of Asian seabass *Lates calcarifer* larvae. Aquac Res 49(11):3606–3613. https://doi.org/10.1111/are.13828

Rasdi NW, Ikhwannuddin M, Syafika CA, Azani N, Ramli A (2021) Effects of using enriched copepod with microalgae on growth, survival, and proximate composition of giant freshwater prawn (*Macrobrachium rosenbergii*). Iran J Fish Sci 20(4):986–1003

Reynolds CS (2006) In: Hudnell KH (ed) The ecology of phytoplankton, vol 619. Cambridge University Press, Stewart

Riaz MN (2023) Extrusion of aquafeeds formulated with alternative protein sources. In: Alternative protein sources in aquaculture diets. CRC Press, Boca Ratón, FL, pp 523–546. https://doi.org/10.1201/9781003421214

Rico-Villa B, Le Coz JR, Mingant C, Robert R (2006) Influence of phytoplankton diet mixtures on microalgae consumption, larval development and settlement of the Pacific oyster *Crassostrea gigas* (Thunberg). Aquaculture 256(1–4):377–388. https://doi.org/10.1016/j.aquaculture.2006.02.015

Rizkallah IR, Ali KAE, Mohamed MA, Said MM (2023) Evaluation of different regimes for seabream (*Sparus aurata*) larval rearing. J Anim Poult Fish Prod 12(1):7–14. https://doi.org/10.21608/JAPFP.2023.337278

Rizwan M, Mujtaba G, Memon SA, Lee K, Rashid N (2018) Exploring the potential of microalgae for new biotechnology applications and beyond: a review. Renew Sust Energ Rev 92:394–404. https://doi.org/10.1016/j.rser.2018.04.034

Robinson CB, Samocha TM, Fox JM, Gandy RL, McKee DA (2005) The use of inert artificial commercial food sources as replacements of traditional live food items in the culture of larval shrimp, *Farfantepenaeus aztecus*. Aquaculture 245(1–4):135–147. https://doi.org/10.1016/j.aquaculture.2004.11.051

Romanova E, Romanov V, Shadyeva L, Lubomirova V, Shlenkina T, Vasiliev A, Turaeva E (2022) Functional biologically active feed additive for breeding stock of fish. In: E3S Web of Conferences, Les Ulis, 363. https://doi.org/10.1051/e3sconf/202236303060

Salin KR, Arun VV, Mohanakumaran Nair C, Tidwell JH (2018) Sustainable aquafeed. In: Sustainable aquaculture. Applied environmental science and engineering for a sustainable future. Springer, Cham, pp 123–151. https://doi.org/10.1007/978-3-319-73257-2_4

Samat NA, Yusoff FM, Rasdi NW, Karim M (2020) Enhancement of live food nutritional status with essential nutrients for improving aquatic animal health: a review. Animals 10(12):2457. https://doi.org/10.3390/ani10122457

Sandeep KP, Sivaramakrishnan T, Sudhin S, Raymond JAJ, Sudheer NS, Raja RA, Ambasankar K (2023) Influence of dietary microalgal concentrates on growth, survival and health status of *Penaeus vannamei*. Aquac Int:1–21. https://doi.org/10.1007/s10499-023-01114-7

Santhanam P, Perumal P (2012) Evaluation of the marine copepod *Oithona rigida Giesbrecht* as live feed for larviculture of Asian seabass *Lates calcarifer Bloch* with special reference to nutritional value. Indian J Fish 59(2):127–134

Sicuro B (2021) World aquaculture diversity: origins and perspectives. Rev Aquac 13(3):1619–1634. https://doi.org/10.1111/raq.12537

Sumaila UR, Tai TC (2019) Ending overfishing can mitigate impacts of climate change. Institute for the Oceans and Fisheries, The University of British Columbia, Working Paper Series 5

Sundarraj DK, Karuppaiya ND, Nagarajan K, Ayyanar SD, Selvakumaran J, Meril D, Pachiappan P (2023) Evaluation of suitability of P-deficient medium-cultured microalga and copepod as an alternative live feed for Pacific whiteleg shrimp *Penaeus vannamei* post-larvae. Biomass Convers Biorefinery:1–15. https://doi.org/10.1007/s13399-023-03887-6

Suzer C, Çoban D, Kamaci HO, Saka Ş, Firat K, Otgucuoğlu Ö, Küçüksari H (2008) *Lactobacillus* spp. bacteria as probiotics in gilthead sea bream (*Sparus aurata*, L.) larvae: effects on growth performance and digestive enzyme activities. Aquaculture 280(1–4):140–145. https://doi.org/10.1016/j.aquaculture.2008.04.020

Tachibana L, Telli GS, de Carla Dias D, Goncalves GS, Ishikawa CM, Cavalcante RB, Ranzani-Paiva MJT (2020) Effect of feeding strategy of probiotic enterococcus faecium on growth performance, hematologic, biochemical parameters and non-specific immune response of Nile tilapia. Aquac Rep 16:100277. https://doi.org/10.1016/j.aqrep.2020.100277

Tacon AG (2020) Trends in global aquaculture and aquafeed production: 2000–2017. Rev Fish Sci Aquac 28(1):43–56. https://doi.org/10.1080/23308249.2019.1649634

Tan K, Wang W (2022) The early life culture and gonadal development of giant freshwater prawn, *Macrobrachium rosenbergii*: a review. Aquaculture 559:738357. https://doi.org/10.1016/j.aquaculture.2022.738357

Tang Y, Wang R, Tan L, Guo L, Duan Y, Yang L, Huang J (2020) Effects of live microalgae and algae powder on microbial community, survival, metamorphosis and digestive enzyme activity of *Penaeus monodon* larvae at different growth stages. Aquaculture 526:735344. https://doi.org/10.1016/j.aquaculture.2020.735344

Tayag CM, Lin YC, Li CC, Liou CH, Chen JC (2010) Administration of the hot-water extract of Spirulina platensis enhanced the immune response of white *shrimp Litopenaeus vannamei* and its resistance against *Vibrio alginolyticus*. Fish Shellfish Immunol 28(5–6):764–773. https://doi.org/10.1016/j.fsi.2010.01.023

Thépot V, Mangott A, Pirozzi I (2016) Rotifers enriched with a mixed algal diet promote survival, growth and development of barramundi larvae, *Lates calcarifer* (Bloch). Aquac Rep 3:147–158. https://doi.org/10.1016/j.aqrep.2016.02.003

Torsabo D, Ishak SD, Noordin NM, Koh ICC, Abduh MY, Iber BT, Abol-Munafi AB (2022) Enhancing reproductive performance of freshwater finfish species through dietary lipids. Aquac Nutr 2022:7138012. https://doi.org/10.1155/2022/7138012

Verschuere L, Rombaut G, Sorgeloos P, Verstraete W (2000) Probiotic bacteria as biological control agents in aquaculture. Microbiol Mol Biol Rev 64(4):655–671. https://doi.org/10.1128/mmbr.64.4.655-671.2000

Wang C, Liu Y, Sun G, Li X, Liu Z (2019) Growth, immune response, antioxidant capability, and disease resistance of juvenile Atlantic salmon (*Salmo salar* L.) fed *Bacillus velezensis* V4 and *Rhodotorula mucilaginosa* compound. Aquaculture 500:65–74. https://doi.org/10.1155/2022/7138012

Wu Y, Duan Y, Wei Y, An D, Liu J (2022) Application of intelligent and unmanned equipment in aquaculture: a review. Comput Electron Agric 199:107201. https://doi.org/10.1016/j.compag.2022.107201

Zarei M, Amirkolaei AK, Trushenski JT, Sealey WM, Schwarz MH, Ovissipour R (2022) Sorghum as a potential valuable aquafeed ingredient: nutritional quality and digestibility. Agriculture 12(5):669. https://doi.org/10.3390/agriculture12050669

Zettl S, Cree D, Soleimani M, Tabil L (2019) Mechanical properties of aquaculture feed pellets using plant-based proteins. Cogent Food Agric 5(1):1656917. https://doi.org/10.1080/23311932.2019.1656917

Zhang X, Jin M, Luo J, Xie S, Guo C, Zhu T, Zhou Q (2022) Effects of dietary carbohydrate levels on the growth and glucose metabolism of juvenile swimming crab, *Portunus trituberculatus*. Aquac Nutr 2022:1–15. https://doi.org/10.1155/2022/7110052

The Importance of Rotifer as Live Feed in Mariculture

3

Sitti Raehanah Muhamad Shaleh, Rizal Ismail, Norfazreena Mohd Faudzi, Rossita Shapawi, and Ching Fui Fui

Abstract

Live feed that consists of phytoplankton and zooplankton plays a vital role in aquaculture, serving as a critical component for the successful seed production of fish and crustaceans. Live feed such as rotifer, artemia and copepod are tiny multicellular zooplankton that are propagated for the first feeding of early larval stage. At the first feeding, the body and mouth or gape size of most larvae are small, limiting the size of food a larvae can capture in this stage; fish larvae require live feed with suitable nutritional value and suitable size range for the success of larviculture. The significance of rotifer being used as a primary live feed is due to the small size, nutritional content, digestibility and high reproductive rate. Rotifers are propagated in a dedicated culture system to provide a consistent and high-quality live feed. The culture methods emphasize maintaining optimal growth parameters that promote the rapid reproduction of rotifers. Apart from that, rotifer enrichment is another key focus in the culture to ensure the live feed meets the larval dietary requirements for optimal growth and survival. Understanding and optimizing rotifer culture techniques remain essential for the success of hatchery operations in the mariculture industry.

Keywords

First feeding · Larviculture · Enrichment · Propagation techniques

S. R. Muhamad Shaleh (✉) · N. Mohd Faudzi · R. Ismail · R. Shapawi · C. F. Fui
Higher Institution Centre of Excellence (HICoE), Borneo Marine Research Institute, Universiti Malaysia Sabah, Kota Kinabalu, Sabah, Malaysia
e-mail: sittirae@ums.edu.my

N. M. Faudzi et al. (eds.), *Essentials of Aquaculture Practices*,
https://doi.org/10.1007/978-981-97-6699-4_3

3.1 Introduction

Fisheries and aquaculture remained an important source of nutritious food for millions of people around the globe. Fish is indeed a healthy and popular food because it is rich in essential nutrients, high protein and low fat. Vitamins and minerals in fish are important in maintaining overall health, while protein is essential for muscle growth and omega-3 lipids are good for our hearts and brain. Because of the health benefits and the taste, the demand for fish is increasing and thus has led to the aquaculture development. The scarcity of fish landing from the sea due to overfishing also makes aquaculture a more reliable option for seafood production.

In Malaysia, the volume of marine fish landings amounted to approximately 1.308 million tonnes in 2022, a slight decrease from the previous year at around 1.328 million metric tonnes. On the other hand, aquaculture production increased to 0.573 million metric tonnes in 2022, while only 0.417 million metric tonnes in the year 2021 (Malaysia Department of Fisheries 2022). In the vein of the global trend, aquaculture sectors in Malaysia are potentially increasing in importance as food fish supply with the target of 0.958 million metric tonnes in 2030. This has led the government to undertake various efforts to improve the aquaculture industry in the country.

Most aquaculture production consists of capturing young or immature wild stocks and culturing them to marketable sizes before selling. This is an unsustainable practice since wild stocks are depleted and sometimes overfished due to the high demand. A viable alternative would involve the establishment of full cycle culture involving the usage of hatchery-produced seeds for aquaculture. Hybrid grouper (Fig. 3.1), for example, has been commercially produced to fulfil the increasing demand. However, larviculture of most species remains the bottleneck in establishing full cycle culture.

Fig. 3.1 Hybrid grouper, cross-breed between male (*Epinephelus lanceolatus* and female *E. fuscoguttatus*) has gained popularity among the fish farmers primarily due to its high growth rate and excellent taste

3.2 First Feeding of Larval Stage

Success in aquaculture is always linked to success in seed production, and it hinges on a deep understanding of the early larval stage of the cultured species. Larviculture, or the first feeding of early larval stages, is apparently the major bottleneck for the professional upscaling of the culture of marine fishes. First feeding is a critical stage in larval development of many fish and aquatic organisms. During this stage, the larvae undergo a transition period from relying on their yolk sac reserves to external food sources. The first feeding is initiated as the nutritious yolk sacs that are rich in essential nutrients for larval development are fully absorbed, and the larvae start to demonstrate feeding behaviour of exogenous consumers.

The first feeding of the early larval stage or larviculture is an important period for optimal development of marine fish larvae. At the first feeding, the body and mouth or gape size of most larvae are small which limits the size of food a larvae can capture to successfully survive. During this stage, fish larvae require live feed with suitable nutritional value and suitable size range for the success of larviculture of many fish species. The size of the initial feed should match the size of the larvae's mouth and should be small enough to be ingested by the larvae. Formulated feed, though nutritionally sufficient, still unable to replace live feed with problems pertaining to the digestibility and palatability. Fish larvae have specific nutritional requirements that are best met by live feed of important phytoplankton and zooplankton.

3.3 Important Live Feed Organisms in Aquaculture

Live feed is a tiny living organism that is used as primary food source for the larval stage of fish, shrimp, crabs and other aquatic organism. They remain essential and play an important role in fish and crustacean hatcheries as they offer a highly nutritious and digestible diet for the rapid growth and development of cultured species. In addition, live feed can contain beneficial microorganisms and natural compounds that contribute to the immune system of the juvenile fish, thus reducing the risk of diseases. Different species of aquatic organisms at the larval stage have specific requirements when it comes to live feed as they vary in terms of size, nutritional content and suitability. In certain conditions, a combination of different live feed may be used to provide a complete diet to support the healthy growth and survival of larval fish. The common types of live feed used in aquaculture include phytoplankton and zooplankton.

3.3.1 Phytoplankton

Phytoplankton are microscopic algae, also known as microalgae, that are the root of the food chain in aquaculture as well as a biological starting point for energy flow through most aquatic ecosystems. They thrived in aerated, liquid cultures where the cells have sufficient access to light, carbon dioxide, nutrients and ideal temperature for growth (Fig. 3.2). Phytoplankton species that were used in aquaculture must meet various criteria such as ease of culturing, lack of toxicity, high nutritional value with correct cell size (2–20 μm) and shape and digestible cell wall to make nutrients available (Patil et al. 2007; Abd Rahman et al. 2018). Microalgae supply essential nutrients to aquatic organisms in the form of proteins, lipids and carbohydrates. *Isochrysis galbana* is known as one of the most commonly used species due to its large amount of essential lipids as reported by Martinez-Fernandez et al. (2004), and *Chaetoceros* sp., a marine diatom, is used in marine shrimp hatcheries due to its essential PUFAs mainly omega 3 and highly unsaturated fatty acids (HUFAs) that are necessary for aquatic organism growth.

Microalgae are applied in aquaculture either as direct consumption or indirectly by small and large organisms. They are directly consumed by larvae of mollusc and crustaceans or indirectly as feed for the live prey to improve their nutritional content before being fed to small fish larvae (Ferreira et al. 2018). Unlike bivalve and crustacean larvae, most marine fish larvae do not feed directly on microalgae. However, a previous study reported that the presence of green water in larval rearing tanks supports better survival, growth and food conversion index of most marine finfish species than in clear conditions (Papandroulakis et al. 2000; Rojo-Cebreros et al. 2017). *Nannochloropsis* sp., a green microalga, is used to culture fish either directly or indirectly to create a green water environment in the larval system. They are fed to live prey before being offered to fish larvae. *Tetraselmis suecica* and *T. tetrathele* also green algae commonly used directly as feed for bivalve molluscs, penaeid shrimp and rotifers (Muller-Feuga et al. 2003).

Fig. 3.2 Microalgal cultures were maintained in a temperature-controlled room, aerated and illuminated

3.3.2 Zooplankton

Zooplankton are small drifting or weakly swimming animals found in aquatic ecosystems and serve as a crucial live feed in aquaculture. Zooplankton represents a high-quality food for fish larvae due to its nutritional content. The most common zooplankton used in aquaculture include rotifer, artemia and copepod.

Artemia is a small crustacean with an elongated segmented body shape which is divided into three parts consisting of a head, thorax and abdomen (Fig. 3.3a). Interest in the usage of brine shrimp in aquaculture started when its commercial values were first discovered. The development of fish and shrimp aquaculture created marketing opportunities for *Artemia* cyst to be sold commercially. *Artemia* used as live feed in hatcheries is obtained by incubating and hatching the cysts and finally enriching the nauplii. These steps are the easiest and most convenient process in preparing live feed in a hatchery (Dhont and Van Stappen 2003; Brown et al. 2022). *Artemia* is known to contain a low nutritional profile especially HUPAs which is essential in fish growth. Therefore, bioencapsulation or enrichment of *Artemia* is important to enhance its nutritional value with essential fatty acids.

Although *Artemia* lacks certain fatty acids, despite the simple production, the price for production and packaging of the cysts is more expensive compared to other live feed. Due to the high price, more effort is required to maximize the hatching rate and quality of *Artemia*. Newly hatched *Artemia* are at their most nutritious as the yolk is present and the dark orange colour of the early nauplii functions as a visual stimulus (Treece 2000; Hepburn 2015). However, the swimming rate and sizes increase as the stage increases, and thus *Artemia* is only essential as feed at its nauplius stage (Drillet et al. 2006; Lubzens et al. 2020).

In the wild, copepod is a natural prey for fish larvae, while in aquaculture, it has been proven as the most appropriate live feed for many marine fish larvae (Payne et al. 2001; Odo et al. 2015). Feeding copepods alone or in combination with other live feeds has repeatedly shown better results in terms of growth, survival and the

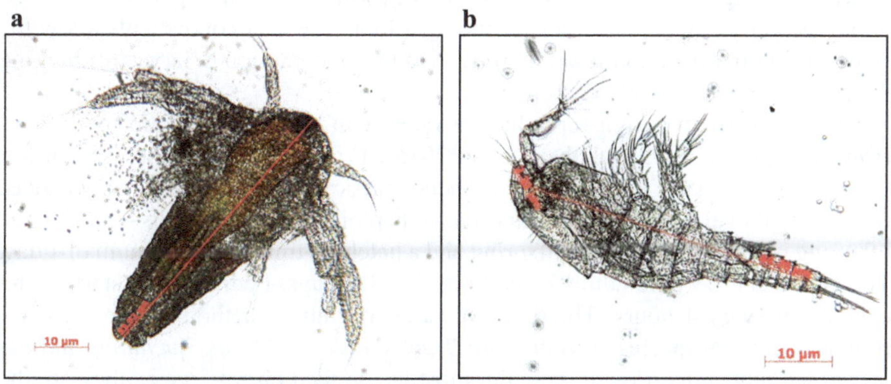

Fig. 3.3 *Artemia* sp. (**a**) and Euterpina *acutifrons* (**b**) are ideal food for marine fish larvae due to its nutritional value, size, trophic ecology and cultivability

overall health of larval fish (Wilcox et al. 2006; Vaidya 2017). Their reputation as an excellent live feed choice is based on their balanced nutritional profile. They contained high levels of protein, omega-3 polyunsaturated fatty acids (PUFAs), carbohydrates and enzymes. Besides being nutritious, the variety of sizes of the copepod is also one of its main potential characteristics as a live feed. They show a wide range of body size between nauplii, copepodite and adults which increases the interest of larvae to feed on them (Kleppel et al. 2005; García-Roger et al. 2019). As the larvae grow, the size of prey that the larvae can ingest may also increase which is an advantage in copepods as they provide a broad range of sizes through their various developmental stages. Thus, they can replace larger live feed such as *Artemia* and rotifers which are commonly used in aquaculture.

The copepod active swimming action can stimulate the feeding response of fish larvae. Copepod has various swimming patterns ranging from continuous to intermittent locomotion (Waggett and Buskey 2007; Vu et al. 2020). Copepod is reported to have a high reproduction rate (50–100 eggs female^{-1} day^{-1}), and of the three orders, calanoid and harpacticoid have been cultured for use in larviculture. The harpacticoid can be grown at high densities and has been proven to have superior nutritional qualities compared to *Artemia* and rotifers (Drillet et al. 2006; Lubzens et al. 2020). The harpacticoid copepod, *Euterpina acutifrons* (Fig. 3.3b), for example, is easily identified from other species because *E. acutifrons* is the only species in the family.

3.4 Rotifer

Rotifers, also known as wheel animalcules, are microscopic zooplanktons that are commonly used in aquaculture as live feeds. Approximately 2000 species of these invertebrates are part of the phylum Rotifera. Rotifers have a broad salinity tolerance that enables them to reside in freshwater, brackish and marine water. Rotifers are 2–3 mm in size, which is smaller than the mouth gape of most larvae and promotes easier digestibility. Figure 3.4 shows the rotifer body that is generally divided into three body components: head, trunk and foot. The foot components allow the rotifer to swim around in the water body to induce locomotion in larvae for hunting the prey.

Rotifers also have high reproductive rates with a life span of up to 15 days (Denekamp et al. 2009; Jakobsen et al. 2015). From Fig. 3.5, in the asexual life cycle, diploid females will produce parthenogenetic diploid amictic eggs, whereas, in the sexual cycle, diploid females produce haploid eggs via meiosis—an average offspring per female of 10–20 offspring and a hatching time of a maximum of 1 day. The larvae can become adults as early as 0.5–1.5 days. Females will start laying eggs again every 4 hours. This indicates a short culture method time needed for rotifers. Common species of rotifers are *Brachionus calyciflorus*, *Euchlanis dilatata* and *Brachionus plicatilis*. In the wild, most larvae feed on rotifers, as they are the optimal choice for early larval feeding due to the size and nutritional profile that can sustain larvae (Ghosh et al. 2016).

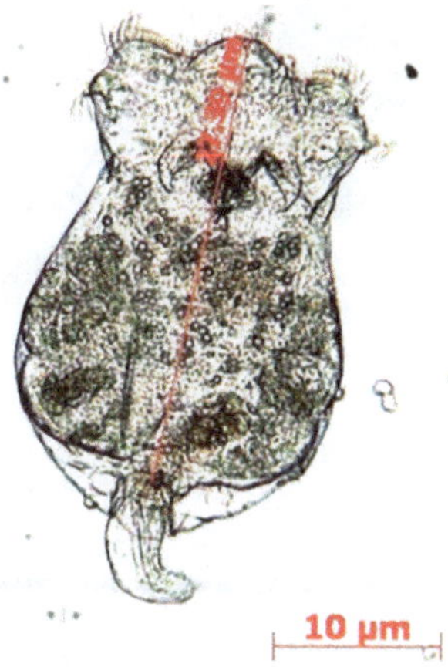

Fig. 3.4 Rotifer, *Brachionus plicatilis*, observed under a light microscope

Rotifer is a minute, unsegmented, pseudocoelomate aquatic invertebrate that has a round and flat spineless body. Rotifers are widely distributed in freshwater lakes and ponds as well as in marine water. Approximately 2000 species have been described (Lubzens and Zmora 2003). The phylum of rotifer is known as Nemathelminthes or Aschelminthes, and it belongs to the class Rotiferia, order Monogononta and family Brachionidae. Initially, rotifer was all classified as one species, *Brachionus plicatilis*, but it was later discovered that *B. plicatilis* comprises two morphologically and ecologically distinct groups, *B. plicatilis* as 'L-type' and *B. rotundiformis* as 'S-type' (Lubzens and Zmora 2003). *Proales similis*, the minute monogonont rotifer, was first reported in 2009 by Wullur et al. as the smallest rotifer with a potential first feed for small mouth marine fish larvae such as humphead wrasse *Cheilinus undulatus*.

The size of the male rotifer is much reduced and less developed than the female rotifers. Lavens and Sorgeloos (1996) noted that the morphology of the rotifer is almost the same as others in the same taxonomic classification. They stated that the body of the rotifer is generally covered by a distinct cuticle and has bilateral symmetry and sexual dimorphism, and the body is comprised of a head with a corona, trunk and foot. The head carries a rotatory organ named the corona, which is easily recognized by its annular ciliation. The corona assures locomotion and a whirling water movement, which facilitates the uptake of small food particles. The trunk contains the digestive tract, a mastax that grinds the ingested particles, the oesophagus, the stomach with gastric glands and the intestine. The foot is an appendage that extends from the body ventrally, with one or four toes.

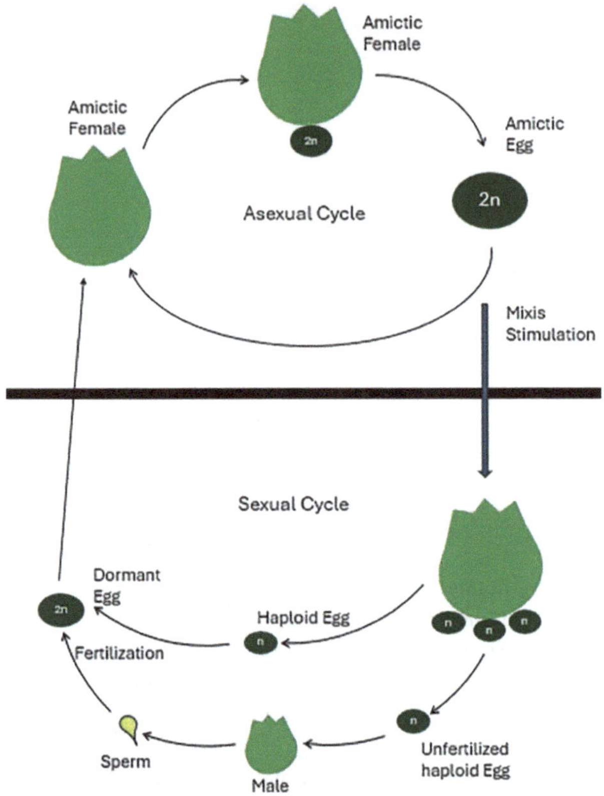

Fig. 3.5 Simplified schematic diagram of the life cycle and reproductive pattern of the rotifer *B. plicatilis* showing the asexual and sexual reproduction (Source: Denekamp et al. 2009)

3.4.1 Reproduction of Rotifer

It is known that rotifer can use two modes of reproduction, asexual and sexual reproduction. The reproduction involves amictic (diploid, 2n chromosomes) and mictic (haploid, n chromosomes) females. Lubzens and Zmora (2003) explained that during asexual reproduction, parthenogenesis happens in the amictic females, which produce amictic eggs. These amictic eggs are later developed and hatch into amictic females. Generally, rotifer undergoes asexual reproduction when there are no environmental threats such as sudden changes in temperature, salinity and feed density. The amictic females will switch to mictic females during sexual reproduction under specific environmental conditions. The mictic females produce mictic eggs and develop into haploid males with a size of about one-quarter of the female. The mictic males and females will undergo sexual reproduction and produce resting eggs. The resting eggs will only develop and hatch into amictic females after exposure to specific environmental conditions, such as a change in temperature or food conditions (Yoshimura et al. 2003; Yuan et al. 2018). Besides the environmental conditions, rotifer population also plays an important role in determining the reproduction mode, where high density can reduce asexual reproduction and promote sexual reproduction (Gilbert 2003, 2017). For propagation sexual reproduction should be avoided since rotifer growth is slower and the resting eggs produced need time to be hatched (Adam 2020).

3.4.2 Nutritional Value of Rotifer

In marine hatcheries, rotifers are the most suitable live feed for rearing marine finfish larvae. They are commercially available and their production as feed for marine fish is not complicated. However, there are also disadvantages to using rotifers as live feed. One of the advantages that make them unsuitable as prey is they are not easily digested. Besides that, their swimming movement does not encourage feeding response in all marine fish due to their steady, random motion that is controlled by their ciliated crown (Chesney 2005; Coutinho et al. 2020). The size of rotifers, with a body width range from 90 to 350 μm, is also larger than the mouth gape of many marine finfish larvae at the first feeding (Ohs et al. 2009; Ogata 2017).

Besides that, the nutritional compositions of rotifers are also poor, and enrichment is needed before offering them to larvae (Sorgeloos et al. 2001; Sun et al. 2013). The nutritional value of rotifer depends on the feed that is given to them. The protein content of rotifer ranges from 28% to 63% and lipid content from 9% to 28% of dry weight. The carbohydrate content ranges from 10.5% to 27.0% of dry weight (Lubzens and Zmora 2003; Lawrence et al. 2012). The lipid content of rotifer unequivocally has the greatest influence on the growth and survival of marine fish larvae. Eicosapentaenoic acid (EPA) and docosahexaenoic acid (DHA) of the fatty acid components in lipids are known to be essential for the survival of marine fish larvae. For this reason, the feed for rotifer is usually enriched to increase the EPA and DHA content. This can be done by feeding rotifer with algae, lipid emulsions, microparticulates or microcapsules containing lipids, or lipids with proteins and carbohydrates (Lubzens and Zmora 2003; Lawrence et al. 2012).

3.4.3 Rotifer Culture Systems

Rotifer was first considered as a pest in eel culture ponds in the 1600s which caused 'Mizukawari' damage or rapid change of water colour (Ito et al. 1981). When the bloom of rotifers occurred, the supply of oxygen became insufficient for the eel that suffered motility due to suffocation. However, they were soon discovered as a potential first feed for freshwater and marine fish (Hirata 1979). Mass production of rotifer was first developed in Japan in the 1960s, and 25 years after the first production of rotifer, several culture systems were being used (Lubzens and Zmora 2003; Lawrence et al. 2012).

In rotifer mass production, a few problems must be considered such as the rotifer strain selected must be appropriate for the larvae to consume. The water quality, feed quality and quantity must be maintained at an adequate level to ensure the survival of the rotifer. In addition, suitable cultural conditions that are convenient for the hatcheries must be considered as well. There are three types of rotifer culture systems: batch, semi-continuous and continuous cultures, which are commonly used in most hatcheries. Each of the culturing systems has advantages and disadvantages. The selection of the culture system is according to the suitability for the hatchery management.

Batch Culture

Most of the efforts made for the culture of rotifer were in the batch culture system, where microalgae raised in a large tank were used as a primary diet for rotifer. Baker's yeast has also been used as food for rotifer, but yeast is lacking in a few nutrients essential for rotifer and fish larvae. Due to this, different direct and indirect methods of rotifer enrichment become necessary before feeding the fish larvae. Now, it is the main challenge for scientists to develop a stable and sustainable high-density culture technique of rotifer, which will reduce the use of hatchery space for live food culture, increase larval fish production and reduce the cost of production.

Batch culture is a system that harvests totally all the rotifer, with part of the harvest used as feed and the other part as inoculum for the next batch cultures. According to Lubzens and Zmora (2003), this culture was initially developed in 1964. This culture method has some disadvantages as it can cause highly variable conditions in growth performance in terms of the biochemical composition of the rotifer population due to unstable physico-chemical water parameters (Dhert et al. 2001; Ogata 2017).

Continuous Culture

Continuous culture controls the feeding, water quality and density of cultured organisms. This culture method is also known as 'feedback' culture where a certain volume of rotifer is harvested daily from the culture tank. Continuous culture offers easy manipulation of rotifer physiological and nutritional quality (Lubzens and Zmora 2003; Lawrence et al. 2012). The system produces a log-phase rotifer that can be harvested continuously. Nutritional quality is maintained by providing adequate feed. However, the initial cost of this culture is high due to the use of an automated control system (Lubzens and Zmora 2003; Lawrence et al. 2012).

The continuous culture methods can be seen to have high daily production with stabilized water parameters that can be achieved by chemostat or turbidostat systems. The chemostat system allows constant flow rate of culture water with a single component controlling the growth rate of culture. This causes the water parameters to be more stable and produce a favourable environment to produce a higher density of rotifers. However, for the turbidostat system, there is no constant flow; however, water is changed in a period and the specific growth rate is controlled internally by the culture reactions. When the culture condition drops slightly, the water is changed immediately. Rotifer culture is successful when microalgae diets provided are sufficient and can sustain the rotifer to avoid a culture crash.

Semi-Continuous Culture

The semi-continuous culture method is a combination of the batch culture and the continuous culture method (Fig. 3.6). A certain volume of rotifer is harvested daily from the culture tank. This is like the continuous culture, but rotifer is harvested according to the batch culture during tank changing, which is called the 'thinning' system. This culture is usually conducted in larger tanks such as 50–100 tonnes and it is always maintained at low densities. The disadvantages of this culture system are the requirement for a larger area in the hatcheries for culturing and the excretory products (solid wastes and nitrogenous products) that accumulate in the tanks that lead to their collapse (Lubzens and Zmora 2003; Alajmi and Zeng 2015).

The selection of a culture method is critical to rotifer mass production. In comparison with batch culture techniques, semi-continuous cultures are considered more promising and reliable for stable and cost-effective rotifer production (Lubzens and Zmora 2003; Alajmi and Zeng 2015). Chemostats are designed for constant input of limiting nutrients by which the growth rate of the culture organisms is maintained constant. In contrast, turbidostats are continuous culture devices where the population density of an organism is held constant by continually adjusting the washout rate or harvest rate to offset the variations. Both chemostat and turbidostat cultures offer easier manipulation of rotifer physiological and nutritional quality. The log-phase rotifers can be harvested continuously, and their nutritional quality is maintained by providing adequate food organisms (Jakobsen et al. 2015). Turbidostats have some advantages over chemostats. Turbidostats can effectively regulate biomass in the system and thus can be safely operated at a higher dilution rate. Subsequently, a relatively steady rotifer population with a higher growth rate could be achieved and maintained (Boraas and Bennett 1988; Jakobsen et al. 2015).

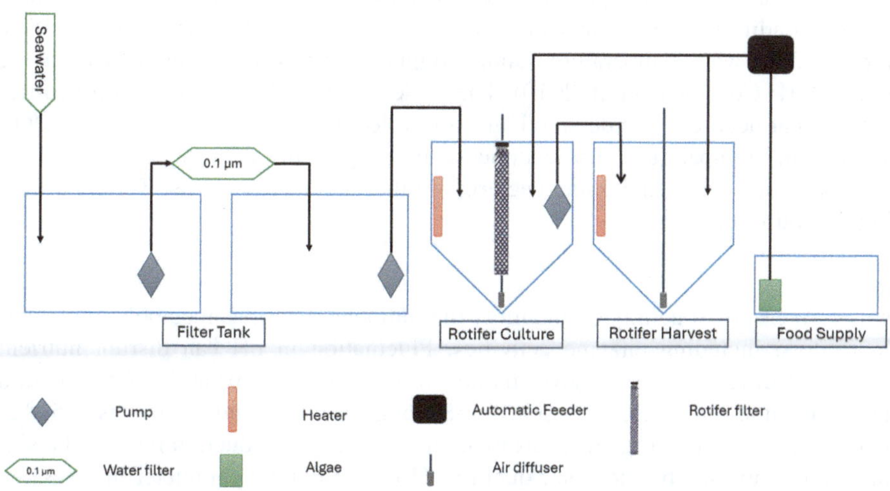

Fig. 3.6 Schematic diagram of the semi-continuous culture system of rotifer

Ultrahigh-Density Culture

Yoshimura et al. (2003) and Yuan et al. (2018) introduced high-density rotifer culture by using concentrated microalgae as food. This technique cultures rotifer in high density of 2000–20,000 individuals/mL. The advantage of this culture system is by eliminating the need for space in the hatchery for microalgae and rotifer culture (Hagiwara et al. 2001). Also, the concentrated microalgae used in the culture have been enriched with fatty acids and vitamin B_{12} (Hagiwara et al. 2001). Enriched algae are commercially available and can be used for mass culture of rotifers that are directly fed to fish larvae, and this can reduce the labour for culture enrichment process.

High-density culture can use batch, semi-continuous or continuous culture. Recently, a new high-density culture, the recirculation system, was introduced to culture rotifer. By using the recirculation system instead of a batch culture system for high-density culture, the disadvantages can be solved. The system can increase rotifer production indirectly and reduce labour cost and water consumption while ensuring stable water quality.

3.4.4 Growth Parameters for Rotifer

Until today, the success of the worldwide aquaculture industry is still quite dependent on the mass culture of rotifer. However, there is a lack of a continuous, stable and reliable supply of nutritionally adequate rotifer to get marine finfish larvae through the first weeks of life. Thus, advancements in the rotifer culture system design as well as the optimum level of physico-chemical parameters and the determination of the ideal feeding rates and feeding regime are needed to be able to effectively move towards adequate commercial production. Rotifer is well suited to mass culture because it is prolific and known to tolerate a wide variety of environmental conditions. The abiotic factors affecting rotifer culture are high densities, types of food, water temperature, salinity, light penetration and water quality (Cheng et al. 2004; Conceicao et al. 2010). The basic water quality parameters for rotifer culture that need to be understood and controlled are pH, dissolved oxygen (DO), salinity and temperature. The chemicals that impact the culturing water such as ammonia, organic particles and bacteria or other organisms are also factors affecting the rotifer culture.

pH

pH is a measure of acidity or alkalinity that influences various metabolic processes in rotifers, including enzyme activities. Fluctuation in pH can disrupt nutrient uptake, leading to decreased growth rates. Rotifer lives at a pH above 7.5 in natural environments. However, the optimal pH range for culturing rotifer is 7.5–8.5. Rotifers are sensitive organisms as their survival and reproduction rate can be significantly impacted by the water quality. pH is an essential parameter in assessing the overall water quality; thus, maintaining pH within the optimal range is important to create a stable and healthy environment for rotifer. As reported by Lubzens and

Zmora (2003), pH can affect the percentage of ammonium nitrate (NH_3-N) in the water, and the acceptable NH_3-N range is 6–10 ppm.

Temperature

Temperature can affect rotifer individuals and populations. Some factors of the individuals are closely related to the temperature, such as the duration of embryonic development, pre-reproductive period, post-embryonic growth and rotifer activities. Generally, cold-adapted rotifer shows faster development in low temperatures but slower development in high temperatures (Kotani et al. 2005; Bielańska-Grajner et al. 2017.

However, the warm-adapted species show opposite results. As in rotifer populations. Stelzer (2006) concluded that *B. plicatilis* and *B. calyciflorus* also show highest growth values at temperatures over 27 °C. However, the maximum reproduction rate of rotifer occurs between 30 and 34 °C (Assavaaree et al. 2003; Sterzelecki et al. 2020). This may be because their metabolism, development and food ingestion rate are high at high temperatures. It is also emphasized that at high temperatures, there is rapid population growth and enhanced grazing, particularly by rotifers that consume the feed faster than at lower temperatures. Moreover, rotifers complete their life cycle faster when temperatures are elevated (Stelzer 2006; Sterzelecki et al. 2020).

Dissolved Oxygen

According to Lubzens and Zmora (2003), although rotifer can tolerate dissolved oxygen (DO) lower than 1 ppm, culture requires aeration or DO that is maintained above 4 ppm. In rotifer culture, adequate aeration is provided using perforated polyvinyl chloride (PVC), air-stones or small-diameter open-ended tubes. Pure oxygen gas should be provided through perforated tubes in a high-density rotifer culture (Yoshimatsu et al. 1997; Brown et al. 2022). Dissolved oxygen and temperature are regarded as two of the most critical environmental factors (potential cues) that could affect the reproductive strategy of rotifers, and both should be maintained relatively stable to achieve culture stability and longevity. Low DO can significantly affect rotifer production. Many factors impact DO levels in the rotifer cultures, and previous reports show that the oxygen consumption rate in rotifer increased when the temperature increased.

Salinity

The salinity of the culture water is one of the important factors that control rotifer distribution and growth. Research has been done to identify the optimum salinity level for rotifer, but there is still little information. There is good evidence to suggest that highly evolved rotifer can regulate the salt concentration of the pseudocoelomic fluid (Lubzens and Zmora 2003). Nevertheless, high salinity results in a reduction in rotifer growth due to a decrease in fecundity and survival (Green 1994; García-Roger et al. 2019). Usually, *B. plicatilis* and *B. rubens* are found at salinity ranges of 0.27–31.00 ppt (Green 1994). However, from previous study, *B. plicatilis* lives in a very broad range of salinities. Even so, the data on the growth rate, for most strains, can be interpreted as an optimum curve with a maximum or plateau located at moderate salinities between 10 and 20 ppt (Lubzens and Zmora 2003; Lawrence et al. 2012).

3.4.5 Diet and Enrichment Types

The nutritional quality of rotifers is crucial for the health and growth of the larvae of marine organisms. Rotifer diets are carefully designated to provide essential nutrients including proteins, lipids, vitamins and minerals for rotifer growth as well as for a highly nutritive rotifer. Rotifer enrichment involves enhancing the nutritional content of rotifers to provide a more balanced diet for larval growth.

Yeast

Baker's yeast is often used as a substitute diet in rotifer culture. However, rotifer culture fed with yeast is sometimes subject to a sudden unexpected decrease or suppressed growth of the population. Vitamin B_{12} is an essential vitamin for rotifer growth. Previous studies show that rotifer grows better at low oxygen levels when fed with yeast because the bacteria in the culture water produce vitamin B_{12} that are facultative anaerobic bacteria (Hirayama 1987; Hamre 2016). The bacteria in the culture water due to the suspension feeds and animal excretion products also become feed in rotifer culture. However, excess food and excretion products tend to pollute culture water by increasing ammonia levels, which limits the population growth of rotifer.

Microalgae

Microalgae, also known as phytoplankton, occur in a vast range of sizes and forms, both as sippet cells and as colonies, which may affect their growth and nutrient storage. Graham (2003) also stated that grazers such as protozoa, rotifers and crustaceans consume small microalgal cells. However, larger cells and colonies above 50 μm in diameter are immune to predation by crustaceans but are more prone to attack by parasites. Sippet cells and colonies can be categorized both as flagellate motile cells and non-motile cells. Microalgae are an important food source and feed additive in the commercial rearing of many aquatic animals, especially the larvae and spat of bivalve molluscs, penaeid prawn larvae and live food organisms such as *Brachionus* sp. which, in turn, are used to feed the larvae of marine finfish and crustaceans (Borowitzka 1997; Khatoon et al. 2010).

Microalgae that are being cultured widely in hatcheries for rotifer feed purposes are *Nannochloropsis*, *Chlorella*, *Tetraselmis* and *Isochrysis* (Patil et al. 2007; Odo et al. 2015). In order to cope with the large demand for an essential fatty acid-fortified food chain during the production of fish larvae, many hatcheries have established *Nannochloropsis* sp. culture as it has high-quality algal biomass. A study was conducted by Abd Rahman et al. (2018) on the effects of microalgae as live food for *B. plicatilis* in an intensive culture system. The five microalgae, *Nannochloris* sp., *Tetraselmis* sp., *Isochrysis* sp., *Chlorella* sp. and *Nannochloropsis* sp., were fed to the rotifer at different densities and culture volumes. Daily algal densities were recorded. *Tetraselmis* sp. showed the significantly highest μ value at 0.1×10^6 cells/ml compared to the other microalgae, leading it to be the optimum species to be used in intensive culture.

Formulated Diets

Reliable and affordable methods for continuous large-scale production are the key barriers in maximizing the use of rotifers. The development of an artificial diet, which eliminates the requirement of an additional enrichment phase for the increase of the rotifers' nutritional value, is a breakthrough in production technology. The most frequently used formulated feed for rotifer cultures is Culture Selco (CS) (INVE N.V., Belgium). Before feeding, the dry product had to be suspended in water. The Culture Selco food suspension can be used in automated feeding for up to 30 h as long as it is kept cold and regularly aerated. Several rotifer strains were evaluated in 100 L tanks using a typical feeding procedure utilizing Culture Selco (Islam et al. 2017).

Prebiotic and Probiotic

Prebiotics and probiotics can enhance the gut health of larvae. Enrichment with prebiotics involves adding substances that promote the growth of beneficial microorganisms, while probiotic enrichment involves adding live beneficial bacteria to the rotifer culture. Prebiotics such as fructooligosaccharides (FOS) and probiotics such as *Bacillus* species may be used for rotifer enrichment. Quy et al. (2018) have reported that both *Artemia* and rotifers enriched with a commercial probiotic increased the survival and metamorphosis rates of mud crab megalopa.

Vitamins and Minerals

Vitamins and minerals are essential for various physiological functions in fish or shrimp larvae. Vitamin enrichment is applied by adding vitamin emulsions or multivitamin solutions to the rotifer culture. Mineral enrichment, including trace elements, involves adding mineral supplements such as marine salt solutions or commercial minerals mixes to the rotifer culture. A study by Srivastava et al. (2011) shows that rotifers can be fortified up to the copepod level with thiamine, vitamin C, vitamin E and iodine.

3.4.6 Culture Setup

Setting up a rotifer culture tank involves creating and maintaining optimal conditions for the growth and reproduction of rotifers. There are many culture techniques from simple to more sophisticated methods that can be set up once the basic has been mastered. Fig. 3.7 is a schematic diagram of the basic rotifer culture setup that typically involves a cylindrical or conical tank equipped with essential systems for aeration and filtration. The aeration system is installed to provide gentle water movement and to ensure adequate oxygenation for the rotifers.

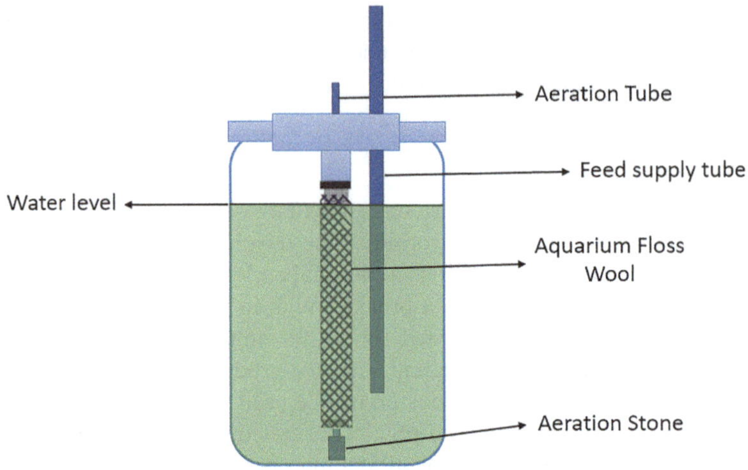

Fig. 3.7 Schematic design of basic requirements for rotifer batch culture that includes filtration, aeration and feed supply tube

The filtration system in a rotifer culture setup plays a crucial role in maintaining water quality and ensuring the productivity of the culture. The purpose of the filter is to remove debris and particulate matter from the culture water. Debris can interfere with the growth of the rotifers and hinder the water clarity. The filtration system for a simple tank consisted of an aquarium floss wool that is a loose weave filter that allows water to easily pass through. Aeration plays a major role in this system, as it will create a water current under the filter which push upwards the suspended debris and waste into the filter. To ensure the filter is functionally well, proper cleaning must be done thrice a week, depending on the filter condition. The filter itself can develop a biological filter within time. The nitrifying bacteria such as nitrosomonas, nitrospira and nitrobacter convert the excreted ammonium from rotifer to non-toxic nitrate adding more effectiveness to the filter in maintaining the proper water quality of the culture tank.

3.5 Conclusion

The use of rotifers as live feed in aquaculture offers numerous advantages that include their nutritional value, digestibility, sustainability and potential benefits for disease management. Incorporating rotifers into aquaculture practices can contribute to the improvement of seed production techniques and enhance the resource efficiency and overall sustainability of the industry. Ongoing research and technological advancements are essential to optimize the mass production and the potential benefits of rotifers in aquaculture.

Acknowledgement The work was supported by the Universiti Malaysia Sabah Flagship Grant (Grant Code: DFM2101) and the Higher Institution Centre of Excellence (HICoE) Research Grant Scheme provided by the Ministry of Higher Education, Malaysia.

References

Abd Rahman AR, Cob ZC, Jamari Z, Mohamed AM, Toda T, Ross O (2018) The effects of microalgae as live food for *Brachionus plicatilis* (rotifer) in intensive culture system. Trop Life Sci Res:127–138

Adam S (2020) Minimizing time fed rotifers maximizes hybrid striped bass larval growth in recirculating aquaculture systems. N Am J Aquac 82:208–214

Alajmi F, Zeng C (2015) Evaluation of Microalgal Diets for the Intensive Cultivation of the Tropical Calanoid Copepod. Parvocalanus crassirostris. Aquaculture Research 46(5):1025–1038

Assavaaree M, Hagiwara A, Kogane T, Arimoto M (2003) Effect of temperature on resting egg formation of the tropical SS-type rotifer Brachionus rotundiformis Tschugunoff. Fish Sci 69:520–528

Bielańska-Grajner I, Mieczan T, Cieplok A (2017) Ecology of moss-dwelling rotifers in a raised bog: differentiation of rotifer communities in microhabitats. Biologia 72(2):175–183

Boraas ME, Bennett WN (1988) Steady-state rotifer growth in a two-stage, computer-controlled turbidostat. J Plankton Res 10(5):1023–1038

Borowitzka MA (1997) Microalgae for Aquaculture: Opportunities and Constraints. Journal of Applied Phycology 9:393–401

Brown MT, Viglia S, Love D, Asche F, Nussbaumer E, Fry J, Haugslien TP, Engeland K, Jonell M, Nussbaumer E, Eide A, Neff R (2022) Quantifying the environmental support to wild catch Alaskan sockeye salmon and farmed Norwegian Atlantic Salmon: an emergy approach. J Clean Prod 369:133379

Cheng SH, Aoki S, Maeda M, Hino A (2004) Competition between the rotifer Brachionus rotundiformis and the ciliate Euplotes vannus fed on two different algae. Aquaculture 241:331–343

Chesney EJ (2005) Copepods as live prey: a review of factors that influence the feeding success of marine fish larvae. In: Lee CS, O'Bryen PJ, Marcus NH (eds) Copepods in aquaculture. Blackwell Publishing, Ames, Iowa, pp 133–150

Conceicao LECD, Yufera M, Makridis P (2010) Live feeds for early stages of fish rearing. Aquac Res 41:613–640

Coutinho P, Ferreira M, Freire I, Otero A (2020) Enriching rotifers with "premium" microalgae: Rhodomonas lens. Mar Biotechnol 22(1):118–129

Denekamp NY, Thorne MA, Clark MS, Kube M, Reinhardt R, Lubzens E (2009) Discovering genes associated with dormancy in the monogonont rotifer Brachionus plicatilis. BMC Genomics 10:108

Dhert P, Rombaut G, Suantika G, Sorgeloos P (2001) Advancement of rotifer culture and manipulation techniques in Europe. Aquaculture 200:129–146

Dhont J, Van Stappen G (2003) Biology, tank production and nutritional value of a. franciscana Artemia. In: Stottrup JG, McEvoy LA (eds) Live feeds in marine aquaculture. Blackwell Science Ltd., Oxford, UK, pp 65–121

Drillet G, Jorgensen NOG, Sorensen TF, Ramlov H, Hansen BW (2006) Biochemical and technical observations supporting the use of copepods as live feed organisms in marine larviculture. Aquac Res 37:756–772

Ferreira M, Cortina-Burgueño Á, Freire I, Otero A (2018) Effect of nutritional status and concentration of *Nannochloropsis gaditana* as enrichment diet for the marine rotifer *Brachionus sp.* Aquaculture 491:351–357

García-Roger EM, Serra M, Carmona MJ, Gómez A, Adcock GJ (2019) Facing adversity: dormant embryos in rotifers. Biol Bull 237(2):119–144

Ghosh S, Megarajan S, Ranjan R, Dash B, Pattnaik P, Edward L, Xavier B (2016) Growth performance of Asian seabass Lates calcarifer (Bloch, 1790) stocked at varying densities in floating cages in Godavari Estuary, Andhra Pradesh, India. Indian J Fish 61(3):146–149

Gilbert JJ (2003) Environment and endogenous control of sexuality in a rotifer life cycle: developmental and population biology. Evol Dev 5(1):19–24

Gilbert JJ (2017) Morphological variation and its significance in a polymorphic Rotifer, environmental, endogenous, and genetic controls. Bioscience 20(10):1–13

Graham JM (2003) Phytoplankton ecology. In: Graham LE, Wilcox LW (eds) Algae. Prentice-Hall, New York, pp 544–602

Green J (1994) The temperate-tropical of planktonic Protozoa and Rotifera. In: Dumont HJ, Green J, Masundire H (eds) Studies on the ecology of tropical zooplankton. Kluwer Academic Publishers, Brussels, pp 13–26

Hagiwara A, Gallardo WG, Assavaaree M, Kotani T, de Araujo AB (2001) Live food production in Japan: recent progress and future aspects. Aquaculture 200:111–127

Hamre K (2016) Nutrient profiles of rotifers (Brachionus sp.) and rotifer diets from four different marine fish hatcheries. Aquaculture 450:136–142

Hepburn J (2015) Taking aquaculture seriously. Organic farming, winter soil association. Int J Res Fish Aquac 63:74–83

Hirata H (1979) Rotifer culture in Japan. In: Styczynska-Jurewicz T, Backiel E, Jaspers J, Persoone G (eds) Cultivation of fish Fry and its live food. European Mariculture Society Special Publication 4, pp 361–375

Hirayama K (1987) A consideration of why mass culture of the rotifer Brachionus plicatilis with Baker's yeast is unstable. Hydrobiologia 147:269–270

Islam MR, Hassan MR, Begum M, Punom NJ, Begum MK, Sultana N, Rahman MS (2017) Effects of feeding zooplankton, Moina macrocopa (Straus, 1820) on the growth of Nile tilapia *Oreochromis niloticus* L. Bangladesh J Sci Ind Res 52(2):81–88

Ito S, Sakamoto H, Hori M, Hirayama K (1981) Morphological characteristics and suitable temperature for the growth of several strains of the rotifer, *Brachionus plicatilis*. Bull Fac Fish Nagasaki University 51:9–16

Jakobsen HH, Jepsen PM, Blanda E, Jørgensen NOG, Novac A, Engell-Sørensen K, Hansen BW (2015) Plankton composition and biomass development: a seasonal study of a semi-intensive outdoor system for rearing of turbot. Aquac Nutr 22:1239–1250

Khatoon N, Chaudhuri A, Sen Roy S, Kundu N, Mukherjee S, Mazumdar D, Homechaudhuri S, Pal R (2010) Algae as feed supplement in fish nutrition. J Bot Soc Bengal 64(2):85–93

Kleppel GS, Hazzard SE, Burkart CA (2005) Maximizing the nutritional values of copepods in aquaculture: managed versus balanced nutrition. In: Lee CS, O'Bryen PJ, Marcus NH (eds) Copepods in aquaculture. Blackwell Publishing, Ames, Iowa, pp 67–72

Kotani T, Hagiwara A, Snell TW, Serra M (2005) Euryhaline Brachionus strains (Rotifera) from tropical habitats: morphology and allozyme patterns. In: Herzig A, Gulati RD, Jersabek CD, May L (eds) Rotifera X: rotifer research: trends, new tools and recent advances. Kluwer Academic Publishers, Willemstad, pp 161–167

Lavens P, Sorgeloos P (1996) Manual on the production and use of live food for aquaculture (no. 361). Food and Agriculture Organization (FAO)

Lawrence C, Sanders E, Henry E (2012) Methods for culturing saltwater rotifers (*Brachionus plicatilis*) for rearing larval zebrafish. Zebrafish 9(3):140–146

Lubzens E, Zmora O (2003) Production and nutritional value of rotifers. In: Stottrup JG, McEvoy LA (eds) Live feeds in marine aquaculture. Blackwell Science, Oxford, UK, pp 17–64

Lubzens E, Hamo R, Blais I, Jeries S, Almog-Gabai O, Assaraf YG (2020) Rotifer resting eggs: an alternative for rotifer mass cultures in fish farms—lessons from a comprehensive study on production, storage and hatching of resting eggs. Aquaculture 528:735–505

Malaysia Department of Fisheries (2022) Annual Report 2022. ATTIN Press Sdn. Bhd, Selangor, p 164

Martinez-Fernandez E, Acosta-Salmon H, Rangel-Davalos C (2004) Ingestion and digestion of 10 species of microalgae by winged pearl oyster Pteria sterna (Gould, 1851) larvae. Aquaculture 230(1–4):417–423

Muller-Feuga A, Robert R, Cahu C, Robin J, Divanach P (2003) Uses of microalgae in aquaculture. In: Stottrup JG, McEvoy LA (eds) Live feeds in marine aquaculture. Blackwell Science Ltd, Oxford, UK, pp 253–299

Odo GE, Agwu JE, Iyaji FO, Madu JC, Ossai NI, Allison LN (2015) Mass production of rotifer (*Brachionus calyciflorus*) for aquaculture in South-Eastern Nigeria. Int J Fish Aquac 7(9):15

Ogata Y (2017) Use of freshwater Brachionus for aquaculture. In: Hagiwara A, Yoshinaga T (eds) Rotifers (pp. 75–85). Fisheries science series. Springer

Ohs CL, Cassiano EJ, Rhodes A (2009) Choosing an appropriate live feed for larviculture of marine fish. FA167. University of Florida

Papandroulakis N, Markakis G, Divanach P, Kentouri M (2000) Feeding requirements of sea bream (Sparus aurata) larvae under intensive rearing conditions. Development of a fuzzy logic controller for feeding. Aquac Eng 21:285–299

Patil V, Källqvist T, Olsen E, Vogt G, Gislerod HR (2007) Fatty acid composition of 12 microalgae for possible use in aquaculture feed. Aquac Int 15:1–9

Payne MF, Rippingale RJ, Cleary JJ (2001) Cultured copepods as food for West Australian dhufish (Glaucosoma hebraicum) and pink snapper (Pagrus auratus) larvae. Aquaculture 194:137–150

Quy MO, Fotedar R, Thy TTH (2018) Effects of extended-rotifers inclusion and live food-enrichment with probiotics on the survival, metamorphosis, development time and growth of mud crab, Scylla paramamosain (Estampador) larvae. Am J Appl Sci 15(7):375–386

Rojo-Cebreros AH, Ibarra-Castro L, Guerrero-Carlock E, Sanchez-Tellez JL, Alvarez-Lajonchere L (2017) Pilot-scale production of the rotifer *Brachionus sp.* under different culture systems. Rev Biol Mar Oceanogr 52(3):539–549

Sorgeloos P, Dhert P, Candreva P (2001) Use of the brine shrimp, Artemia spp. in marine fish larviculture. Aquaculture 200:147–159

Srivastava A, Stoss J, Hamre K (2011) A study on enrichment of the rotifer Brachionus "Cayman" with iodine and selected vitamins. Aquaculture 319(3–4):430–438

Stelzer CP (2006) Competition between two planktonic rotifer species at different temperatures: an experimental test. Freshw Biol 51(12):2187–2199

Sterzelecki FC, dos Santos Cipriano F, Vasconcelos VR, Sugai JK, Mattos JJ, Derner RB, Cerqueira VR (2020) Minimum rotifer density for best growth, survival and nutritional status of Brazilian sardine larvae, *Sardinella brasiliensis*. Aquaculture:736–264

Sun S, Chen L, Ge X, Qin J (2013) Examination of a practical method for copper enrichment of euryhaline rotifers (*Brachionus plicatilis*) as diet of *Eriocheir sinensis* zoea larvae. Aquac Nutr 19:809–817

Treece GD (2000) Artemia production for marine larval fish culture (SRAC Publ. No. 702)

Vaidya SR (2017) Role of zooplankton in aquaculture. Int J Adv Res Dev 2:951–954

Vu UN, Pham TH, Huynh PV, Huynh TG (2020) Importance of the freshwater rotifer Brachionus angularis for improved survival rate of early life history stages of pangasius catfish, *Pangasianodon hypophthalmus*. Aquac Res:1–10

Waggett RJ, Buskey EJ (2007) Copepod escape behavior in non-turbulent and turbulent hydrodynamic regimes. Mar Ecol Prog Ser 334:193–198

Wilcox JA, Tracy PL, Marcus NM (2006) Improving live feeds: effect of a mixed diet of copepod nauplii (Acartia tonsa) and rotifers on the survival and growth of first feeding larvae of the southern flounder, Paralichthys lethostigma. J World Aquacult Soc 37:113–120

Yoshimatsu T, Imoto H, Hayashi M, Toda K, Yoshimura K (1997) Preliminary Results in Improving Essential Fatty Acids Enrichment of Rotifer Cultured in High Density. Hydrobiologia 358:153–157

Yoshimura K, Tanaka K, Yoshimatsu T (2003) A novel culture system for the ultra-high-density production of the rotifer, Brachionus rotundiformis—a preliminary report. Aquaculture 227:165–172

Yuan D, Zhan X, Wang M, Wang X, Feng W, Gong Y, Hu Q (2018) Biodiversity and distribution of microzooplankton in Spirulina (Arthrospira) platensis mass cultures throughout China. Algal Res 30:38–49

Nutritional Requirements and Alternative Ingredients in Feed for Marine Finfish

4

Norfazreena Mohd Faudzi, Rita Rostika, Ching Fui Fui, Shigeharu Senoo, Rossita Shapawi, and Annita Yong Seok Kian

Abstract

The increasing demand for fish around the world in recent years has led to a significant expansion of the aquaculture industry. The need for improvement in fish nutrition to enable efficient and sustainable production is highlighted by this growth. Marine finfish, with their unique biological and physiological needs, require a balanced feed in order to maintain good health and growth. Proteins and amino acids, lipids and fatty acids, carbohydrates and vitamins and minerals are important nutritional elements for marine finfish. Proteins and essential amino acids are necessary for tissue repair and growth, while lipids supply energy and are needed for cell membrane function. Although carbohydrates are used less, they help to spare proteins for growth. Sufficient amounts of vitamins and minerals are required to promote bone formation, metabolic activities and general health. Traditional feed ingredients such as fish meal and fish oil have been used widely to produce commercial artificial feed due to their high nutritional value, yet their use raises issues with sustainability and economics. The chapter reviews several alternative feed ingredients that encourage solving the issues. By understanding and addressing the specific dietary requirements of marine finfish and incorporating alternative feed ingredients, the aquaculture industry can be expanded while reducing the environmental impact and cost of production.

N. Mohd Faudzi (✉) · C. F. Fui · S. Senoo · R. Shapawi · A. Y. Seok Kian
Higher Institution Centre of Excellence (HICoE), Borneo Marine Research Institute, Universiti Malaysia Sabah, Kota Kinabalu, Sabah, Malaysia
e-mail: fazreenafaudzi@ums.edu.my

R. Rostika
Department of Fisheries, Faculty of Fisheries and Marine Science, Universitas Padjadjaran, Bandung, Indonesia

N. M. Faudzi et al. (eds.), *Essentials of Aquaculture Practices*,
https://doi.org/10.1007/978-981-97-6699-4_4

Keywords

Nutrition · Dietary requirements · Alternative ingredients · Marine finfish

4.1 Introduction

Fish provides necessary amino acids, unsaturated fatty acids (mainly omega-3 fatty acids) and minerals that are vital for upholding a healthy lifestyle, making it a significant source of protein in human diets (Tacon and Metian 2018; FAO 2020). However, there is a limited number of wild fish in the world which is insufficient to meet the growing demand (Tacon and Metian 2018). As a result, the aquaculture industry has seen its fastest growth in producing fish worldwide, especially in Asia (FAO 2020). However, a significant obstacle to the aquaculture industry's growth is the demand for reasonably priced and high-quality feed that is nutrient-balanced. It is reported that about 50–60% of all production expenditures are related to fish feed (Gasco et al. 2018; FAO 2020). Harvest quality, growth, disease risk and total yield of the fish are all directly impacted by the feed quality (Hardy and Lee 2010). Therefore, it is critical to give cultured fish all the nutrients that are required in the right amounts, according to their specific requirement for the development stage, efficiently (NRC 2011). The majority of fish species cultured or selected for the aquaculture industry are well-documented, depending on their species, age, size and production level (NRC 2011; Craig et al. 2017; Prabu et al. 2017).

4.2 Nutritional Requirement of Marine Finfish

4.2.1 Protein and Amino Acids

Protein, the important organic material in the fish body tissue, is formed by amino acids consisting of carbon, hydrogen, oxygen and nitrogen (Wilson 2002). Protein is essential in fish feed to build new tissues during growth and reproduction or repair the existing body tissues (Kpogue et al. 2013). A previous study on tiger grouper, *Epinephelus fuscoguttatus*, revealed that the fish is utilizing protein as a primary source of energy (Giri et al. 2004). Other than that, protein is required as a source of essential amino acids such as arginine, histidine, isoleucine, leucine, lysine, methionine, phenylalanine, threonine, tryptophan and valine that cannot be synthesized by fish (Miles and Chapman 2006). Due to the expensive price of protein sources, providing excessive protein in fish feed beyond the requirement of fish led to increased feed cost (Jiang et al. 2015) and more nitrogen loss to the environment (Wu and Gatlin III 2014).

The requirement for protein is varied among fish species. Generally, the protein requirement of fish is affected by different trophic levels (Craig 2009). Carnivorous fish require more protein compared to omnivorous and herbivorous fish. As groupers are carnivorous marine fish, the requirement for dietary protein is high which is within the range of 40–55% to achieve maximum growth (Teng and Chua 1978; Chen and Tsai 1994; Shiau and Lan 1996; Luo et al. 2004, 2005). This differs from other species such as Nile tilapia (*Oreochromis niloticus*) which require 20–30% of dietary protein for optimum growth (El-Sayed 2006). Meanwhile, Ahmad (2008) indicates that African catfish (*Clarias gariepinus*) (initial BW, 9.5–10.3 g) required 30% of dietary protein for optimum growth and effective protein utilization. Furthermore, Craig (2009) stated that the life stage of fish influences protein requirements. The protein requirement is high during the early life stage and decreases as the fish grow larger (Craig 2009). According to Lupatsch and Kissil (2005), white grouper (*E. aeneus*) with a size of 2–10 g required 55% dietary protein, while 500–750 g required only 40% dietary protein for optimum growth performance. Besides that, the culture environment such as stocking density and water temperature also contributes to the variation of protein requirement by the fish (Craig 2009).

Studies have proven that providing an optimum dietary protein resulted in the highest growth. In the previous study, tiger grouper (initial body weight, BW, 8.8 g) achieved the highest growth when fed with 50% dietary protein (Shapawi et al. 2014). The study also showed that providing a higher level of protein (55%) than the optimum level did not further increase fish growth. It has been suggested that low dietary protein (40%) is not sufficient to promote better growth for the hybrid grouper (*E. fuscoguttatus* × *E. lanceolatus*) (initial BW, 4.7 and 2.6 g) in 8-week duration, respectively (Jiang et al. 2015; Rahimnejad et al. 2015). In the 8-week feeding experiment, Malabar grouper (*E. malabaricus*) (initial BW, 17.0 g) attained the highest growth when fed with 55% of dietary protein (Tuan and Williams 2007). Other than that, Japanese flounder (*Paralichthys olivaceus*) (initial BW, 22.7 g) required 45–50% of dietary protein for optimum growth in 9-week duration (Lee et al. 2002).

In addition to that, the dietary protein in fish feed influenced the utilization of nutrients. A recent study on hybrid grouper showed that the protein efficiency ratio (PER) value decreased when the dietary protein increased from 40% to 55% (Jiang et al. 2015). However, the PER value increased when the dietary protein in the feed for gilthead sea bream (*Sparus aurata*) (initial BW, 0.8 g) increased from 40% to 55% (Vergara et al. 1996). However, the PER value decreased when the dietary protein increased to 65%. Providing sufficient protein resulted in the highest PER and net protein utilization (NPU), indicating that the protein was efficiently utilized for protein synthesis (Kim et al. 2016). However, providing protein beyond the optimal level in the feed resulted in a reduction of PER and NPU (Kim and Lee 2009; Kpogue et al. 2013) which indicates that the fish did not utilize the protein.

The dietary protein level in fish feed also reflects the nutrient retention and body protein composition. Similar to the PER and NPU values, the protein retention and body protein increased with the increase of dietary protein. However, the value of protein retained in the body decreased when the protein intake was not utilized for protein synthesis, and higher energy was required to deaminate and discharge extra nitrogenous load into the environment (Siddiqui and Khan 2009). In certain species, a combination of optimum dietary protein with appropriate lipid levels increased the nutrient utilization, body protein and lipid, indicating that the dietary lipid provided allows protein to be spared. Shapawi et al. (2014) reported that tiger grouper tends to utilize non-protein energy sources. Similarly, higher protein was retained in juvenile Pacific bluefin tuna (*Thunnus orientalis*) (initial BW, 0.3 g) when dietary lipid in the feed increased from 9.2% to 17.9% which reflects a better growth and feed utilization (Biswas et al. 2009).

Apart from the utilization of protein for growth, using an optimum protein in the feed reduced nitrogenous waste in the environment (Catacutan and Coloso 1995; Tibbetts et al. 2000). Excessive dietary protein in the feed is commonly reported to increase ammonia content proportionately in the water, leading to eutrophication and changes in the aquatic ecosystem (Lazzari and Baldisserotto 2008; Oishi et al. 2010). Other than that, excessive dietary protein in fish feed increases production cost. Thus, providing an optimum dietary protein in fish feed can substantially reduce the feed cost (Kim et al. 2003; Kpogue et al. 2013).

4.2.2 Lipid and Fatty Acids

Lipid is a group of organic compounds which comprises fats, oils and triglycerides (Sargent et al. 2002). Lipid is hydrophobic, where it is insoluble in water but dissolvable in non-polar natural solvents. Lipid plays a few roles in fish feed particularly as vitality sources (Ali et al. 2000; Lee et al. 2002). It contributes twofold vitality contrasted with protein and lessens feed costs by mostly saving protein (Ali et al. 2000). In addition, lipid is essential for ordinary development and tissue maintenance (Watanabe 1982). Further, lipid acts as a transporter for fat-solvent nutrients (A, D, E and K) and a source of essential fatty acids such as eicosapentaenoic acid (EPA) and docosahexaenoic acid (DHA), which cannot be synthesized by fish (Ali et al. 2000; Jiang et al. 2015; Rahimnejad et al. 2015).

However, it has been reported that inappropriate lipid combination with protein leads to a negative impact on fish. High lipids in fish feed led to suppressed fish appetite and consequently reduced the feed intake (Giri et al. 1999; Lin and Shiau 2003; Williams et al. 2004; Yong et al. 2015). Moreover, excess lipid in the feed is commonly related to fat deposition in the liver or flesh, which affects the fish's health and market quality (Regost et al. 2001; Cheng et al. 2006; Wang et al. 2016). On the other hand, insufficient lipid supply in fish feed may lead to a deficiency of essential fatty acids, which can cause an increase in mortality and suppress the growth and abnormal development of the fish.

The lipid requirement of different fish species varies. Luo et al. (2004) reported that orange-spotted grouper (*E. coioides*) (initial BW, 10.9 g) cultured in floating cages required 10% of lipid level with fish oil as the sole lipid source for maximum growth. Similar to that, Malabar grouper (initial BW, 4.4 g) required 9% of lipid level with a combination of fish oil and corn oil (1:1) to obtain the best growth and feed efficiency (Lin and Shiau 2003). However, contradicting results have been reported on the optimum level of dietary lipid for hybrid grouper (Jiang et al. 2015; Rahimnejad et al. 2015). Jiang et al. (2015) recommended 7% lipid level for the hybrid grouper, while Rahimnejad et al. (2015) recommended doubling the lipid level at 14% for the same hybrid grouper.

Several studies reported that increasing dietary lipids in fish feed can improve fish performances, feed utilization and nutrient retention (Mohanta et al. 2013; Yong et al. 2015). An appropriate combination level of dietary lipids allows the protein to be spared for fish growth. In the previous study, increasing dietary lipid levels from 8% to 16% was reported to maximize the growth of juvenile tiger grouper (Shapawi et al. 2014). A similar observation was reported by Li et al. (2016) where the growth and FCR of giant grouper improved when the lipid level increased from 13% to 15.1%. The hybrid grouper also improved weight gain, FCR and PRE when dietary lipid increased from 8% to 16% in fish feed regardless of the protein level (Yong et al. 2019). Other than that, Pacific bluefin tuna and olive flounder also showed a tendency to perform well when dietary lipid increased in the feed (Biswas et al. 2009; Kim et al. 2010). On the other hand, no protein-sparing effect also has been reported in several species such as white sea bream (*Diplodus sargus*) and tiger puffer (*Takifugu rubripes*) (Ozorio et al. 2006; Kikuchi and Furuta 2009).

4.2.3 Carbohydrate

Carbohydrates are the least expensive energy source in the formulation of aquaculture feed (Zamora-Sillero et al. 2013). Adding carbohydrates to aquaculture feeds is important to lower feed costs and improve the binding qualities when the feed is manufactured. As feed costs account for significant total production costs, producing a cost-effective feed in large-scale aquaculture operations is very important (NRC 2011). Although fish do not specifically require carbohydrates, incorporating carbohydrates in the feed can help spare protein and lipids for other vital physiological functions such as growth and tissue repair (Ismail et al. 2018). Carbohydrates are stored as glycogen in the liver and muscle tissues. When there is a lot of activity or stress, the stored glycogen can be quickly released to meet the energy needs by the fish (Wilson 1994). Generally, carnivorous fish (Enes et al. 2010; Cui et al. 2010), herbivorous fish (Ren et al. 2015) and omnivorous fish (Tan et al. 2006; Sá et al. 2008) utilized carbohydrates from polysaccharide groups such as starch and dextrin efficiently than mono- and disaccharides. According to previous studies, the ability of fish to use various sources of carbohydrates is species-dependent (Wilson 1994; Ismail et al. 2018). Other than that, carbohydrates also mainly function to improve binding activity during the feed manufacturing process. This will reduce waste and maximize the feed intake of the fish as the structural integrity of water is maintained (NRC 2011).

4.2.4 Vitamins and Minerals

Vitamins are essential organic compounds that fish need for normal growth, development and overall health. Since most of the vitamins cannot be synthesized by fish, they must be supplied in the feed. Vitamins are categorized into two main groups, which are water-soluble vitamins and fat-soluble vitamins. Water-soluble vitamins include B vitamins (thiamine, riboflavin, niacin, pantothenic acid, pyridoxine, biotin, folic acid and cobalamins), inositol, choline and vitamin C (ascorbic acid) (Craig et al. 2017). Among other vitamins, ascorbic acid is the most important as it is a powerful antioxidant and promotes growth and normal skeletal development (Ebi et al. 2018). Other than that, most water-soluble vitamins serve as coenzymes in the body for biochemical reactions (Parker 2012). Meanwhile, there are four fat-soluble vitamins, which are vitamins A (retinol, beta-carotene), D (cholecalciferol), E (tocopherols) and K (phylloquinone). Similar to vitamin C, vitamin E plays an important role as an antioxidant. Both vitamins prevent the oxidation of dietary lipids, thus extending their shelf life (Craig et al. 2017). Each vitamin deficiency has its specific symptoms, but reduction in growth is the most common. Deficiency in ascorbic acid and folic acid may cause scoliosis (a fish condition characterized by a bent backbone) and dark colouration (Fig. 4.1) (Ebi et al. 2018).

Minerals are inorganic elements that are required in the feeds for regular body functions of fish, and they can be classified into macrominerals and microminerals. The groups are formed based on the quantity required in the feed and the amount present in the fish. Fish can partially make up for the shortage of minerals in the feed by directly absorbing the minerals from the water through their gills and skin (Craig et al. 2017). The macrominerals are calcium, sodium, chloride, potassium, chlorine, sulphur, phosphorous and magnesium, which are commonly found in the feeds. These minerals support the growth and integrity of bones and control osmotic balance. Meanwhile, iron, copper, chromium, iodine, manganese, zinc and selenium are the microminerals that are considered as trace minerals required in small quantities for enzyme and hormone systems (Parker 2012).

4.3 Fish Meal and Fish Oil as Main Ingredients in Marine Fish Feed

Fish meal (Fig. 4.2) has been used as the main protein source in formulating feeds for marine fish (Kanashiro et al. 2015; Quinto et al. 2017). Generally, fish meal can be made from wild-caught pelagic fish, and a small percentage of fish meal is rendered from the by-products or fish trimmings during processing of seafood products (Miles and Chapman 2006). Fish meal carries large quantities of energy per unit weight and is a nutrient-rich ingredient with a balanced amount of essential amino acids, fatty acids and minerals (Miles and Chapman 2006). Incorporation of fish meal into marine fish feed increases the feed efficiency and fish growth due to the high palatability and improved nutrient digestion and absorption compared to other protein sources (Lim and Lee 2009; Li et al. 2015). However, the fish meal has been reported to have a stagnant supply due to the decline of wild stock landing amount (Macusi et al. 2023) which led to a limitation on fish meal availability (Ween et al. 2017). The fish meal is currently priced at USD 1604/tonne, making it an expensive ingredient for feed production (Indexmundi 2018). In the global fish meal production, about 90% of fish meal is produced by oily pelagic fish species such as anchovy, menhaden and sardine (Barlow 2000). Meanwhile, less than 10% of that is white

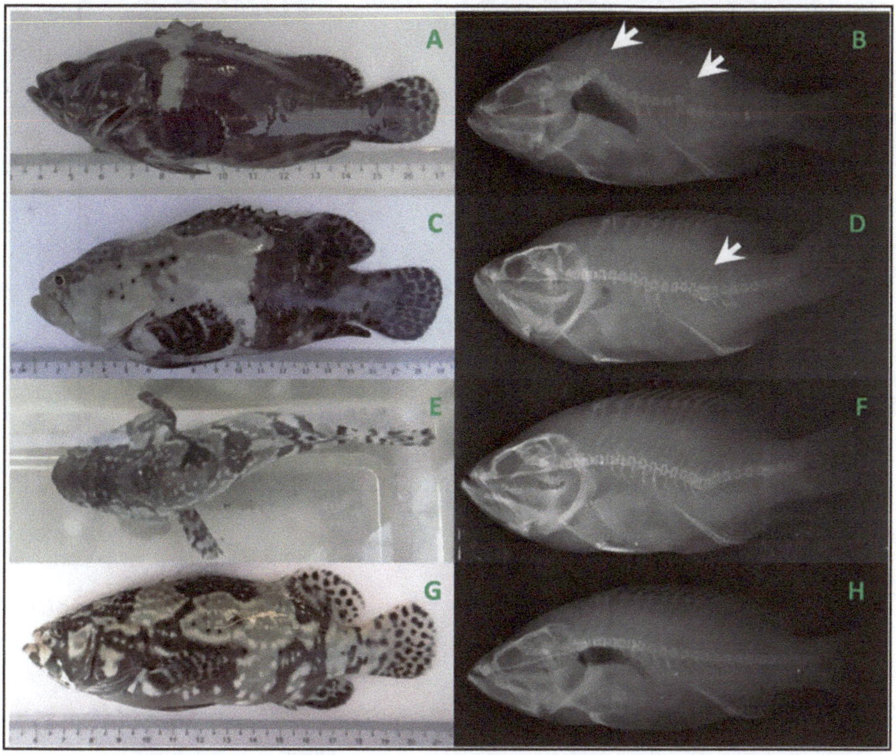

Fig. 4.1 Photos of deformed and healthy fish. (**a**) Fish exhibited a dark skin colour and suffered from kyphosis. (**b**) X-ray image of the fish suffering from kyphosis. Arrowheads point to the affected areas. (**c**) Fish exhibited a half-black skin colour and suffered from fused vertebrae. (**d**) X-ray image of the fish suffering from fused vertebrae. Arrowhead points to the affected area. (**e**) Fish develops a permanent S-shape even when at rest. (**f**) X-ray image of the fish suffering from scoliosis. Arrowhead points to the affected area. (**g**) Healthy fish. (**h**) X-ray image of a normal fish skeleton

fish meal from white fish offal such as cod and haddock (Barlow 2000). The main producers of fish meal are Peru (one-third of the total global fish meal supply), Chile, China, Norway and Denmark (Miles and Chapman 2006). The percentage of fish meal used as a protein source for different applications is shown in Table 4.1. Approximately 63% of fish meal production is used as production in aquaculture feed (Boyd 2015).

Fish oil (Fig. 4.3) has long been an important ingredient in the formulation of feeds for marine finfish due to its high content of essential fatty acids, particularly omega-3 fatty acids such as eicosapentaenoic acid (EPA) and docosahexaenoic acid (DHA) (Turchini et al. 2011; Tocher 2015). These fatty acids support an improved immune system and reproductive function and are essential for the better growth and development of marine finfish. Fish oil is typically extracted from oily fish such as anchovy, menhaden and sardine, similar to the main source of fish meal (Olsen

Fig. 4.2 Fish meal in powder form

Table 4.1 The use of global fish meal supply (Boyd 2015)

Application	Percentage (%)
Aquaculture feed	63
Pig feed	25
Poultry feed	8
Other	4

and Hasan 2012). According to Glencross (2009), fish oil is a very nutritious and energy-dense lipid source that can enhance feed efficiency and accelerate fish growth. Adding fish oil to aquaculture feed increases feed intake and utilization as the palatability improves. Similar to fish meal, the supply of fish oil is restricted due to the declining fisheries stocks and landings (FAO 2020). The price becomes more expensive due to restricted supply and rising demand (Indexmundi 2018). The cost of fish oil is currently around USD 2500 per tonne, indicating its economic importance and the difficulties in ensuring a consistent supply (FAO 2020).

The process of fish meal and fish oil production are started with the freshly caught feed-grade fish were cooked at a temperature of 85–90 °C to coagulate protein and allow oil to be released. The microorganisms in the fresh-caught fish were killed during this process. The low temperature was used to reduce the fish enzyme activity that can cause spoilage. The cooked fish was then passed into a screw-pressed chamber to press liquor and the solids to the drier. Furthermore, the liquor was decanted to remove solids and processed into fish oil, a by-product of fish meal production. Meanwhile, the drier used to dry the cooked fish contains coils that raise the temperature to 90 °C for drying to around 10% moisture after cooling. The fish meal has been ground into powder form and stored for further use in the aquaculture

Fig. 4.3 Fish oil as an ingredient in the fish feed

feed industry. Meanwhile, the fish meal by-product, fish oil, was passed through the purifier to remove solid impurities. More sophisticated refining is used to produce clear odourless fish oil.

4.4 Alternative Ingredients in Marine Fish Feed

Concerns about sustainability and the economy have been highlighted as aquaculture feeds depend on fish meal and fish oil as the main ingredients. Investigating replacement components can lessen the pressure on wild fish populations and reduce feed costs.

4.4.1 Alternative Protein Source

Animal-Based Protein

Animal-based proteins are a crucial element in aquaculture feed formulation, especially for carnivorous and omnivorous species. These proteins are essential because their superior amino acid profiles closely align with the nutritional requirements. Other than fish meal, poultry by-product meal and various rendered animal products such as meat and bone meal are frequently used as sources of animal-based proteins in aquaculture feed. The rendered portions of chicken carcasses are used to make poultry by-product meal, which is a more affordable option than fish meal (Shapawi et al. 2007; Irm et al. 2020). It offers a balanced amino acid profile appropriate for fish feeds and a high protein level (55–65%) (Chaklader et al. 2020). It also contains high lipids, vitamins and minerals, which add to the overall nutritional content of the fish feed (Galkanda-Arachchige et al. 2019). Previous studies revealed that fish meal can be substituted completely or partially with poultry by-product meal in the feeds of a variety of marine fish species including tiger grouper (Gunben et al. 2014), humpback grouper (*Cromileptes altivelis*) (Shapawi et al. 2007) and black sea bream (*Acanthopagrus schlegelii*) (Irm et al. 2020).

Meat and bone meal is a by-product of the animal rendering industry and less expensive than fish meal. This protein source also contains high protein content, well-balanced amino acids and a good source of minerals, especially phosphorus and calcium. Although it has good digestibility, previous studies reported that the value varies among fish species. This is due to its high ash content that limits its aquaculture feed use. However, it has been reported to have successfully replaced partially in the feeds for various species such as Japanese flounder (Kikuchi et al. 1997), olive flounder (Lee et al. 2012) and gilthead seabream (Moutinho et al. 2017).

Even though animal-based proteins are one of the potential options to replace fish meal as protein sources in aquaculture feeds, there are issues and challenges that have been highlighted. Variation of nutritional content in the meals is commonly reported due to variability in raw material composition and processing method for each batch or production. Such variation resulted in an imbalance of amino acid content and variability in the ability of fish to digest the protein sources. Applying standard processing methods and controlling the raw materials used can play a good role in guaranteeing product quality and help promote sustainable aquaculture without affecting fish growth and health.

Plant-Based Protein

As a more affordable and environmentally friendly alternative to fish meal, plant-based protein such as soybean meal, corn gluten meal, canola meal and others plays a more significant role in aquaculture feed composition. Compared to animal-based protein, plant-based protein sources offer several advantages such as being readily available, less expensive and having a less environmental effect. Soybean meal is a plant-based protein commercially used in the aquaculture feed industry (FAO 2016). Compared to other plant proteins, soybean meal contains high protein, a better amino acid profile, lower carbohydrate and fibre content and better digestibility of nutrients (Gatlin et al. 2007; Yang et al. 2011). However, using soybean meal and other plant-based protein as protein replacements for carnivorous marine finfish remains a challenge (Kumar et al. 2012). This is due to the anti-nutritional factors (ANFs) such as phytic acid, trypsin inhibitors and saponins that can affect nutrient digestibility and fish health (Gatlin et al. 2007). Other than that, lacking certain essential amino acids such as methionine and lysine can be a factor that limits the level of replacement in the fish feed (Biswas et al. 2017; Mohd Faudzi et al. 2022).

Few studies have been conducted on protein replacement by soybean meal in several marine finfish species. Firdaus et al. (2016) revealed that feeding the hybrid grouper up to 60% replacement level with defatted soybean meal did not affect the palatability of the feed. However, the study demonstrated that increasing the level of protein replacement with defatted soybean meal from 0% to 80% reduced the growth performances of hybrid grouper. In other marine fish such as Asian seabass

(*Lates calcarifer*) and tiger puffer, increasing the level of defatted soybean meal beyond the optimal level of protein replacement in the feeds reduced the growth performances and feed utilization (Tantikitti et al. 2005; Kikuchi and Furuta 2009). Other than soybean meal, it has been reported that green peas (*Pisum sativum*) also can be another potential alternative protein source (Ganzon-Naret 2013). Partial replacement of green pea meal did not have an adverse effect on the growth and feed utilization of Atlantic salmon (*Salmo salar* L.) (Carter and Hauler 2000).

Feeding marine finfish with plant-based protein feed is commonly associated with inflammation and changes in the morphology of the intestine. The increasing level of DSM in the feeds for orange-spotted grouper led to a reduction in villus height and inflammation in the intestines (Wang et al. 2017). A similar finding was obtained when fish meal protein was partially replaced with defatted soybean meal in the feed for tiger grouper juveniles (Shapawi et al. 2013a, b). However, the hybrid grouper showed an interesting finding, whereby replacement of fish meal protein with defatted soybean meal at 30% level did not affect the morphology of the fish intestine (Mohd Faudzi et al. 2022). The changes in villus height in fish intestines influence the absorption area of nutrients, thus reflecting the growth performance and nutrient digestibility (Wang et al. 2017).

Proteins derived from plants have a great potential to replace fish meal in aquaculture feeds sustainably and economically. The nutritional constraints of plant-based proteins can be overcome by adding necessary amino acids, using enzyme treatments and applying sophisticated processing methods. The strategies addressed will enable aquaculture feed production to lessen the demand for fish meal, which then promotes sustainable and cost-effective feeds while at the same time preserving the growth and health of cultured fish.

Insect Meal

As the nutritional composition is more similar to fish meal, insect meal is a promising alternative protein source for aquaculture feed production. While insects have been studied and tested as a possible food source for animals since 1969 (Alfiko et al. 2022), there has been little focus on incorporating them into aquaculture feeds until the 2000s (Ogunji et al. 2008). Research on the impact of replacing insect meal for fish meal on the growth and health of aquaculture species has been conducted for the past 20 years (Daniel 2018). Numerous additional study areas exist, such as large-scale insect farming and breeding, safety concerns, nutritional value and identification of insect species (Alfiko et al. 2022). Recently, insect meal has been used to partially replace a fish meal as a protein source in feeding trials for numerous aquaculture species such as rainbow trout (*Oncorhynchus mykiss*) (St-Hilaire et al. 2007), gilthead seabream (Karapanagiotidis et al. 2014) and hybrid grouper (Mohamad-Zulkifli et al. 2019). Although it depends on fish and insect species, the feeding trial showed promising results of insect meal as feed ingredients.

The insect meal production is expanding quickly in North China, Southeast Asian countries, North America, Europe and Australia. There are about eight promising insect species used for aquaculture feed production, which are silkworms (*Bombyx mori*) (Duan et al. 2010), black soldier fly (*Hermetia illucens*) (Barry

2004), housefly (*Musca domestica*) (Malik et al. 2007), yellow mealworm (*Tenebrio molitor*) (Li et al. 2013), lesser mealworm (*Alphitobius diaperinus*) (Rumbos et al. 2019), house cricket (*Acheta domesticus*) (Hessler Frelinckx 2019), banded cricket (*Gryllodes sigillatus*) (Vandeweyer et al. 2018) and Jamaican field cricket (*Gryllus assimilis*) (Masson et al. 2020). Table 4.2 showed the nutritional composition of fish meal, soybean meal and insect meal.

There are several procedures to be followed in order to produce the insect meal. The continuous supply of biomass insects should be identified first as the nutrient composition is greatly influenced (Tschirner and Simon 2015). The insect biomass was decontaminated through a thermal or radiation process and dried with convection. The insect pupa was broken into small pieces using the grinding method. It is necessary to undergo a defatted process in some insect species by aqueous and solvent-based methods. Many methods are available to extract protein from the insects, such as enzymes, microwave and high-pressure extraction (Pojić et al. 2018). It is also important to remove chitin from some species containing chitin through fermentation with microorganisms and enzymes (Rumpold et al. 2017).

Since insect meal has a high possibility of becoming a viable alternative protein, it is crucial to improve its growing, harvesting and processing into mass quantities in order to make it profitable for the feed industry (Hawkey et al. 2021). Though no best practices have been established, mass rearing and breeding of insects will not require a complicated infrastructure, but environmental conditions must be managed to maximize insect growth. High-quality animal protein will be generated due to the development and technology application throughout insect farming industries.

4.4.2 Alternative Lipid Source

Plant-Based Oil

Plant oils are lipid materials derived from the plants, either from fruits, flowers or seeds. Plant oils are the most common oil used as an alternative to replace fish oil in aquaculture feed. There are several types of plant oils, such as palm oil, canola oil and soybean oil. Palm oil, canola oil and soybean oil are the most abundant plant oils in the world and their production has steadily increased at lower prices than fish oil (Turchini et al. 2011). Most plant oils are poor sources of omega-3 fatty acids compared to fish oil (Table 4.3). However, plant oils are rich in omega-3 and omega-9 fatty acids, mainly linoleic acid and oleic acid. They also have relatively higher saturated fatty acids and monounsaturated fatty acids, which can be catabolized by the fish as energy sources (Bell et al. 2001; Regost et al. 2003; Ng et al. 2007).

As the demand for fish oil continues to increase and exceeds the supply, finding an alternative to fish oil as a lipid source in aquaculture feed is crucial. Several studies have proven that plant oils can replace fish oil partially and even completely depending on the type of plant oil and aquaculture fish species. The previous studies suggested that palm oil can replace 40% of fish oil in the feed for Japanese seabass (*Lateolabrax japonicus*) (Gao et al. 2012) and red sea bream (*Pagrus major*)

Table 4.2 Nutritional components (%) in fish meal, soybean meal and eight insect species (Amza and Tamiru 2017; Makkar et al. 2014; Mlček et al. 2018; Tran et al. 2015; Yu et al. 2018; Alfiko et al. 2022)

			Species of insect for insect meal							
Nutritional components	Fish meal	Soybean meal	Silkworm pupae meal	Black soldier fly larvae	Housefly maggot meal	Yellow mealworm	Lesser mealworm	House cricket	Banded cricket	Jamaican field cricket
Crude protein	70.6	51.8	60.7	42.1	50.4	52.8	57.3	63.3	59.8	55.5
Crude lipid	9.9	2.0	25.7	26.0	18.9	36.1	8.5	17.3	13.3	11.8
Calcium	4.34	0.39	0.38	7.56	0.47	0.27	0.13	1.01	0.2	NA
Phosphorus	2.79	0.69	0.60	0.90	1.6	0.78	0.11	0.79	1.04	NA
Eicosapentaenoic acid (EPA)	0.3–11.1	–	0.21–0.79	–	–	–	–	–	–	1.16
Docosahexaenoic acid (DHA)	3.7–29.1	–	1.96–5.52	–	–	–	–	–	0.07	0.15

NA data is not available

Table 4.3 Fatty acid composition (% total fatty acids) of fish oil and plant oils (NRC 1993; Gunstone et al. 1994)

Oil	SFA	MUFA	LA	AA	Omega-6	ALA	EPA	DHA	Omega-3
Fish oil	19.4	46.0	1.4	1.6	3.0	0.6	14.4	12.0	27.0
Palm oil	48.8	37.0	9.1	NA	9.1	0.2	NA	NA	0.2
Soybean oil	14.2	23.2	51.0	NA	51.0	6.8	NA	NA	6.8
Canola oil	4.6	62.3	20.2	NA	20.2	12.0	NA	NA	12.0

SFA saturated fatty acid, *MUFA* monounsaturated fatty acid, *LA* linoleic acid, *AA* arachidonic acid, *ALA* a-linolenic acid, *EPA* eicosapentaenoic acid, *DHA* docosahexaenoic acid, *NA* data is not available

(Komilus et al. 2008) without affecting survival, feed utilization and health of the fish. A complete replacement of fish oil with palm oil can be applied to barramundi feed without affecting growth performance (Wan Ahmad et al. 2013). Similar to that, soybean oil and linseed oil can replace fish oil 100% in the feed for sharpsnout seabream (*Diplodus puntazzo*) (Piedecausa et al. 2007). Other than that, previous study suggests that replacing fish oil with plant oil has enhanced immune functions in certain fish species. The leucocyte superoxide anion production was significantly higher in Malabar grouper fed with plant oil-based feed than fish oil-based feed (Lin and Shiau 2007).

4.5 Conclusion

Understanding and fulfilling the dietary needs of marine fish is essential for sustainable aquaculture practices. The industry has significantly benefited from using traditional feed ingredients such as fish meal and fish oil. However, there are also substantial advantages to exploring and incorporating alternative ingredients in terms of sustainability, cost-effectiveness and environmental impact. The aquaculture industry can support ecological balance and economic viability while improving the health and growth of marine finfish by carrying out more research and development of these alternatives.

Acknowledgement We are grateful to the Higher Institution Centre of Excellence (HICoE), Borneo Marine Research Institute, Universiti Malaysia Sabah, for providing the necessary resources and support throughout the completion of the chapter.

References

Ahmad MH (2008) Response of African catfish, *Clarias gariepinus* to different dietary protein and lipid levels in practical diets. J World Aquac Soc 39(4):541–548

Alfiko Y, Xie D, Astuti RT, Wong J, Wang L (2022) Insect as a feed ingredient for fish culture: status and trends. Aquac Fish 7(2):166–178

Ali A, Al-Ogaily SM, Al-Asgah NA, Ali S (2000) Effect of dietary lipid source on the growth performance and body composition of *Oreochromis niloticus*. Pak Vet J 20(2):57–63

Amza N, Tamiru M (2017) Insects as an option to conventional protein sources in animal feed: a review paper. Glob J Sci Front Res 17:12

Barlow SM (2000) Fishmeal and fish oil: sustainable feed ingredients for aquafeeds. Glob Aquac Advocate 4:85–88

Barry T (2004) Evaluation of the economic, social, and biological feasibility of bioconverting food wastes with the black soldier fly (*Hermetia illucens*). University of North Texas, Denton, Texas

Bell JG, McEvoy J, Tocher DR, McGhee F, Campbell PJ, Sargent JR (2001) Replacement of fish oil with rapeseed oil in diets of Atlantic salmon (*Salmo salar*) affects tissue lipid compositions and hepatocyte fatty acid metabolism. J Nutr 131:1535–1543

Biswas BK, Ji SC, Biswas AK, Seoka M, Kim YS, Kawasaki KI, Takii K (2009) Dietary protein and lipid requirements for the Pacific bluefin tuna, *Thunnus orientalis* juvenile. Aquaculture 288:114–119

Biswas A, Araki H, Sakata T, Nakamori T, Kato K, Takii K (2017) Fish meal replacement by soy protein from soymilk in the diets of red sea bream, *Pagrus major*. Aquac Nutr 23(6):1379–1389

Boyd CE. (2015) Overview of aquaculture feeds. Feed and Feeding Practices in Aquaculture. 3–25

Carter CG, Hauler RC (2000) Fish meal replacement by plant meals in extruded feeds for Atlantic salmon, *Salmo salar* L. Aquaculture 185(3–4):299–311

Catacutan MR, Coloso RM (1995) Effect of dietary protein to energy ratios on growth, survival, and body composition of juvenile Asian seabass, *Lates calcarifer*. Aquaculture 131(1):125–133

Chaklader MR, Siddik MAB, Fotedar R (2020) Total replacement of fishmeal with poultry by-product meal affected the growth, muscle quality, histological structure, antioxidant capacity and immune response of juvenile barramundi, *Lates calcarifer*. PLoS One 15(11):e0242079

Chen HY, Tsai JC (1994) Optimal dietary protein level for the growth of juvenile grouper, *Epinephelus malabaricus* fed semipurified diets. Aquaculture 119:265–271

Cheng AC, Chen CY, Liou CH, Chang CF (2006) Effects of dietary protein and lipid on blood parameters and superoxide anion production in the grouper, *Epinephelus coioides* (Serranidae: Epinephelinae). Zool Stud 45(4):492–502

Craig S (2009) Understanding fish nutrition, feeds and feeding. Virginia State University. Pp:1–4

Craig S, Kuhn D, Schwarz M (2017) Understanding fish nutrition, feeds and feeding. Virginia Coop Ext:1–6

Cui XJ, Zhou QC, Liang HO, Yang J, Zhao LM (2010) Effects of dietary carbohydrate sources on the growth performance and hepatic carbohydrate metabolic enzyme activities of juvenile cobia (*Rachycentron canadum* Linnaeus.). Aquac Res 42:99–107

Daniel N (2018) A review on replacing fish meal in aqua feeds using plant protein sources. Int J Fish Aquat Stud 6(2018):164–179

Duan J, Li R, Cheng D, Fan W, Zha X, Cheng T (2010) SilkDB v2.0: a platform for silkworm (*Bombyx mori*) genome biology. Nucleic Acids Res 38(2010):D453–D456

Ebi I, Yong ASK, Lim LS, Shapawi R (2018) Dietary ascorbic acid requirement for the optimum growth performances and normal skeletal development in juvenile hybrid grouper, *Epinephelus fuscoguttatus* × *Epinephelus lanceolatus*. J King Saud Univ Sci 30(4):493–499

El-Sayed AFM (2006) Tilapia culture. CABI Publishing, Wallingford, p 274

Enes P, Peres H, Couto A, Oliva-Teles A (2010) Growth performance and metabolic utilization of diets including starch, dextrin, maltose or glucose as carbohydrate source by gilthead sea bream (*Sparus aurata*) juveniles. Fish Physiol Biochem 36:903–910

FAO (2016) The State of World Fisheries and Aquaculture. Food and Agriculture Organization of the United Nations, Rome

FAO (2020) The state of world fisheries and aquaculture 2020. Sustainability in action. Food and agriculture organization of the United Nations, Rome, Italy

Firdaus RF, Lim LS, Kawamura G, Shapawi R (2016) Assessment on the acceptability of hybrid grouper, *Epinephelus fuscoguttatus* ♀ × *Epinephelus lanceolatus* ♂ to soybean meal-based diets. AACL Bioflux 9(2):284–290

Galkanda-Arachchige HSC, Wilson AE, Davis DA (2019) Success of fish meal replacement through poultry by-product meal in aquaculture feed formulations: a meta-analysis. Rev Aquac:1–13

Ganzon-Naret ES (2013) The use of green pea (*Pisum sativum*) as alternative protein source for fish meal in diets for Asian sea bass, *Lates calcarifer*. AACL Bioflux 6(4):399–406

Gao J, Koshio S, Ishikawa M, Yokoyama S, Ren T, Komilus CF, Han Y (2012) Effects of dietary palm oil supplements with oxidized and non-oxidized fish oil on growth performances and fatty acid compositions of juvenile Japanese seabass, *Lateolabrax japonicus*. Aquaculture 324-325:97–103

Gasco L, Gai F, Maricchiolo G, Genovese L, Ragonese S, Bottari T, Caruso G (2018) Fishmeal alternative protein sources for aquaculture feeds. In: Feeds for the aquaculture sector. Springer briefs in molecular science. Springer, Cham, pp 1–28

Gatlin DM, Barrows FT, Brown P (2007) Expanding the utilization of sustainable plant products in aquafeeds. Aquac Res 38:551–579

Giri NA, Suwirya K, Marzuqi M (1999) Effect of dietary protein and energy on growth of juvenile humpback grouper (*Cromileptes altivelis*). Indonesian Fish Res J 8:1

Giri NA, Suwirya K, Marzuqi M (2004) Optimum level of dietary protein and lipid for rearing juvenile tiger grouper (*Epinephelus fuscoguttatus*). In: Rimmer MA, McBride S, Williams KC (eds) Advances in grouper aquaculture. ACIAR Monograph, Canberra, Australia, pp 92–94

Glencross BD (2009) Exploring the nutritional demand for essential fatty acids by aquaculture species. Rev Aquac 1:71–124

Gunben EM, Senoo S, Yong ASK, Shapawi R (2014) High potential of poultry by-product meal as a main protein source in the formulated feeds for a commonly cultured grouper in Malaysia (*Epinephelus fuscoguttatus*). Sains Malaysiana 43(3):399–405

Gunstone FD, Harwood JL, Padley FB (1994) The lipid handbook. The University Press, Cambridge

Hardy RW, Lee C (2010) Aquaculture feed and seafood quality. Bull Fish Res Agen 31:43–50

Hawkey KJ, Lopez-Viso C, Brameld JM, Parr T, Salter AM (2021) Insects: a potential source of protein and other nutrients for feed and food. In: Lewin HA, Roberts RM (eds) Annual review of animal biosciences, pp 333–354

Hessler Frelinckx JC (2019) Behavioural study of the house cricket *(Acheta domesticus)*: behavioural study of the house cricket *(Acheta domesticus)*. Swedish University of Agricultural Science, Uppsala

Indexmundi (2018) The price of fish meal and fish oil. Available from URL: http://www.indexmundi.com/commodities

Irm M, Taj S, Jin M, Luo J, Andriamialinirina HJT, Zhou Q (2020) Effects of replacement of fish meal by poultry by-product meal on growth performance and gene expression involved in protein metabolism for juvenile black sea bream (*Acanthoparus schlegelii*). Aquaculture 528:735544

Ismail R, Yong ASK, Lim LS, Kawamura G, Shapawi R (2018) Utilization of different dietary carbohydrate sources in hybrid grouper, tiger grouper (*Epinephelus fuscoguttatus*) x giant grouper (*E. lanceolatus*) juveniles. Int J Aquat Sci 9(2):85–92

Jiang S, Wu X, Li W, Wu M, Luo Y, Lu S, Lin H (2015) Effects of dietary protein and lipid levels on growth, feed utilization, body and plasma biochemical compositions of hybrid grouper, *Epinephelus lanceolatus* X *Epinephelus fuscoguttatus* juveniles. Aquaculture 446:148–155

Kanashiro E, Díaz-Rosales P, Castro C, Dió́genes A, Oliva-Teles A, Peres H (2015) Effect taurine and hydroxyproline supplementation in low fish meal diets for European seabass (*Dicentrarchus labrax*). In: Aquaculture Europe 2014 San Sebastian, Spain, pp 15–17

Karapanagiotidis I, Daskalopoulou E, Vogiatzis I, Rumbos C, Mente E, Athanassiou C (2014) Substitution of fishmeal by fly *Hermetia illucens* prepupae meal in the diet of gilthead seabream (*Sparus aurata*). HydroMediT, November 13–15, Volos, Greece, pp. 110–114

Kikuchi K, Furuta T (2009) Use of defatted soybean meal and blue mussel meat as substitute for fish meal in the diet of tiger puffer, *Takifugu rubripes*. J World Aquacult Soc 40:472–482

Kikuchi K, Sato T, Furuta T, Sakaguchi I, Deguchi Y (1997) Use of meat and bone meal as a protein source in the diet of juvenile Japanese flounder. Fish Sci 63:29–32

Kim SS, Lee KJ (2009) Dietary protein requirement of juvenile tiger puffer, (*Takifugu rubripes*). Aquaculture 287:219–222

Kim K, Wang X, Bai SC (2003) Re-evaluation of the dietary protein requirement of Japanese flounder (*Paralichthys olivaceus*). J World Aquac Soc 34(2):133–139

Kim KD, Kang YJ, Lee HYM, Kim KW, Jang MS, Choi SM, Lee SM, Cao SH (2010) Effects of dietary protein and lipid levels on growth and body composition of subadult olive flounder, *Paralichthys olivaceus*, at a suboptimal water temperature. J World Aquacult Soc 41(S2):263–269

Kim KW, Moniruzzaman M, Kim KD, Han HS, Yun H, Lee S, Bai SC (2016) Re-evaluation of the optimum dietary protein level for maximum growth of juvenile barred knifejaw, *Oplegnathus fasciatus* reared in cages. Fish Aquat Sci 19:24

Komilus CF, Shichi N, Koshio S, Ishikawa M, Yokoyama S, Michael FR, Gao J, Makita C (2008) Influences of palm oil blended with fish oil on growth performances and lipid profiles of red sea bream, *Pagrus major*. Aquac Sci 56:317–326

Kpogue D, Gangbazo H, Fiogbe E (2013) A preliminary study on the dietary protein requirement of *Parachanna obscura* (Günther, 1861) larvae. Turk J Fish Aquat Sci 13:111–117

Kumar V, Sinha AK, Makkar HPS, Boeck GD, Becker K (2012) Phytate and phytase in fish nutrition. J Anim Physiol Anim Nutr 96:335–364

Lazzari R, Baldisserotto B (2008) Nitrogen and phosphorus waste in fish farming. Boletim Instituto Pesca, Sao Paulo 34(4):591–600

Lee SM, Jeon IG, Lee JY (2002) Effects of digestible protein and lipid levels in practical diets on growth, protein utilization and body composition of juvenile rockfish (*Sebastes schlegel*). Aquaculture 211:227–239

Lee J, Choi IC, Kim KT, Cho SH, Yoo JY (2012) Response of dietary substitution of fishmeal with various protein sources on growth, body composition and blood chemistry of olive flounder (*Paralichthys olivaceus*, Temminck and Schlegel, 1846). Fish Physiol Biochem 38:735–744

Li L, Zhao Z, Liu H (2013) Feasibility of feeding yellow mealworm (*Tenebrio molitor* L.) in bioregenerative life support systems as a source of animal protein for humans. Acta Astronaut 92(2013):103–109

Li PY, Wang JY, Song ZD, Zhang LM, Zhang H, Li XX, Pan Q (2015) Evaluation of soy protein concentrate as a substitute for fishmeal in diets for juvenile starry flounder, *Platichthys stellatus*. Aquaculture 448:578–585

Li W, Wu X, Lu S, Jiang S, Luo Y, Wu M, Wang J (2016) Effects of different dietary carbohydrate/lipid ratios on growth, feed utilization and body composition of early giant grouper, *Epinephelus lanceolatus* juveniles. J Aquac Res Dev 7:415

Lim SJ, Lee KJ (2009) Partial replacement of fish meal by cottonseed meal and soybean meal with iron and phytase supplementation for parrot fish Oplegnathus fasciatus. Aquaculture 290(3):283–289

Lin XY, Shiau SY (2003) Dietary lipid requirement of grouper, *Epinephelus malabaricus* and effects on immune responses. Aquaculture 225:243–250

Lin YH, Shiau SY (2007) Effects of dietary blend of fish oil with corn oil on growth and non-specific immune responses of grouper, *Epinephelus malabaricus*. Aquac Nutr 13:137–144

Luo Z, Liu YJ, Mai KS, Tian LX, Liu DH, Tan XY (2004) Optimal dietary protein requirement of grouper *Epinephelus coioides* juveniles fed isoenergetic diets in floating net cages. Aquac Nutr 10:247–252

Luo Z, Liu YJ, Mai KS, Tian LX, Yang HJ, Liu D (2005) Dietary L-methionine requirement of juvenile grouper *Epinephelus coloides* at a constant dietary cystine level. Aquaculture 249:409–418

Lupatsch I, Kissil GW (2005) Feed formulations based on energy and protein demands in white grouper, *Epinephelus aeneus*. Aquaculture 24:83–95

Macusi ED, Cayacay MA, Borazon EQ, Sales AC, Habib A, Fadli N, Santos MD (2023) Protein Fishmeal Replacement in Aquaculture: A Systematic Review and Implications on Growth and Adoption Viability. Sustainability. 15(16):12500

Makkar HPS, Tran G, Henze V, Ankers P (2014) State-of-the-art on use of insects as animal feed. Anim Feed Sci Technol 197(2014):1–33

Malik A, Singh N, Satya S (2007) House fly (*Musca domestica*): a review of control strategies for a challenging pest. J Environ Sci Health B 42(2007):453–469

Masson MV, de Souza Tavares W, Alves JM, Ferreira-Filho PJ, Barbosa LR, Wilcken CF (2020) Bioecological aspects of the common black field cricket, *Gryllus assimilis* (Orthoptera: Gryllidae) in the laboratory and in Eucalyptus (Myrtaceae) plantations. J Orthoptera Res 29(2020):83

Miles RD, Chapman FA. (2006) The benefits of fish meal in aquaculture diets. Fisheries and Aquatic Sciences Department, UF/IFAS Extension. Pp: 1–6

Mlček J, Adámková A, Adámek M, Borkovcová M, Bednarová M, Kourimska L (2018) Selected nutritional values of field cricket (*Gryllus assimilis*) and its possible use as a human food. Indian J Tradit Knowl 17(2018):518–524

Mohamad-Zulkifli NFN, Yong ASK, Kawamura G, Lim LS, Senoo S, Devic E, Mustafa S, Shapawi R (2019) Apparent digestibility coefficient of black soldier fly (*Hermetia illucens*) larvae in formulated diets for hybrid grouper (*Epinephelus fuscoguttatus* ♀ x *Epinephelus lanceolatus* ♂). AACL Bioflux 12(2):513–522

Mohanta KN, Subramanian S, Korikanthimath VS (2013) Effect of dietary protein and lipid levels on growth, nutrient utilization and whole-body composition of blue gourami, *Trichogaster trichopterus* fingerlings. J Anim Physiol Anim Nutr 97:126–136

Mohd Faudzi N, Shapawi R, Senoo S, Biswas A, Yong ASK (2022) The performances of hybrid grouper, *Epinephelus fuscoguttatus* × *E. lanceolatus* fed with defatted soybean meal-based feeds with supplementation of phytase. Malays Appl Biol 51(6):73–83

Moutinho S, Martınez-Llorens S, Tomas-Vidal A, Jover-Cerda M, Oliva-Teles A, Peres H (2017) Meat and bone meal as partial replacement for fish meal in diets for gilthead seabream (*Sparus aurata*) juveniles: growth, feed efficiency, amino acid utilization, and economic efficiency. Aquaculture 468(1):271–277

Ng WK, Tocher DR, Bell JG (2007) The use of palm oil in aquaculture feeds for salmonid species. Eur J Lipid Sci Technol 109:394–399

NRC (1993) Nutrient requirements of fish. National Research Council, National Academy Press, Washington

NRC (2011) Nutrient requirements of fish and shrimp. National Research Council. The National Academies Press, Washington, DC

Ogunji J, Schulz C, Kloas W (2008) Growth performance, nutrient utilization of Nile tilapia *Oreochromis niloticus* fed housefly maggot meal (magmeal) diets. Turk J Fish Aquat Sci 8(2008):141–147

Oishi CA, Nwanna LC, Filho MN (2010) Optimum dietary protein requirement for Amazonian Tambaqui, *Colossoma macropomum* Cuvier, fed fish meal free diets. Acta Amazon 40(4):757–762

Olsen RL, Hasan MR (2012) A limited supply of fishmeal: impact on future increases in global aquaculture production. Trends Food Sci Technol 27:120–128

Ozorio ROA, Valente LMP, Pousao-Ferreira P, Oliva-Teles A (2006) Growth performance and body composition of white seabream (*Diplodus sargus*) juveniles fed diets with different protein and lipid levels. Aquac Res 37:255–263

Parker R (2012) Fundamentals of nutrition in aquaculture. In: Aquaculture Science. Delmar Cengage Learning, USA, pp 252–269

Piedecausa MA, Mazón MJ, García García B, Hernández MD (2007) Effects of total replacement of fish oil by vegetable oils in the diets of sharpsnout seabream (*Diplodus puntazzo*). Aquaculture 263:211–219

Pojić M, Misan A, Tiwari B (2018) Eco-innovative technologies for extraction of proteins for human consumption from renewable protein sources of plant origin. Trends Food Sci Technol 75:93–104

Prabu E, Felix S, Felix N, Ahilan B, Ruby P (2017) An overview on significance of fish nutrition in aquaculture industry. Int J Fish Aquat 5(6):349–355

Quinto BPT, Albuquerque JV, Bezerra RS, Peixoto S, Soares R (2017) Replacement of fish meal by two types of fish protein hydrolysates in feed for postlarval shrimp, *Litopenaeus vannamei*. Aquac Nutr 24:2

Rahimnejad S, Bang IC, Park JY, Sade A, Choi J, Lee SM (2015) Effects of dietary protein and lipid levels on growth performance, feed utilization and body composition of juvenile hybrid grouper, *Epinephelus fuscoguttatus* × *E. lanceolatus*. Aquaculture 446:283–289

Regost C, Arzel J, Cardinal M, Robin J, Laroche M, Kaushik SJ (2001) Dietary lipid level, hepatic lipogenesis and flesh quality in turbot (*Psetta maxima*). Aquaculture 193:291–309

Regost C, Arzel J, Robin JH, Rosenlund G, Kaushik SJ (2003) Total replacement of fish oil by soybean or linseed oil with a return to fish oil in turbot (*Psetta maxima*): growth performance, flesh fatty acid profile and lipid metabolism. Aquaculture 217:465–482

Ren M, Habte-Tsion HM, Xie J, Liu B, Zhou Q, Ge X, Pan L, Chen R (2015) Effects of dietary carbohydrate source on growth performance, diet digestibility and liver glucose enzyme activity in blunt snout bream, *Megalobrama amblycephala*. Aquaculture 438:75–81

Rumbos CI, Karapanagiotidis IT, Mente E, Athanassiou CG (2019) The lesser mealworm *Alphitobius diaperinus*: a noxious pest or a promising nutrient source? Rev Aquac 11(2019):1418–1437

Rumpold BA, Klocke M, Schluter O (2017) Insect biodiversity: underutilized bioresource for sustainable applications in life sciences. Reg Environ Chang 17:1445–1454

Sá R, Ferreira-Pousão P, Oliva-Teles A (2008) Effect of dietary starch source (normal versus waxy) and protein levels on the performance of white sea bream *Diplodus sargus* (Linnaeus) juveniles. Aquac Res 39:1069–1076

Sargent JR, Tocher DR, Bell JG (2002) The lipids. In: Halver JE, Hardy RW (eds) Fish nutrition, 3rd edn. Elsevier (Academic Press), San Diego, CA, pp 181–257

Shapawi R, Ng WK, Mustafa S (2007) Replacement of fish meal with poultry by-product meal in diets formulated for the humpback grouper, *Cromileptes altivelis*. Aquaculture 273:118–126

Shapawi R, Ebi I, Yong ASK (2013a) Soybean meal as a source of protein in formulated diets for tiger grouper, *Epinephelus fuscoguttatus* juvenile. Part I: effects on growth, survival, feed utilization and body compositions. Agric Sci 4:317–323

Shapawi R, Ebi I, Yong ASK, Chong M, Lu KC, Sade A (2013b) Soybean meal as a source of protein in formulated diets for tiger grouper, *Epinephelus fuscoguttatus* juvenile. Part II. Improving diet performances with phytase supplementation. Agric Sci 4:19–24

Shapawi R, Ebi I, Yong ASK, Ng WK (2014) Optimizing the growth performance of brown-marbled grouper, *Epinephelus fuscoguttatus* (Forskal), by varying the proportion of dietary protein and lipid levels. Anim Feed Sci Technol 191:98–105

Shiau SY, Lan CW (1996) Optimum dietary protein level and protein to energy ratio for growth of grouper (*Epinephelus malabaricus*). Aquaculture 145:259–266

Siddiqui TQ, Khan MA (2009) Effects of dietary protein levels on growth, feed utilization, protein retention efficiency and body composition of young *Heteropneustes fossilis* (Bloch). Fish Physiol Biochem 35:479–488

St-Hilaire S, Cranfill K, McGuire M, Mosley E, Tomberlin J, Newton L, Sealey W, Sheppard C, Irving S (2007) Fish offal recycling by the black soldier fly produces a foodstuff high in omega-3 fatty acids. J World Aquacult Soc 38:309–313

Tacon AG, Metian M (2018) Food matters: fish, income and food supply a comparative analysis. Rev Fish Sci 26(1):15–28

Tan Q, Xie S, Zhu X, Lei W, Yang Y (2006) Effect of dietary carbohydrate sources on growth performance and utilization for gibel carp (*Carassius auratus gibelio*) and Chinese longsnout catfish (*Leiocassis longirostris* Günther). Aquac Nutr 12:61–70

Tantikitti C, Sangpong W, Chiavareesajja S (2005) Effects of defatted soybean protein levels on growth performance and nitrogen and phosphorus excretion in Asian seabass (*Lates calcarifer*). Aquaculture 248:41–50

Teng SK, Chua TE (1978) Effect of stocking density on the growth of estuary grouper *Epinephelus salmoides* Maxwell, cultured in floating net-cages. Aquaculture 15:273–287

Tibbetts SM, Lall SP, Anderson DM (2000) Dietary protein requirement of juvenile American eel (*Anguilla rostrata*) fed practical diets. Aquaculture 186:145–155

Tocher DR (2015) Omega-3 long-chain polyunsaturated fatty acids and aquaculture in perspective. Aquaculture 449:94–107

Tran G, Heuze V, Makkar H (2015) Insects in fish diets. Anim Front 5(2015):37–44

Tschirner M, Simon A (2015) Influence of different growing substrates and processing on the nutrient composition of black soldier fly larvae destined for animal feed. J Insects Food Feed 1(2015):249–259

Tuan LA, Williams KC (2007) Optimum dietary protein and lipid specifications for juvenile malabar grouper, *Epinephelus malabaricus*. Aquaculture 267:129–138

Turchini GM, Ng WK, Tocher DR (2011) Fish oil replacement and alternative lipid sources in aquaculture feeds. Taylor and Francis, CRC Press, Boca Raton, p 533

Vandeweyer D, Wynants E, Crauwels S, Verreth C, Viaene N, Claes J (2018) Microbial dynamics during industrial rearing, processing, and storage of tropical house crickets (*Gryllodes sigillatus*) for human consumption. Appl Environ Microbiol 84(2018):e00255–e0e218

Vergara JM, Fernandez-Palacios H, Robaina L, Jauncey K, Higuera MDL, Izquierdo M (1996) The effects of varying dietary protein level on the growth, feed efficiency, protein utilization, and body composition of gilthead sea bream fry. Fish Sci 62(4):620–623

Wan Ahmad WA, Stone DAJ, Schuller KA (2013) Dietary fish oil replacement with palm or poultry oil increases fillet oxidative stability and decreases liver glutathione peroxidase activity in barramundi, *Lates calcarifer*. Fish Physiol Biochem 39:1631–1640

Wang L, Lu Q, Luo S, Zhan W, Chen R, Low B, Xu D (2016) Effect of dietary lipid on growth performance, body composition, plasma biochemical parameters and liver fatty acids content of juvenile yellow drum, *Nibea albiflora*. Aquac Rep 4:10–16

Wang YR, Wang L, Zhang CX, Song K (2017) Effects of substituting fishmeal with soybean meal on growth performance and intestinal morphology in orange-spotted grouper (*Epinephelus coioides*). Aquac Rep 5:52–57

Watanabe T (1982) Lipid nutrition in fish. Comp Biochem Physiol 73B:3–15

Ween O, Stangeland JK, Fylling TS, Aas GH (2017) Nutritional and functional properties of fishmeal produced from fresh by-products of cod (*Gadus morhua* L.) and saithe (*Pollachius virens*). Heliyon 3(7):e00343

Williams KC, Irvin S, Barclay M (2004) Polka dot grouper, *Cromileptes altivelis* fingerlings require high protein and moderate lipid diets for optimal growth and nutrient retention. Aquac Nutr 10:125–134

Wilson RP (1994) Utilization of dietary carbohydrate by fish. Rev Aquac 124:67–80

Wilson RP (2002) Amino acids and protein. In: Halver JE, Hardy RW (eds) Fish nutrition, 3rd edn. Academic Press, an imprint of Elsevier Science, California, pp 144–171

Wu XY, Gatlin DM III (2014) Effects of altering dietary protein content in morning and evening feedings on growth and ammonia excretion of red drum, *Sciaenops ocellatus*. Aquaculture 434:33–37

Yang YH, Wang YY, Lu Y, Li QZ (2011) Effect of replacing fish meal with soybean meal on growth, feed utilization and nitrogen and phosphorus excretion on rainbow trout (*Oncorhynchus mykiss*). Aquac Int 19:405–419

Yong ASK, Ooi SY, Shapawi R, Biswas AK, Takii K (2015) Effects of dietary lipid increments on growth performance, feed utilization, carcass composition and intraperitoneal fat of marble goby, *Oxyeleotris marmorata* juveniles. Turk J Fish Aquat Sci 15(3):653–660

Yong ASK, Mohd Faudzi N, Senoo S, Shapawi R (2019) Optimum level of dietary protein and lipid for hybrid grouper, tiger grouper (*Epinephelus fuscoguttatus*) x giant grouper (*E. lanceolatus*). J Sustain Sci Manag 14(5):1–15

Yu XB, Shen YY, Cui QM, Chen Y, Sun W, Huang XZ (2018) Silkworm (*Bombyx mori*) has the capability to accumulate C20 and C22 polyunsaturated fatty acids. Eur J Lipid Sci Technol 120(2018):1700268

Zamora-Sillero J, Ramos LRV, Romano LA, Monserrat JM, Tesser MB (2013) Effect of dietary dextrin levels on the growth performance, blood chemistry, body composition, hepatic triglycerides and glycogen of Lebranche mullet juveniles (*Mugil liza* Valenciennes 1836, Mugilidae). J Appl Ichthyol 29:1342–1347

Feeding Stimulants in Finfish Aquaculture

5

Leong-Seng Lim, Isabella Ebi, Kianann Tan, Hon Jung Liew, Sharifah Rahmah, Roslianah Asdari, Rovina Kobun, and Rossita Shapawi

Abstract

Aquaculture is a fast-growing industry in the world, and finfish culture has been accounted for the largest share of it. Feed cost in fish farming is very high; hence, the strategy for preventing feed wastage is crucial. Uneaten feed or feed rejection is one of the causes to feed wastage in finfish culture, while this phenomenon can

L.-S. Lim (✉) · I. Ebi · R. Shapawi
Higher Institution Centre of Excellence (HICoE), Borneo Marine Research Institute, Universiti Malaysia Sabah, Kota Kinabalu, Sabah, Malaysia
e-mail: leongsen@ums.edu.my

K. Tan
Guangxi Key Laboratory of Beibu Gulf Marine Biodiversity Conservation, College of Marine Sciences, Beibu Gulf University, Qinzhou, Guangxi, China

H. J. Liew
Higher Institution Centre of Excellence (HICoE), Institute of Tropical Aquaculture and Fisheries (AKUATROP), Universiti Malaysia Terengganu, Kuala Nerus, Terengganu, Malaysia

S. Rahmah
Higher Institution Centre of Excellence (HICoE), Institute of Tropical Aquaculture and Fisheries (AKUATROP), Universiti Malaysia Terengganu, Kuala Nerus, Terengganu, Malaysia

Faculty of Fisheries and Food Science, Universiti Malaysia Terengganu, Kuala Nerus, Terengganu, Malaysia

R. Asdari
Department of Aquatic Sciences, Faculty of Resource Science and Technology, Universiti Malaysia Sarawak, Kota Samarahan, Sarawak, Malaysia

R. Kobun
Faculty of Food Science and Nutrition, Universiti Malaysia Sabah, Kota Kinabalu, Sabah, Malaysia

N. M. Faudzi et al. (eds.), *Essentials of Aquaculture Practices*,
https://doi.org/10.1007/978-981-97-6699-4_5

be due to poor feed palatability. Therefore, dietary inclusion of feeding stimulant, the taste substance that can promote ingestion in fish, is necessary to enhance fish intake and minimize uneaten feed. This chapter aimed to provide an overview on the types of taste substances that have been commonly used as feeding stimulants in finfish aquaculture. Generally, amino acids, betaine, organic acids, and nucleotides have been reported as the functional feeding stimulants for fish. However, fish taste preference for these substances is species-specific. Therefore, before the taste substance was used as the feeding stimulant, a preliminary confirmation of the fish preference for it through behavioral assay is strongly recommended.

Keywords

Feeding stimulants · Taste substances · Nucleotides · Amino acids

5.1 Introduction

Aquaculture is a fast-growing industry in the world, and it has been identified as an efficient protein generator in the grand challenge of feeding more than nine billion people by 2050 (Godfray et al. 2010; FAO 2022). In 2020, the global total aquaculture production has reached approximately 122 million tons (worth about USD 265 billion), while finfish has been accounted for the largest share of it (FAO 2022). In finfish aquaculture, feed cost is very high as it can reach up to more than 60% of the total operational expenses (Sathishkumar et al. 2021). To prevent feed wastage, feed palatability is an important criterion to be taken seriously (feed will be rejected by fish and wasted otherwise) (Coloso 2014). In the present time, however, alternative protein sources, especially plant proteins, are commonly used to replace fishmeal protein in fish diets for cost reduction, and such practice often deteriorates the feed palatability and leads to feed wastage. The impact of waste products from aquaculture, including feed wastage, has raised public concern and threatens the sustainability of aquaculture practices (Dauda et al. 2019). Feeding stimulant (FS) is, therefore, needed in the feed formulation to improve its palatability and promote fish ingestion (Dias et al. 1997; Kubitza et al. 1997; Papatryphon and Soares 2000a, b; Lim et al. 2015a, b). FS is defined as the substance that promotes ingestion in fish (Kasumyan and Døving 2003). In other words, FS is the substance that suits the fish taste preference. As the taste preference in fish is species specific (Kasumyan and Døving 2003), the knowledge of using the suitable FS for effective feeding in fish farming is essential. This chapter aimed to provide a comprehensive overview on the feeding stimulants that are commonly used in finfish aquaculture. Although there are studies reported on the utilization of marine- and terrestrial animal-based ingredients or extracts as an effective FS in finfish farming (Kubitza et al. 1997; Kikuchi and Furuta 2009; Tusche et al. 2011), the chemical compositions of these ingredients can vary when they are obtained from different sources and/or been processed by different methods. This circumstance may cause variants in the functionality degree of these ingredients for being FS. Therefore, this chapter do not intend to include information on the feeding stimulants that are made of animal protein sources.

5.2 Amino Acids

5.2.1 General Chemistry

In general, amino acids consist of carbon, hydrogen, oxygen, and nitrogen atoms with the amine ($-NH_2$) and carboxylic acid ($-COOH$) as the functional groups along with the side chain specifically in each amino acid, although there are also other components that can be found in the functional group, such as sulfur. There are 20 standard amino acids that are proteinogenic (meaning protein building) as they combine into the peptide chains or polypeptides that form proteins. These amino acids are known as the "standard amino acids" because they are encoded directly by the codons of the universal genetic code (Jakubke and Sewald 2008). Among all of them, nine amino acids cannot be synthesized from other compounds in the human body (including animals) and must be obtained from food. These amino acids are known as the "essential" one, while the others are the "nonessential" (Reeds 2000; Wu et al. 2013). Other than this categorization, amino acids are also classified into six main groups (aliphatic, hydroxyl or sulfur containing, acidic and amide containing, cyclic, aromatic, and basic) based on the structure and general chemical characteristics of their side chains. Table 5.1 shows the 20 standard amino acids based on their classifications. All of these amino acids, except glycine, can exist in two enantiomers (mirror images of each other), which are known as the L- and D-types (Fleck and Petrosyan 2014). Figure 5.1 shows the molecular structures of the L- and D-types of alanine.

Table 5.1 Names of amino acids based on their classifications

	Essential	Nonessential
Aliphatic	Isoleucine, leucine, valine	Alanine, glycine
Hydroxyl or sulfur-containing	Methionine, threonine	Serine, cysteine
Acidic and amide- containing	–	Asparagine, aspartic acid, glutamic acid, glutamine
Cyclic	–	Proline
Aromatic	Phenylalanine, tryptophan	Tyrosine
Basic	Histidine, lysine	Arginine

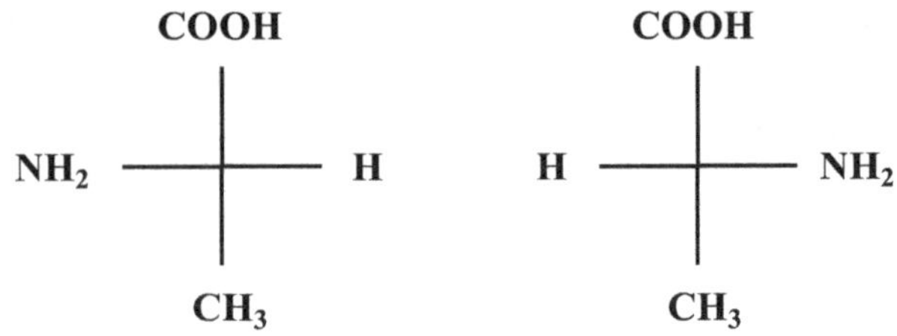

Fig. 5.1 Molecular structures of the D type (left) and L type (right) of alanine

5.2.2 Amino Acid Supplementation in Animal Feeds

Amino acids have been commonly supplemented into the feeds for many animals including swine, poultry, and ruminants, due to their biological significance to the animals' performance (D'Mello 1993; Oldham 1993; Brooks et al. 2011; Wu et al. 2013). Fish amino acid nutrition also has drawn the attention of aqua nutritionists (Li et al. 2009) and recommended that amino acid supplementation to fish diets is necessary, especially for those formulated with alternative protein sources that are usually lacking in some essential amino acids (Nunes et al. 2014). In fact, interesting findings revealed that amino acid dietary supplementation, particularly methionine, can be practiced as an approach to reduce fishmeal protein in the diets for shrimp and fish. Huai et al. (2010) reported that dietary crude protein could be reduced from 41.26% to 35.52% in the diets of *L. vannamei* when the synthetic amino acids (methionine plus cysteine, lysine, and threonine) were supplemented. Moreover, Browdy et al. (2012) have demonstrated that the *L. vannamei* growth, body weight, survival, yield, and FCR were supported by 2-hydroxy-4-(methylthio) butanoic acid (HMTBa) supplementation when 150.0 g kg^{-1} of fish meal was replaced by soybean meal and other ingredients, at 50% and 100%. For fishes, Zhao et al. (2010) reported that fishmeal could be completely replaced by the soy protein concentrate (SPC) without negative effect on the growth of Nile tilapia (*Oreochromis niloticus* GIFT strain, less than 2 g) by methionine hydroxy analogue (MHA) supplementation and increasing feeding frequency. Besides that, dietary supplementation of MHA also has been reported to affect antioxidant status of immune organs and promoted immune response in the Jian carp (*Cyprinus carpio* var. Jian) juveniles (Kuang et al. 2012). Other than those, Figueiredo-Silva et al. (2015) also have proven that supplementing SBM-based diet with graded levels of DL-methionine is an effective strategy to reduce fishmeal content in practical diets for hybrid tilapia.

5.2.3 Amino Acids as Feeding Stimulant in Fish Feeds

Amino acid supplementation has also been utilized as a feeding stimulant to promote feed intake in fish instead of only fulfilling their nutrition requirement. According to Kasumyan and Døving (2003), the nonessential amino acids are usually more palatable than the essential amino acids, while the L-isomer of amino acids (L-amino acids) are usually palatable but the D-isomer of amino acids

(D-amino acids) are often identified as a deterrent or indifferent taste substance. Table 5.2 shows successful examples of the use of amino acids as feeding stimulants in different fish species to promote intake of the diets formulated with different alternative protein sources. Nevertheless, there is also an exceptional case where amino acid supplementation did not promote feed intake in fish. For example, Kubitza et al. (1997) reported the failure of utilizing the mixture of amino acids (alanine, glycine, proline, serine, leucine, valine, histidine, and tryptophan) to promote intake of soybean meal diet in the largemouth bass (*Micropterus salmoides*). Such failure could be mainly due to the lack in knowledge on the taste preference of the target fish before the supplementation was made because the fish preference for amino acids is species specific (Kasumyan and Døving 2003). Table 5.3 shows the acceptance level of several amino acids by several different species of fish species based on behavioral assays. To avoid supplementing the unsuitable amino acids (deterrents) as FS in fish feeds, it is therefore highly recommended to identify the fish preferable amino acids before the supplementation is done.

5.3 Betaine

5.3.1 General Chemistry and Its Usage in Animal Feeds

Betaine is a chemical compound that was first discovered in sugar beets (*Beta vulgaris*) and then later in other plants and animals (Craig 2004). It is a neutral compound because of its "zwitterionic" structure that contains a cationic (positive charge) functional group such as a quaternary ammonium (N-R4 with R being an alkyl- or an aryl-group) or phosphonium (P-R4) and an anionic (negative charge) functional group such as a carboxylate group—[R-COO]*n*. Figure 5.2 shows the molecular structure of betaine.

Table 5.2 Successful examples of using amino acids as feeding stimulants to promote fish intake of diets with different alternative protein sources

Fish species	Basal diets	Amino acids	References
European seabass *Dicentrarchus labrax*	Soy protein concentrates	A mixture of 20 amino acids (detail see Dias et al. 1997)	Dias et al. (1997)
Striped bass *Morone saxatilis*	Mixture of plant proteins	A mixture of L-alanine, L-serine, and other non-amino acid substances	Papatryphon and Soares (2000b)
Gibel carp *Carassius auratus gibelio*	Fish meal + meat and bone meal	Glycine, L-lysine, L-methionine, L-phenylalanine	Xue and Cui (2001)
Japanese flounder *Paralichthys olivaceus*	Soy protein concentrates	Mixture of L-alanine, DL-methionine. L-leucine, L-valine, L-threonine	

Table 5.3 Acceptance levels of several amino acids by different fish species as determined behaviorally (modified from Kasumyan and Døving 2003)

	Fish species			
Amino acids	Common carp *Cyprinus carpio*	Tench *Tinca tinca*	Navaga *Eleginus nawaga*	Brown trout *Salmo trutta caspius*
Cysteine	High	High	Reject	High
Lysine	Low moderate	High	Reject	High moderate
Serine	Low	High	High	Low
Alanine	High moderate	High	High	Low moderate

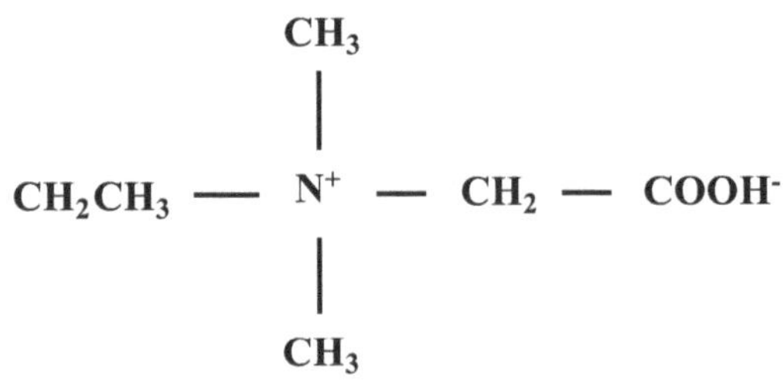

Fig. 5.2 Molecular structure of betaine

Betaine has been commonly supplemented as an additive into the feeds for many farmed terrestrial animals (such as poultry, cows, horses) because such practice has been proven to be friendly to both consumers and the environment (EFSA FEEDAP Panel 2013). In addition, the functions of betaine and its optimum supplementation rate in the feed for each animal species have been well studied and reviewed (Eklund et al. 2005; Ratriyanto et al. 2009). The standard maximum supplementation rate in the feeds for all animal species has been decided at 2000 mg kg^{-1} (EFSA FEEDAP Panel 2013). Betaine also has been widely used as an additive in fish feeds. The general maximum supplementation rate of betaine in the feeds for fish and crustaceans has also been decided at 1500 mg kg^{-1} (EFSA FEEDAP Panel 2013). The functions of betaine in fish performance have been evaluated by various authors on many fish species in different studies. Dietary betaine supplementation can be used to reduce the osmoregulation-related stress in fish (Virtanen et al. 1989; Clarke et al. 1994; Ayec Genc et al. 2006). Studies also demonstrated that betaine as a methyl donor is potent in mitigating the endosulfan and elevated temperature-induced stress in *Labeo rohita* fingerlings (Kumar et al. 2012a, b). The other advantageous function of dietary betaine supplementation to fish is being a taste enhancer.

5.3.2 Betaine as a Taste Enhancer in Fish Feeds

Dietary betaine supplementation generally has a positive effect in improving feed intake. Fredette et al. (2000) demonstrated that 1% of betaine supplementation in the fish meal-based diet can significantly improve the feed consumption of the 1-year-old juvenile winter flounder (*Pleuronectes americanus*). The supplementation of 0.25% dietary betaine also significantly improved the feed intake of Indian major carp (*L. rohita*) fry on the basal diet (Shankar et al. 2008). Xue and Cui (2001) reported that the feed intake of a diet with meat and bone meal in juvenile gibel carp (*Carassius auratus gibelio*) can be improved with 0.5% of dietary betaine supplementation. Tiril et al. (2008) also reported that supplementation of 1.5% dietary betaine significantly improved the intake of plant protein-based diets in rainbow trout (*O. mykiss*). Zakipour Rahimabadi et al. (2012) and Fattahi et al. (2013) confirmed that 1% supplementation dietary betaine can significantly improve the feed intake of pike perch (*Sander lucioperca*) and Caspian roach (*Rutilus rutilus caspicus*) fingerlings on the basal diets, respectively. Indeed, Lim et al. (2015a, b) also reported that betaine supplementation can promote ingestion of the soybean meal-based diets in the brown-marbled grouper (*Epinephelus fuscoguttatus*). Nonetheless, there are exceptional cases. Betaine dietary supplementation was reported to be insignificant in promoting the feed intake of largemouth bass (*M. salmoides*) (Kubitza et al. 1997), fish piauçu (*Leporinus macrocephalus*) (Normandes et al. 2006), and cobia (*R. canadum*) (Trushenski et al. 2011). According to Kasumyan and Døving (2003), betaine can work as a feeding stimulant, taste enhancer, taste-indifferent substance, or deterrent in fish, and the effect is species specific. Feed enhancer is defined as a substance that is tasteless in itself, but it has synergistic properties to enhance the flavor of amino acids; hence, it can be used to improve the palatability of fish diets. Indeed, Lim et al. (2016a) have demonstrated that the *E. fuscoguttatus* rejected the agar gel pellet with betaine, but it showed significantly higher ingestion of the agar gel pellet with betaine + amino acid mixture than the one without betaine. Table 5.4 shows the taste effects of betaine on different species of fish as determined behaviorally in the previous studies. Apparently, dietary betaine supplementation with the aim to promote feed intake does not work universally to all fish species. Behavioral assays should be conducted to confirm the taste effect of betaine on the fish before the dietary supplementation is done.

5.4 Organic Acids

Organic acids are organic compounds with acidic properties, while the most common type of organic acid is carboxylic acid that contains the carboxylic group (COOH). Organic acids are commonly used as a preservative in food industry due to its antimicrobial capabilities that can contribute to reduce food spoilage (Anyasi et al. 2017), and it has been proven that dietary organic acid supplementation can benefit the farmed terrestrial and aquatic animals' performance, including swine (Partanen and Mroz 1999; Suiryanrayna and Ramana 2015), poultry (Khan and Iqbal 2016), beef cattle and dairy cows (Castillo et al. 2004; Gheller et al. 2020), fish, and shrimp (Ng and Koh 2017). Although not many studies have reported organic acids as a feeding stimulant, Adams et al. (1988) have demonstrated the preference of the red belly tilapia (*Tilapia zillii*) for acidic food, while Kasumyan and Prokopova (2001) have reported a taste preference of the tench (*Tinca tinca*) for 19 types of organic acids. According to Kasumyan and Døving (2003), the

Table 5.4 Taste effects of betaine in several fish species

Fish species	Categories	References
Pigfish *Orthopristis chrysoptera*	Taste enhancer of amino acids; Feeding stimulant	
Pinfish *Lagodon rhomboides*	Taste enhancer of amino acids; Feeding stimulant	
Puffer *Fugu pardalis*	Taste enhancer of amino acids; Indifferent taste substance	Ohsugi et al. (1978)
Red sea bream *Chrysophrys major*	Taste enhancer of amino acids; Indifferent taste substance	
Dover sole *Solea solea* (L.)	Taste enhancer of amino acids; Feeding stimulant	Mackie and Mitchell (1985)
Rainbow trout adult *Salmo gairdneri*	Feeding stimulant	Mearns et al. (1987)
Atlantic salmon parr *Salmo salar*	Deterrent	
Brown-marbled grouper *Epinephelus fuscoguttatus*	Taste enhancer of amino acids; Indifferent taste substance	Lim et al. (2016a)
Marble goby *Oxyeleotris marmorata*	Indifferent taste substance	Lim et al. (2016b)

relationship between food acidity level and fish taste preference is species specific. Indeed, the marble goby (*Oxyeleotris marmorata*) was found to highly prefer the highly acidic (low pH) organic acids (Lim et al. 2017a; Teoh et al. 2018), similar to *T. tinca* (Kasumyan and Prokopova 2001), but opposite results (prefer the low acidic or high pH organic acids) were found on the bitterling (*Rhodeus sericeus amarus*) and dace (*Leuciscus leuciscus*) (Kasumyan and Døving 2003). Apparently, fish preference for food acidity should be confirmed if organic acids are intended to be used as a feeding stimulant.

5.5 Nucleotides

Nucleotides consist of a nitrogenous base, a ribose or 2′-deoxyribose sugar, and a phosphate group, while those without the phosphate group are known as the nucleosides (Li and Gatlin 2006). Both nucleotides and nucleosides are the building blocks of deoxyribonucleic acid (DNA) and ribonucleic acid (RNA); hence, they are naturally present in all plants and animals. Fish nucleotide nutrition has been well studied, and it is proven that nucleotide dietary supplementation can enhance fish immunity for disease resistance and stimulate feed intake (Li and Gatlin 2006;

Varghese et al. 2015; Hossain et al. 2020). In fact, nucleotides and nucleosides are efficient feeding stimulants, but fish preference for them is species specific. For instance, Mackie and Adron (1978) investigated 47 types of nucleotides and nucleosides and discovered that inosine and inosine-5′-monophosphate (IMP) were the best feeding stimulants for turbot (*Scophthalmus maximus*). Both inosine and IMP are also effective feeding stimulants for brill (*Scophthalmus rhombus*) (Mackie and Mitchell 1985), rainbow trout (*Oncorhynchus mykiss*) (Mearns et al. 1987), marble goby (*Oxyeleotris marmorata*) (Lim et al. 2016b), and hybrid grouper (*Epinephelus fuscoguttatus* ♀ × *E. lanceolatus* ♂) (Lim et al. 2017b). However, the largemouth bass (*Micropterus salmoides*) and jack mackerel (*Trachurus japonicus*) preferred only the IMP, but not inosine (Ikeda et al. 1988; Kubitza et al. 1997). Therefore, preliminary evaluation to confirm the fish's preference for nucleotides is necessary before it can be supplemented as a dietary feeding stimulant for the fish. Besides inosine and IMP, other nucleotides and nucleosides including adenosine-5′-monophosphate (AMP), guanosine-5′-monophosphate (GMP), uridine-5′-monophosphate (UMP), uridine-5′-diphosphate (UDP), and uridine-5′-triphosphate (UTP) were also reported to be taste stimulative to fish (Ikeda et al. 1988, 1991; Takaoka et al. 1990).

5.6 Method to Confirm Fish Preference for Taste Substances

In the study of searching suitable FS for fish, the electrophysiological technique—electrogastrogram (EGG)—has been conventionally used to examine the fish gustatory neural response to taste substances. In general, the EGG can be performed on the different branches of the facial nerve (including the facial-trigeminal complex and the palatine nerve), glossopharyngeal nerve, or vagal nerve. The facial nerve innervates the taste buds within the rostral oral cavity, lips, and possibly the external body surface, depending on fish species (Kohbara and Carpio 2001). The glossopharyngeal nerve innervates those taste buds on the anterior part of the oral cavity, while the vagal nerve innervates those on the pharynx. In marine fish species, recordings were usually done on the palatine nerve, a descending branch of the facial nerve (Hara 1994a, b). Nevertheless, the EGG recording technique is complex, and it usually requires a lot of preparation time and practice to excel especially for beginners. In the EGG procedure, a cautious surgery of removing the fish eyeball is necessary to obtain the branches of the facial-trigeminal complex or the palatine nerve (Carpio 1978; Kiyohara et al. 1985; Hara et al. 1994; Kohbara and Carpio 2001; Yacoob et al. 2001; Yamashita et al. 2006), while removal of the operculum is needed to reach to the glossopharyngeal nerve (Kohbara and Carpio 2001). Besides that, the gustatory neural response reflects only the fish gustatory sensitivity to the taste substance, but not its preference. A highly gustatory stimulative taste substance could be a FS but also, oppositely, a deterrent to the fish.

Behavioral assay is the more suitable method to confirm fish preference or acceptance level to various taste substances. In behavioral assay, generally, the taste substance to be tested is delivered to the fish using a solid matrix carrier. Subsequently,

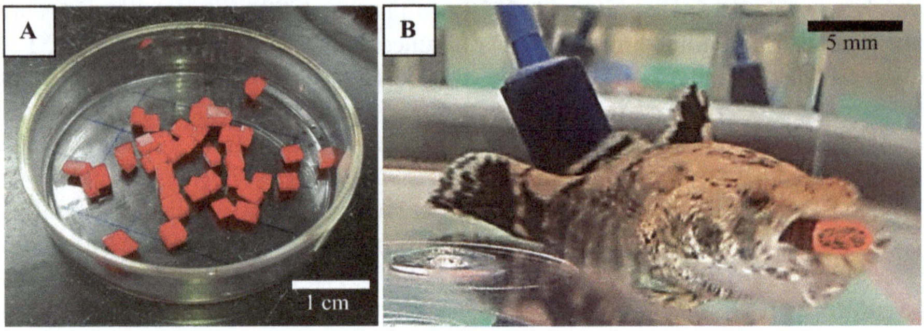

Fig. 5.3 (**a**) Agar gel pellet with red food dye; (**b**) *O. marmorata* captured the agar gel pellet once it was introduced into the aquarium during the behavioral assay

the fish feeding response (capture the solid matrix, manipulate it in the mouth, then ingest or reject) is observed and evaluated quantitatively (e.g., scoring method) or qualitatively (descriptive). In the study of the Atlantic salmon taste preference, Sutterlin and Sutterlin (1970) used a polyurethane disc as a carrier, coated it with the taste substances, and delivered it to the fish. Jones (1989, 1990) used the taste solution-soaked cotton pellets, while Mackie and Adron (1978) incorporated the test substances into the freeze-dried pellets to study the taste preference in the rainbow trout and turbot (*Scophthalmus maximus*), respectively. On the contrary, some researchers used starch pellets as the medium to deliver taste substances to the fish (Hidaka et al. 1978; Ohsugi et al. 1978; Murofushi and Ina 1981). In a study of the rainbow trout adult and Atlantic salmon parr taste preference by Mearns et al. (1987), agar gel pellet was used as the medium to deliver the taste substances to the fish. This method was then commonly used by other researchers in the more recent studies on fish taste preference (Kasumyan and Prokopova 2001; Goli et al. 2015; Lim et al. 2016a, b, 2017a, b; Teoh et al. 2018; Jamil et al. 2019). Figure 5.3 shows an example of the agar gel pellet used in the study of taste preference in the juvenile marble goby (*O. marmorata*) by Lim et al. (2017a).

5.7 Conclusion

In summary, amino acids, betaine, organic acids, and nucleotides (including nucleosides) are functional feeding stimulants for fish. However, fish taste preference for these substances is species specific. Preliminary confirmation of the fish preference for these substances through behavioral assays is highly recommended before they are used as feeding stimulants in fish farming.

Acknowledgments This work was supported by the Higher Institution Centre of Excellence (HICoE) Research Grant Scheme (Grant Code: HIC2403), provided by the Ministry of Higher Education, Malaysia.

References

Adams MA, Johnsen PB, Zhou HQ (1988) Chemical enhancement of feeding for the herbivorous fish *Tilapia zillii*. Aquaculture 72(1–2):95–107

Anyasi TA, Jideani AIO, Edokpayi JN, Anokwuru CP (2017) Application of organic acids in food preservation. In: Vargas C (ed) Organic acids: characteristics, properties and synthesis. Nova Publishers, New York, pp 1–46

Ayec Genc M, Tekelioglu N, Yilmaz E, Hunt AO, Yanar Y (2006) Effect of betaine on growth performance and body composition of *Oreochromis aureus* reared in fresh and sea water: a comparative study. J Anim Vet Adv 5:1185–1188

Brooks MA, Porter JH, Kerley MS (2011) Assessment of amino acid supplementation in rumen microbial efficiency and nitrogen metabolism using a continuous culture system. Prof Anim Sci 27:152–159

Browdy CL, Bharadwaj AS, Venero JA, Nunes AJP (2012) Supplementation with 2-hydroxy-4-(methylthio) butanoic acid (HMTBa) in low fish meal diets for the white shrimp, *Litopenaeus vannamei*. Aquac Nutr 18:432–440

Carpio J (1978) Olfaction and taste in the channel catfish: an electrophysiological study of the responses to amino acids and derivatives. J Comp Physiol 123:357–371

Castillo C, Benedito JL, Méndez J, Pereira V, Lopez-Alonso M, Miranda M, Hernández J (2004) Organic acids as a substitute for monensin in diets for beef cattle. Anim Feed Sci Technol 115(1):101–116

Clarke WC, Virtanen E, Blackburn J, Higgs DA (1994) Effects of a dietary betaine/ amino acid additive on growth and seawater adaptation in yearling chinook salmon. Aquaculture 121:137–145

Coloso RM (2014) Feed formulation for sustainable aquaculture. In: Proceedings of international workshop on resource enhancement and sustainable aquaculture practices in Southeast Asia, pp 223–230

Craig SAS (2004) Betaine in human nutrition. Am J Clin Nutr 80:539–549

D'Mello JPF (1993) Amino acid supplementation of cereal-based diets for non-ruminants. Anim Feed Sci Technol 45:1–8

Dauda AB, Ajadi A, Tola-Fabunmi AS, Akinwole AO (2019) Waste production in aquaculture: sources, components and managements in different culture systems. Aquac Fish 4(3):81–88

Dias J, Gomes EF, Kaushik SJ (1997) Improvement of feed intake through supplementation with an attractant mix in European seabass fed plant-protein rich diets. Aquat Living Resour 10:385–389

EFSA FEEDAP Panel (European Food Safety Authority Panel on Additives and Products or Substances used in Animal Feed) (2013) Scientific opinion on the safety and efficacy of betaine (betaine anhydrous and betaine hydrochloride) as feed additive for all animal species based on a dossier submitted by VITAC EEIG. EFSA J 11:3210–3232

Eklund M, Bauer E, Wamatu J, Mosenthin R (2005) Potential nutritional and physiological functions pf betaine in livestock. Nutr Res Rev 18:31–48

FAO (2022) The state of world fisheries and aquaculture 2022. Towards blue transformation. FAO, Rome

Fattahi S, Hosseini SA, Sudagar M (2013) The effects of dietary betaine on the growth performance and carcass synthesis of Caspian roach (*Rutilus rutiluscaspicus*) fingerlings. Sci J Anim Sci 2:132–137

Figueiredo-Silva C, Lemme A, Sangsue D, Kiriratnikom S (2015) Effect of DL-methionine supplementation on the success of almost total replacement of fish meal with soybean meal in diets for hybrid tilapia (*Oreochromis niloticus* × *Oreochromis mossambicus*). Aquac Nutr 21(2):234–241

Fleck M, Petrosyan AM (2014) Salts of amino acids: crystallization, structure and properties. Springer, Cham

Fredette M, Batt J, Castell J (2000) Feeding stimulant for juvenile winter flounders. N Am J Aquac 62:157–160

Gheller LS, Ghizzi LG, Marques JA, Takiya CS, Grigoletto NTS, Dias MSS, Silva TBP, Nunes AT, da Silva GG, Fernandes LGX, Rennó LN, Rennó FP (2020) Effects of organic acid-based products added to total mixed ration on performance and ruminal fermentation of dairy cows. Anim Feed Sci Technol 261:114406

Godfray HCJ, Beddington JR, Crute IR, Haddad L, Lawrence D, Muir JF, Pretty J, Robinson S, Thomas SM, Toulmin C (2010) Food security: the challenge of feeding 9 billion people. Science 327:812–818

Goli S, Jafari V, Ghorbani R, Kasumyan A (2015) Taste preferences and taste thresholds to classical taste substances in the carnivorous fish, kutum *Rutilus frisii kutum* (Teleostei: Cyprinidae). Physiol Behav 140:111–117

Hara TJ (1994a) The diversity of chemical stimulation in fish olfaction and gustation. Rev Fish Biol Fish 4:1–35

Hara TJ (1994b) Olfaction and gustation in fish: an overview. Acta Physiol Scand 152:207–217

Hara TJ, Kitada Y, Evans RE (1994) Distribution patterns of palatal taste buds and their responses to amino acids in salmonids Journal of Fish Biology 45(3) 453-465. 10.1111/jfb.1994.45.issue-3 https://doi.org/10.1111/j.1095-8649.1994.tb01328.x

Hidaka I, Ohsugi T, Kubomatsu T (1978) Taste receptor stimulation and feeding behaviour in the puffer, *Fugu pardalis* I. Effect of single chemicals. Chem Senses 3(4):341–354

Hossain MS, Koshio S, Kestemont P (2020) Recent advances of nucleotide nutrition research in aquaculture: a review. Rev Aquac 12(2):1028–1053

Huai MY, Liu YJ, Tian LX, Deng SX, Xu AL, Gao W, Yang HJ (2010) Effect of dietary protein reduction with synthetic amino acids supplementation on growth performance, digestibility and body composition of juvenile Pacific white shrimp, *Litopenaeus vannamei*. Aquac Int 18:255–269

Ikeda I, Hosokawa H, Shimeno S, Takeda M (1988) Identification of feeding stimulants for jack mackerel in its muscle extract. Bull Jpn Soc Sci Fish 54:229–233

Ikeda I, Hosokawa H, Shimeno S, Takeda M (1991) Feeding stimulant activities of nucleotides, tryptophan, and their related compounds for jack mackerel. Bull Jpn Soc Sci Fish 57:1539–1542

Jakubke HD, Sewald N (2008) Peptides from A to Z: a concise encyclopedia. Wiley-VCH, Weinheim

Jamil AZR, Lim LS, Tuzan AD, Shapawi R, Kawamura G (2019) Acceptability to betaine as a feed enhancer in the brown-marbled grouper (*Epinephelus fuscoguttatus*) at grow-out stage. Songklanakarin J Sci Technol 41(3):490–493

Jones KA (1989) The palatability of amino acids and related compounds to rainbow trout, *Salmo gairdneri* Richardson. J Fish Biol 34(1):149–160

Jones KA (1990) Chemical requirements of feeding in rainbow trout, *Oncorhynchus mykiss* (Walbaum); palatability studies on amino acids, amides, amines, alcohols, aldehydes, saccharides, and other compounds. J Fish Biol 37(3):413–423

Kasumyan AO, Døving KB (2003) Taste preferences in fish. Fish Fish 4:289–347

Kasumyan AO, Prokopova OM (2001) Taste preferences and the dynamics of behavioral taste response in the tench *Tinca tinca* (Cyprinidae). J Ichthyol 41(8):640–653

Khan SH, Iqbal J (2016) Recent advances in the role of organic acids in poultry nutrition. J Appl Anim Res 44(1):359–369

Kikuchi K, Furuta T (2009) Inclusion of blue mussel extract in diets based on fish and soybean meals for tiger puffer *Takifugu rubripes*. Fish Sci 75:183–189

Kiyohara S, Hidaka I, Kitoh J, Yamashita S (1985) Mechanical sensitivity of the facial nerve fibers innervating the anterior palate of the puffer, *Fugu pardalis*, and their central projection to the primary taste center. J Comp Physiol A 157:705–716

Kohbara J, Carpio J (2001) Taste responses of the facial and glossopharyngeal nerves to amino acids in the rainbow trout. J Fish Biol 58:1062–1072

Kuang SY, Xiao WW, Feng L, Liu Y, Jiang J, Jiang WD, Hu K, Li SH, Tang L, Zhou XQ (2012) Effects of graded levels of dietary methionine hydroxyl analogue on immune response and antioxidant status of immune organs in juvenile Jian carp (*Cyprinus carpio* var. Jian). Fish Shellfish Immunol 32:629–636

Kubitza F, Lovshin LL, Lovell RT (1997) Identification of feed enhancers for juvenile largemouth bass *Micropterus salmoides*. Aquaculture 148:191–200

Kumar N, Jadhao SB, Chandan NK, Kumar K, Jha AK, Bhushan S, Kumar S, Rana RS (2012a) Dietary choline, betaine and lecithin mitigates endosulfan-induced stress in *Labeo rohita* fingerlings. Fish Physiol Biochem 38:989–1000

Kumar N, Jadhao SB, Jha AK, Kumar K, Chandan NK, Shahbaz Akhtar M, Aklakur M, Kumar S, Rana RS (2012b) Methyl donors potentials growth, metabolic status and neurotransmitter enzyme in *Labeo rohita* fingerlings exposed to endosulfan and temperature. Fish Physiol Biochem 38:1343–1353

Li P, Gatlin DM (2006) Nucleotide nutrition in fish: current knowledge and future applications. Aquaculture 251(2):141–152

Li P, Mai KS, Trushenski J, Wu GY (2009) New developments in fish amino acid nutrition: towards functional and environmentally oriented aquafeed. Amino Acids 37:43–53

Lim LS, Ebi I, Chor WK, Kawamura G, Shapawi R (2015a) Determination of the possibility of dietary betaine supplementation to improve feed intake of soybean meal-based diet in the juvenile grouper (*Epinephelus fuscoguttatus*): a pilot study. Malays Appl Biol 44(2):137–141

Lim LS, Ebi I, Chor WK, Lu KC, Chong M, Sade A, Shapawi R (2015b) Optimizing betaine supplementation level in soybean-meal based diets to enhance feed intake and growth performance of juvenile grouper, *Epinephelus fuscoguttatus*. Adv Environ Biol 9(17):12–17

Lim LS, Chor WK, Tuzan AD, Shapawi R, Kawamura G (2016a) Betaine is a feed enhancer for juvenile grouper (*Epinephelus fuscoguttatus*) as determined behaviourally. J Appl Anim Res 44:415–418

Lim LS, Lai SKJ, Yong ASK, Shapawi R, Kawamura G (2016b) Evaluation on the potential of betaine, taurine, nucleotide and nucleoside as feeding stimulant for juvenile marble goby *Oxyeleotris marmoratus* through behavioural assays. Int Aquat Res 8:161–167

Lim LS, Lai SKJ, Yong ASK, Shapawi R, Kawamura G (2017a) Feeding response of marble goby (*Oxyeleotris marmorata*) to organic acids, amino acids, sugars and some classical taste substances. Appl Anim Behav Sci 196:113–118

Lim LS, Firdaus RF, Shapawi R, Kawamura G (2017b) Taste preference of hybrid grouper (*Epinephelus fuscoguttatus* ♀ × *Epinephelus lanceolatus* ♂) for nucleoside and nucleotides. Borneo J Mar Sci Aquac 1:39–43

Mackie AM, Adron JW (1978) Identification of inosine and inosine 5′-monophosphate as the gustatory feeding stimulants for the turbot, *Scophthalmus maximus*. Comp Biochem Physiol A Physiol 60(1):79–83

Mackie AM, Mitchell AI (1985) Identification of gustatory feeding stimulants for fish applications in aquaculture. In: Nutrition and feeding in fish, pp 177–189

Mearns KJ, Ellingsen OF, Døving KB, Helmer S (1987) Feeding behaviour in adult rainbow trout and Atlantic salmon parr, elicited by chemical fractions and mixtures of compounds identified in shrimp extract. Aquaculture 64(1):47–63

Murofushi S, Ina K (1981) Survey of feeding stimulants for the sea bream present in the dried pupae of silkworms. Agric Biol Chem 45(6):1501–1504

Ng WK, Koh CB (2017) The utilization and mode of action of organic acids in the feeds of cultured aquatic animals. Rev Aquac 9(4):342–368

Normandes EB, Barreto RE, Varvalho RF, Delicio HC (2006) Effects of betaine on the growth of the fish Piauçu, *Leporinus macrocephalus*. Braz Arch Biol Technol 49:757–762

Nunes AJP, Sa MVC, Browdy CL, Vazquez-Anon M (2014) Practical supplementation of shrimp and fish feeds with crystalline amino acids. Aquaculture 431:20–27

Ohsugi T, Hidaka I, Ikeda M (1978) Taste receptor stimulation and feeding behaviour in the puffer, *Fugu pardalis*. II. Effects produced by mixtures of constituents of clam extracts. Chem Senses 3(4):355–368

Oldham JD (1993) Recent progress towards matching feed quality to the amino acid needs of ruminants. Anim Feed Sci Technol 45:19–34

Papatryphon E, Soares JH (2000a) Identification of feeding stimulants for striped bass, Morone saxatilis. Aquaculture 185(3):339–352

Papatryphon E, Soares JH (2000b) The effect of dietary feeding stimulants on growth performance of striped bass, *Morone saxatilis*, fed a plant-feedstuff based diet. Aquaculture 185:329–338

Partanen KH, Mroz Z (1999) Organic acids for performance enhancement in pig diets. Nutr Res Rev 12(1):117

Ratriyanto A, Mosenthin R, Bauer E, Eklund M (2009) Metabolic, osmoregulatory and nutritional functions of betaine in monogastric animals. Asian Aust J Anim Sci 22:1461–1476

Reeds PJ (2000) Dispensable and indispensable amino acids for humans. J Nutr 130:1835S–1840S

Sathishkumar G, Bhavatharaniya U, Felix N, Ranjan A, Prabhu E (2021) Strategies to reduce feed cost by improving gut health and nutrient utilisation of fish in aquaculture. In: Wilkinson S (ed) Aquaculture Asia, vol 25, no. 1 January–March 2021. Network of Aquaculture Centres in Asia-Pacific (NACA), Bangkok, pp 14–16

Shankar R, Shivananda Murthy H, Pavadi P, Thanuja K (2008) Effect of betaine as a feed attractant on growth, survival, and feed utilization in fingerlings of the Indian major carp, *Labeo rohita*. Isr J Aquac 60:95–99

Suiryanrayna MVAN, Ramana JV (2015) A review of the effects of dietary organic acids fed to swine. J Anim Sci Biotechnol 6:45

Sutterlin AM, Sutterlin N (1970) Taste responses in Atlantic salmon (*Salmo salar*) parr. J Fish Board Can 27(11):1927–1942

Takaoka O, Takii K, Nakamura M, Kumai H, Takeda M (1990) Identification of feeding stimulants for marbled rockfish. Nippon Suisan Gakkaishi 56:345–351

Teoh CF, Lim LS, Kawamura G (2018) Remarkably high ingestion ratio of acidic food in juvenile marble goby *Oxyeleotris marmorata*. Int Aquat Res 10(1):95–100

Tiril SU, Alagil F, Yagci FB, Aral O (2008) Effects of betaine supplementation in plant protein-based diets on feed intake and growth performance in rainbow trout (*Oncorhynchus mykiss*). Isr J Aquac 60:57–64

Trushenski J, Laporte J, Lewis H, Schwarz M, Delbos B, Takeuchi R, Sampaio LA (2011) Fish meal replacement with soy-derived protein in feeds for juvenile cobia: influence of replacement level and attractant supplementation. J World Aquac Soc 42:435–443

Tusche K, Berends K, Wuertz S, Susenbeth A, Schulz C (2011) Evaluation of feed attractions in potato protein concentrate based diets for rainbow trout (*Oncorhynchus mykiss*). Aquaculture 321:54–60

Varghese T, Mishal P, Sandeep KP, Pal AK (2015) Nucleotides in aquafeeds: a right step towards sustainable nutrition. Fish Chimes 35:49–55

Virtanen E, Junnila M, Soivio A (1989) Effects of food containing betaine/ amino acid additive on the osmotic adaptation of young salmon, *Salmo salar* L. Aquaculture 83:109–122

Wu G, Wu ZL, Dai ZL, Yang Y, Wang WW, Liu C, Wang B, Wang JJ, Yin YL (2013) Dietary requirements of "nutritionally nonessential amino acids" by animals and humans. Amino Acids 44:1107–1113

Xue M, Cui Y (2001) Effect of several feeding stimulants on diet preference by juvenile gibel carp (*Carassius auratus gibelio*), fed diets with or without partial replacement of fish meal by meat and bone meal. Aquaculture 198:281–292

Yacoob SY, Anraku K, Marui T, Matsuoka T, Kawamura G, Archdale MV (2001) Gustatory sensitivity of the external taste buds of *Oreochromis niloticus* L. to amino acids. Aquac Res 32:217–222

Yamashita S, Yamada T, Hara TJ (2006) Gustatory responses to feeding- and non-feeding-stimulant chemicals, with an emphasis on amino acids, in rainbow trout. J Fish Biol 68:783–800

Zakipour Rahimabadi E, Akbari M, Arshadi A, Effatpanah E (2012) Effect of different levels of dietary betaine on growth performance, food efficiency and survival rate of pike perch (*Sander lucioperca*) fingerlings. Iran J Fish Sci 11:902–910

Zhao H, Jiang R, Xue M, Xie S, Wu X, Guo L (2010) Fish meal can be completely replaced by soy protein concentrate by increasing feeding frequency in Nile tilapia (*Oreochromis niloticus* GIFT strain) less than 2 g. Aquac Nutr 16:648–653

Diseases in Aquaculture

6

Hiam Elabd, Rafidah Othman, and Heba Mahboub

Abstract

The rapidly growing food sector of aquaculture is facing many fish diseases and disease outbreak challenges. Aquaculture may face disease infections due to different pathogenic and nonpathogenic causes. Disease outbreaks are well known to result in manifested economic losses. Disease conditions vary from pathogenic to nonpathogenic. Pathogenic conditions include bacterial, viral, fungal, and parasitic. While nonpathogenic include environmental and nutritional. Among nonpathogenic causes comes the overdensities that render aquatic animals more susceptible to infection and predispose them to the occurrence of various diseases conditions including pathogenic diseases. From the pattern of mortality occurrence, we can identify the causative disease conditions. The intermittent course of mortalities may refer to a pathogenic condition, sudden mortalities may be due to toxins, uprising pattern could be related to nutritional conditions, and morning mortalities are mostly suggested to be related to oxygen content. Therefore, developing aquatic animals' immunity that involves specific and nonspecific immune responses is crucial to prevent those challenging conditions rather than searching for a cure. In this chapter, we will review different diseases of economic importance and their impacts on the aquaculture industry.

H. Elabd (✉)
Aquatic Animals' Medicine Department, Faculty of Veterinary Medicine, Benha University, Moshtohor, Toukh, Egypt
e-mail: hiam.elabd@fvtm.bu.edu.eg

R. Othman
Higher Institution Centre of Excellence (HICoE), Borneo Marine Research Institute, Universiti Malaysia Sabah, Kota Kinabalu, Sabah, Malaysia

H. Mahboub
Department of Aquatic Animal Medicine, Faculty of Veterinary Medicine, Zagazig University, Zagazig, Egypt

N. M. Faudzi et al. (eds.), *Essentials of Aquaculture Practices*,
https://doi.org/10.1007/978-981-97-6699-4_6

Keywords

Infections · Pathogenic and nonpathogenic · Mortality · Immune responses

6.1 Introduction

Aquaculture industries have recently become the world's fastest-growing food sector. Aquaculture plays an important role in securing food and job opportunities for several millions of people all over the world (de Bruijn et al. 2018). The term "disease" is considered a disturbed condition where the physiological functions deviate from normal, causing disease conditions, which may be acute, per-acute, or chronic. Fish face a variety of challenging infections that are capable of causing disease outbreaks that threaten the whole aquaculture sector because they can result in significant economic losses. The overstocking densities and poor water quality parameters can predispose to the occurrence of various diseases (Rohani et al. 2022). Various stressful conditions can accompany aquaculture, which will represent a challenge to the immune system and render aquatic animals more susceptible to diseases. Many factors could collaborate in causing those stressful compromising situations (Elabd et al. 2017; Kuebutornye et al. 2019; Rohani et al. 2022). Those conditions can result from inadequate husbandry and/or management, poor water quality, and intensive farming densities, regardless if it is a mono or poly fish production facility (Marana et al. 2022; Mohd Ali et al. 2023).

Aquaculture is subjected to many pathogenic causes, among which are bacterial, viral, mycotic, and parasitic pathogens that can arise in the environment or as asymptomatic carriers that make the aquaculture facility more exposed to catching a disease and cause severe economic losses and mortalities (Jørgensen 2020; Avila et al. 2022). The exaggeration of the disease condition depends on the interrelationship between the pathogen, environment, and host (Fig. 6.1). High stocking density, suboptimum temperature, hypoxia, and ammonia can predispose to the occurrence and spread of pathogenic diseases and can cause disease outbreaks (Kuebutornye et al. 2019). Prevention is better than cure; this is a good strategy to follow to ensure sustaining proper environmental conditions. Integration of good husbandry practices, such as water quality optimization, quarantine application, supplying high-quality diets, breeding brood stocks that are disease resistant, and vaccination program establishment, has acted as alternative protocols to combat fish diseases. This, in turn, will create a safe and stress-free environment for successful aquaculture practices and industry (Ben Hamed et al. 2018; Elabd et al. 2017, 2020a, b; Othman et al. 2022).

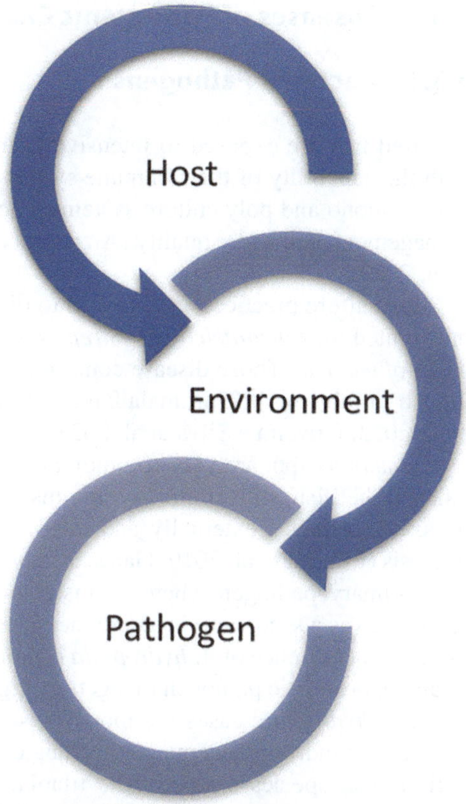

Fig. 6.1 The interrelationship between the pathogen, environment, and host

6.2 Fish Immune Response

A fish's immune system is quite different from that of mammals. Mainly, fish lack bone marrow and lymph nodes. Immune response involves specific and nonspecific immune responses. Nonspecific immunity has been considered the essential component in combating pathogens due to limitations of the specific immune system, and it includes the epithelial/mucosal barrier, the humoral response, and the cellular response. Fish health status depends on the interrelationship between fish and their environment. Various environmental factors such as temperature, photoperiod, salinity, and pH can greatly influence fish immune response and expression of related genes. This review aims to provide an idea about fish immune response and environmental factors, mainly the temperature that affects fish immune response and the expression of various genes, mainly alpha 2-macroglobulins (α 2-M), cytochrome, major histocompatibility (MH), and heat shock proteins (HSPs) genes, especially the 90-kDa heat shock protein (HSP90) (Elabd et al. 2016a, 2017, 2020a, b).

6.3 Diseases of Pathogenic Causes

6.3.1 Bacterial Pathogens

Cultured fish are exposed to intensive farming stress, which interferes negatively with the capability of their immune systems to fight various pathogenic infections in both mono and poly culture systems. Factors may include biosecurity, nutrition, management, and water quality (Alzahrani et al. 2023; Hegde et al. 2023; Mohd Ali et al. 2023).

Aquaculture practices are exposed to different bacterial diseases, including, but not limited to, *Edwardsiellosis*, *Streptococcosis*, *Aeromonas septicemia*, *Vibriosis*, and *Columnaris*. Those disease conditions result in great losses to the aquaculture industry (Table 6.1) (Thirumalaikumar et al. 2021; LaFrentz et al. 2022; van Doan et al. 2022; Oliveira e Silva et al. 2024).

Aeromonas spp. are very common and represent major sources of mortalities in fish and shellfish, with clinical symptoms varying from anorexia, inflammation, redness, hemorrhages externally and internally, enlarged internal organs, and tissue necrosis (El Latif et al. 2019; Ganesan et al. 2023; Hegde et al. 2023). It is classified as a primary pathogen when it causes the disease in stressed fish by itself and referred to as a secondary invader when it relies on the presence of a primary infection. The occurrence of *A. hydrophila* is more common as a secondary pathogen and as an opportunistic pathogen that gets exaggerated when exposed to stressful conditions. *A. hydrophila* cases are mostly associated with red body disease, epizootic ulcerative syndrome, edema, hemorrhagic septicemia, and enteritis and can infect various fish species such as Nile tilapia, catfish, carps, goldfish, and eel fishes (Thirumalaikumar et al. 2021; Oliveira e Silva et al. 2024).

A. salmonicida is well known to infect the salmonids causing furunculosis disease that is manifested with septicemia, external ulcerations, and hemorrhages. It is a nonmotile aeromonad present worldwide affecting both fresh and marine fishes

Table 6.1 Bacterial diseases of marine water fishes in Egypt

Bacterial disease	Causative agent	Fish	Province/ region	References
Vibriosis	*V. anguillarum, V. alginolyticus, V. alginolyticus, V. parahaemolyticus, V. harveyi, V. anguillarum*	*Sparus aurata and Dicentrarchus labrax*	Damietta, Port Said, Alexandria	Abdel-Aziz et al. (2013), Aly et al. (2020), Muniesa et al. (2020)
Motile *Aeromonas septicemia*	*Aeromonas hydrophila* and *A. sobria*	*S. aurata D. labrax*	Port Said Damietta	Haraz et al. (2015)
Streptococcosis	*Streptococcus agalactiae, S. dysgalactiae,* and *S. iniae*	*Red tilapia* (*Oreochromis* spp.)	Damietta, Northern Egypt, Alexandria	Moustafa et al. (2015), Abdelsalam et al. (2017)

and commonly causes high mortalities among rainbow trout *Oncorhynchus mykiss* and Atlantic salmon *Salmo salar*. The *A. salmonicida* subspecies are classified into "typical" and "atypical" strains according to the phenotypic characteristics and biochemical reactions (Pradhan et al. 2023).

A. sobria is a pathogenic bacterial disease that results in higher mortalities in *Cyprinus carpio* and Nile catfish. It can infect fish Rohu (*Labeo rohita*), one of the Indian major carps (IMC), and fish exhibit loss of escape and darkness of the skin linked with hemorrhages. The infected fish were isolated from a mismanagement pond, which has higher pollution and anthropogenic activity (Abdel Rahman et al. 2022a, b).

Edwardsiellosis is caused by *Edwardsiella tarda* and leads to manifested economic aquaculture losses all over the world. It affects all freshwater, brackish, and marine fishes. It is regarded as a widely spread fish pathogenic infection to a wide range of hosts, exceeding fishes to reptiles, birds, amphibians, and mammals globally. Recent studies have included both *E. piscicida* and *E. anguillarum* species as new strains that possess many phenotypic properties similar to *E. tarda*. Plus, the most common species *E. tarda*, *E. hoshinae,* and *E. ictaluri* have been reported as causative agents of Edwardsiellosis. Edwardsiellosis is usually subsequent to stressful environmental conditions, such as suboptimum water temperature and poor water parameters including high ammonia levels. The disease's clinical symptoms include abnormal swimming, abdominal swelling, depigmentation, and representing a health hazard for both fish and humans, causing severe economic losses. The species *E. ictaluri* results in enteric septicemia in reared channel catfish, tilapia, and additional warm-water fishes. *E. hoshinae* was reported in birds and reptiles. *E. anguillarum* is manifested to infect eels and characterized by arabinose fermentation and acetone production from glucose. The genes for the core oligosaccharide of LPS are mainly organized into clusters of contiguous genes (Pandey et al. 2021; Goh et al. 2023; Jordán et al. 2023).

Yersinia ruckeri causes disease conditions among aquaculture farms across the world. It leads to manifested economic losses in salmon industry. It causes the disease enteric red mouth affecting fish fauna and primarily isolated from rainbow trout. The enteric red mouth disease starts with reduced mortalities that increases with time. The disease is usually predisposed with stressful environmental conditions. According to outer membrane protein (OMP), serotype, and biotype profiles, *Y. ruckeri* is divided into clonal types. The serovars II and I cause the greatest epizootic outbreaks in cultured salmonids (Zaheen et al. 2022; Kumar et al. 2023; Nissa et al. 2023).

The opportunistic *Pseudomonas anguilliseptica* affects a wide range of fish species all over the world. It causes red spot disease in eel, sea bream, Atlantic salmon, rainbow trout, and grouper (Fadel et al. 2018; Hegde et al. 2023). *P. anguilliseptica* is regarded as an opportunistic emergent pathogen of fishes of a significant clinical importance and is incriminated to cause mortalities outbreaks in several fish species such as eel, cod, gilthead sea bream, sea bass, and turbot. It is characterized by pale liver, abdominal distension, inflamed kidney, and yellowish exudate in the intestine (Fadel et al. 2018; Alzahrani et al. 2023).

Vibriosis is also a disease condition that hinders aquaculture production. Vibrio are Gram-negative, motile, asporogenous rods, mostly having single polar flagellum when grown in a liquid medium. Most of them produce oxidase and catalase and ferment glucose without producing gas. *Vibrio* spp. are ubiquitous to marine environments clinical disease outbreaks which only occur when fish get exposed to infectious agent with the presence of stressful environmental conditions. *V. alginolyticus* is frequently isolated from outbreaks and mortalities in many marine fish species including gilt-head sea bream. In addition, *V. alginolyticus* causes many epizootic outbreaks among the gilt-head sea bream and European seabass populations in Mediterranean countries and some Egyptian coastal provinces (Abdel-Aziz et al. 2013; Tanekhy 2013; Khalil et al. 2019; Farag et al. 2022). The main causative agents of vibriosis are halophilic *Vibrio anguillarum* and *V. harveyi*, which are present in the aquatic environment and infect a wide variety of fishes. *V. anguillarum* causes hemorrhagic septicemia in rainbow trout *Oncorhynchus mykiss*, Atlantic salmon *Salmo salar*, and groupers. It is characterized by causing large mortalities and economic losses in the aquaculture industry. Vibrio fish pathogens also include *V. alginolyticus*, *V. harveyi*, and *V. parahaemolyticus* (Manchanayake et al. 2023). *Vibrio* spp. are of zoonotic importance that includes *V. parahaemolyticus* and *V. vulnificus* well-documented human pathogens. *V. mimicus* is a recognized pathogen with similar characteristics to *V. cholerae* except the ability to ferment sucrose. Other species within the genus such as *V. alginolyticus*, *V. fluvialis*, and *V. hollisae* are occasional human pathogens. *Vibrio* spp. account for a significant proportion of human infections from the consumption of raw or undercooked fishes (Abdel-Aziz et al. 2013; Tanekhy 2013; Khalil et al. 2019; Farag et al. 2022).

6.3.2 Fungal Pathogens

Temperate fishes are mostly affected by fungal infections because of the ability of the fungal to spread via water. Stressful conditions such as poor water quality render fishes more susceptible to fungal diseases (Sarkar et al. 2022). Pathogenic fungi such as *Aphanomyces* spp., *Dermocystidium* spp., *Branchiomyces* spp., *Aspergillus* spp., *Saprolegnia* spp., *Ichthyophonus* spp., and *Fusarium* spp. can infect several fishes including carps, rainbow trout, salmonids, catfishes, tilapia, and eels by affecting mainly the epidermis and may cause systemic conditions (Tartor et al. 2018; Magray et al. 2020). Fungal diseases could be superficial that attack fish skin (Sarjito et al. 2019; Younis et al. 2019; Mahboub and Shaheen 2020) or systemic infections that invade internal organs (Tartor et al. 2018; Mahboub and Shaheen 2021).

Superficial Fungi

Saprolegniasis is the most important fungal disease that affects the skin. It is also called winter kill syndrome or winter fungus. It includes almost 14 genera and around 146 species, with *S. parasitica* and *S. invaderis* as the most common species. It infects both freshwater and marine fish industries. It is an opportunistic invasive pathogen exaggerated by poor environmental conditions that suppress the immune system. It is characterized by excessive mucus secretion and damage to the epidermal tissues (Özcan 2022). *S. parasitica* causes extensive fish mortalities and decaying of eggs in salmon and trout fishes. It mainly affects the skin, gills, and eggs in the form of a cotton wool growth (As et al. 2018).

Branchiomycosis is a superficial fungus that affects the gills and spreads throughout the world as a fungal disease that mainly affects carp farms. It is caused by *Branchiomyces demigrans* in pike fishes and *Branchiomyces sanguinis* in carps. *Branchiomyces demigrans* can proliferate from the blood vessels to the neighboring gill tissues. It occurs mostly in warm waters. Branchiomycosis is a filamentous fungus blocking the gill filaments of blood vessels. It appears on the gill filaments in late stages (grayish-white) and may fall off leaving the cartilaginous arches exposed. Both fungi are preceded with environmentally stressful conditions, such as decreased pH levels, reduced dissolved oxygen, and excessive algal growth. The fungal spores are the main source of infection that is transported through water and grows at temperatures between 57 and 95 °F. *B. demigrans* and *B. sanguinis* infect the fish gills (As et al. 2018; Mishra et al. 2022).

B Branchiomyces. sanguinis is isolated for the first time and *B. demigrans* in Nile tilapia in Egypt. Both species are identified depending on the outline and diameter of hyphae via compression methods from infected gills and the diameter and localization of spores. This study reveals that *Branchiomyces* spp. exhibits various growth stages in gill tissues, and the tissues look as if it is congested, paled, white-colored, or marbling in appearance based on the development of infection. A crucial diagnostic sign of the genus *Branchiomyces* is the branching of hyphae from tips, which is clear in compression techniques from gills. Moreover, a microscopic inspection of fungal culture assists in the identification (Mahboub and Shaheen 2021). Branchiomycosis mostly negatively influences fish and induces respiratory problems, lethargy, and rapid mortalities. Diagnosis of branchiomyces relies on macroscopic inspection of necrotic areas on the injured gills and compression technique from the affected gill tissues (As et al. 2018).

Systemic Fungi

Currently, systemic fungal diseases have developed as substantial sources of fish morbidity and mortalities. There are some virulent systemic fungal infections that induce numerous lesions in the hepatic and renal tissues of African catfish. They include yeast (*Candida albicans*) and molds like *Aspergillus ochraceous*, *A. flavus*, and *Cladosporium herbarum*. Such fungi result in specific postmortem lesions. Many techniques are employed to recognize the fungi, like gross examination, microscopic inspection of tissue squashes, in vitro culturing, and histopathological assessment. Koch's postulate is another tool to verify the pathogenicity (Mahboub et al. 2022).

Ichthyophonus spp. is another systemic significant pathogen that belongs to protistan parasites but hastily looks like a fungus and releases hyphae and spores. *Ichthyophonus* spp. causes outbreaks in herring, rainbow trout, Atlantic salmon, and chinook. The disease affects the heart by causing necrosis with inflammatory cellular infiltration and, later on, causes fibrous encapsulations. The disease is characterized by the formation of granulomatous lesions externally and grayish-white lesions internally. The causative obligate parasite *I. hoferi* is of spherical or oval shape. The infective agent is in the form of fungal cysts that are produced in fecal matter. Infective spores' ingestion leads to the occurrence of infection. In chronic conditions, the infected skin reveals a sandpaper look with a swinging movement (swinging disease) (Chong 2022a; Neethu and Jyothis 2023).

Recent report has identified *Ichthyophonus* sp. and is distinguished by a circular double-walled multinucleate schizonts that are identified in squash preparations from infected systemic organs (spleen, liver, and kidneys), tissue sections, and histopathological sections. In Nile tilapia, clinical signs of *Ichthyophonus* sp. represent hemorrhages, congestion of the body surface, and dark discoloration. The postmortem lesions include enlargement of the liver, congested kidneys, and white raised nodules mainly in the liver, kidneys, and spleen. The frequency of infection is most common in winter. Moreover, it settled two various forms of growth on MEM media based on culture medium pH (Mahboub and Shaheen 2021).

Dermocystidiosis is a systemic infection by *Dermocystidium* spp. in the internal organs (spleen, liver, stomach, kidneys, and intestine) of Nile tilapia (*Oreochromis niloticus*). The total prevalence of infection is 35.5%, and this infection is higher in the wild (45%) compared with cultured species (27.8%). The greatest infection rates are found in large-sized fish and in the winter season. Numerous protocols are used to investigate and recognize *Dermocystidium* sp. from the internal organs of infected fish involving macroscopic inspection, microscopic examination of squashes from tissues, histological assessment, and in vitro culture. Infected fish are characterized by sluggish movement and ulcers with dark skin discoloration. The postmortem lesions include black nodules on internal organs, especially the liver. Squash swaps from the internal organs show circular spores (signet-ring shape) that distinguish *Dermocystidium* sp. The spores can multiply especially in the freshly dead tissue. In vitro culture of *Dermocystidium* sp. requires Eagle's Minimum Essential Medium at pH 3.5 or on Sabouraud's dextrose agar media provided with 10% duck decoction. Experimental challenge demonstrates that intragastric route was more virulent compared with the immersion one. Histopathological assessments from diseased tissues after natural and experimental infections exhibit spherical *Dermocystidium* spores embedded in the diseased tissue (Mahboub and Shaheen 2020).

6.3.3 Parasitic Pathogens

Parasites affect fishes through shifting the host's feeding and behavioral patterns. They also affect the host's neurological system, digestion, and metabolism and destruct gills and skin, leading to ulcer formation and tissue necrosis. Parasites may alter the fish population. The condition depends on the number of parasites and the size of the fish as a host. Sometimes, the parasite infestation is not fatal but affects the aquaculture production system negatively. The most common parasites affecting fish species include monogeneans, protozoans, and trematodes (Zhi et al. 2020;

Adeshina et al. 2021). Protozoan parasites are crucial, and they include *Ichthyobodo necator*, *Ichthyophthirius multifiliis*, *Trichodina*, and *Hexamita*. They can cause a mortality rate of almost 100% (Adeshina et al. 2021).

Fish trematodes infect humans causing serious conditions if improperly consumed. They result in economic losses, and they belong to the Platyhelminthes phylum. The adult stages are obligatory parasites and can complete their life cycle in four hosts (Chi et al. 2009; Muñoz et al. 2023). *Heterophyidae* trematodes include 22 zoonotic species that represent hazard to humans worldwide. Zoonotic species *Haplorchis pumilio* and *Centrocestus formosanus* were isolated from catfish species. Heterophyids are intestinal trematodes with mammals and birds as final hosts (Ogbon et al. 2023).

Trematode metacercariae can cause retardation of young fishes, and immune suppression predisposing to the occurrence of secondary infections. The *Clinostomum complanatum* metacercarial stage is of zoonotic importance and reported to cause Halzoun or laryngopharyngitis disease, which could be fatal causing mortalities from asphyxia and breathing blockage (Mahdy et al. 2023). Several other disease conditions include black spot disease, inflammation of internal organs, necrotic changes, displacements, malformations, and morbidity (Adeshina et al. 2021; Muñoz et al. 2023).

Fish monogenea include a wide range of fresh and marine fish ectoparasites. Monogeneans cause anemia, respiratory distress, and osmoregulatory failure, and they also exaggerate the secondary pathogenic infections and result in great financial losses. The monogenean mainly attaches to the external body including the skin, gills, fins, and eyes. They fed on mucus, blood, and tissue debris of the host. Their size varies from small to medium in a single host. Different members of monogenean species can infect Nile tilapia and cause high mortalities levels in both the wild and cultured fishes. *Gyrodactylidae*, *Ancyrocephalidae*, and *Dactylogyridea* members are investigated in both the cultured and wild fishes. The host-to-host interaction is very significant in the transmission process of those parasites on, plus drifting with water flow that enables them to occupy a new host according to some environmental factors like water quality and temperatures. In polluted water, parasitic infections commonly increase, and they provide an indication of water quality (Kamchoo and Chai 2023; Ogbon et al. 2023; Shafiq et al. 2023).

6.4 Diseases of Nonpathogenic Causes

6.4.1 Nutritional Diseases

When it comes to dietary elements and diet composition, the disease conditions can arise and leave the fish more susceptible to catching other diseases. Fish nutritional diseases develop as a consequence of dietary essential nutrient deficiency. Excessive dietary element increase can also result in disease conditions. The key condition is to maintain the dietary composition at optimum level and condition. Essential ingredients to meet the fish growth and developmental needs usually include proteins, lipids, carbohydrates, minerals, and vitamin salts (Abd El-Hamid et al. 2021; Farag et al. 2022). Moreover, in the case of switching diets, nutritional diseases may arise that will in turn manifest with loss of appetite, slower or suppressed growth rate, decreased fertility, susceptibility to diseases, histopathological alterations, and mortalities (Table 6.2) (Chong 2022b; Syanya et al. 2023).

Fish scurvy is one of the noninfectious diseases caused by a deficiency of ascorbic acid condition and poor preparation of diets. The grow-out stage is the most affected, and spinal deformities arise among post-larval stages. The clinical picture of affected fishes includes loss of appetite, emaciation, retarded growth, external hemorrhages, exophthalmia, abdominal distension, fin erosion, and malformations in fins, snout, skull, spinal column, and eye. The condition can be treated with supplementary vitamin C that helps to improve protection against diseases including pathogenic infections (Muniesa et al. 2020; Désiré Adéyèmi et al. 2023). Lipidosis is another noninfectious disease of fishes. It is usually caused by intake of rancid diets or poor diet storage. It mostly affects grow-out fish. It is manifested in poor growth, low mortality rate, ocular opacity, lethargy, abdominal distention, and pale liver appearance (Abdel-Tawwab et al. 2018a, b; Lin et al. 2022; Désiré Adéyèmi et al. 2023).

Hyponatremia is a noninfectious disease of marine fish. It is caused by deficiency of salts and characterized with decreased appetite, weakness, lethargy, and loss of balance. The condition can be treated with infusion with sodium chloride and freshwater supply (Désiré Adéyèmi et al. 2023; Qian et al. 2024).

Thiamin deficiency is related to nutritional conditions that can also affect fish as it plays a role in carbohydrate metabolism as a coenzyme. It is important for supporting digestion, nervous integrity, and reproduction of normal functions and activities. The thiamin deficiency is related to the enzyme thiaminase loss as a result of the presence of some dietary anti-thiaminic substances. Affected fishes with thiamin deficiency exhibit anorexia, nervous manifestations, pale body, external hemorrhages, abnormal swimming movement, and mortalities (Harder et al. 2018; Jackson et al. 2024).

Table 6.2 Different vitamin deficiency conditions in fishes

Deficient vitamin	Signs
Vitamin K	Anemia, paleness
Thiamine	Anorexia, poor appetite and growth, atrophy, paleness, unbalanced movement
Vitamin C	Anorexia, hemorrhages, darkness, loss of balance
Folic acid	Anemia, retarded growth
Riboflavin	Hemorrhages, darkness, poor appetite, and growth
Niacin	Anemia, retarded growth
Vitamin D	Anorexia, poor appetite, and growth

6.4.2 Environmental Diseases

The immune system and response of fish can be greatly influenced by various external factors like temperature, light, water quality, salinity, and different stress inducers (Abdel-Tawwab et al. 2018a, b; Zhi et al. 2020; Abu-Elala et al. 2021).

Photoperiod in addition to temperature is considered the prime controller of living organisms. Photoperiod has been shown to impact the immune response through a reduction in the number of circulating leukocytes that is caused by increasing daylight hours. Moreover, an increase in the photoperiod can increase the activity of lysozyme and the circulating IgM. Photoperiod is important to change the breeding cycle and allow the production of eggs/fry within the normal breeding cycle. In salmon farming, photoperiod changes can be used to induce the animals into molting, allowing farmers to have more than one input into seawater each year and thus stretch the production cycle (Bowden 2008).

Oxygen levels in the aquatic environment can modulate the immune response; hypoxia depresses the respiratory burst activity of macrophages and lowers the levels of circulating antibodies, which, in turn, are elevated by hyperoxia (Abdel-Tawwab et al. 2018a, b; Abarike et al. 2020). Channel catfish showed that a reduction in oxygen from 6 mg/L to less than 2 mg/L resulted in a drop of complement hemolytic activity (Welker et al. 2007). In our recent study, yellow perch (*Perca flavescens*) subjected to hypoxia stress showed significant decreases ($P < 0.05$) in both aspartate aminotransferase (AST) and alanine transaminase (ALT) and antioxidant enzymes (superoxide dismutase (SOD) and catalase (CAT)) (Elabd et al. 2017, 2020a).

Natural changes in salinity can occur as a consequence of rainwater diluting seawater or mixing of estuarine waters or seawater into freshwater areas. Increasing salinity can increase phagocytosis and respiratory burst. In addition, lysozyme and IgM levels can increase with increasing salinity (Bowden 2008). A recent study on rainbow trout showed that fishes which were transferred from freshwater to seawater had an enhanced head kidney macrophage respiratory burst (Taylor et al. 2007). Moreover, brown trout (*Salmo trutta*) transferred from freshwater to seawater showed an increase in the phagocytic activity of head kidney leucocytes. A further study in brown trout found an increase in lysozyme after seawater transfer (Khalil et al. 2019; Othman et al. 2022).

Changes in the pH affect mainly the freshwater systems with the surrounding water than seawater, which has a significant buffering capacity to overcome pH changes. In freshwater lakes this level of buffering and so can be impacted by various changes in pH (Abdel-Tawwab et al. 2018a, b; Othman et al. 2022). Moreover, a study on rainbow trout showed that reducing the pH to 4.5 resulted in a reduction in the levels of plasma lysozyme (Yada et al. 2006). Another study on Nile tilapia maintained at pH 7.9 and pH 4 for 2 weeks showed a highly significant ($P < 0.001$) increase in lysozyme levels at the lower pH after 2 weeks (Dominguez et al. 2005). In the study we previously performed on a giant freshwater prawn (*Macrobrachium rosenbergii*), juveniles exposed to high pH showed significantly higher total hemocyte count (THC), hemocyte viability (HV), differential hemocyte count (DHC), phagocytic rate (PR), and phagocytic index (PI) compared with the control group (Elabd et al. 2012).

Temperature is a crucial environmental parameter and has been reported to affect growth, survival, physiological function, and immune function in teleost (Bowden 2008; Chen et al. 2022). Changes in ambient temperature affect all kinds of biological processes; it has been demonstrated that an increase in water temperature stimulates the basal release of the pituitary gland gonadotrophins (Tort and Balasch 2022); in addition, growth is enhanced with rising temperatures. In addition, low environmental temperatures have been shown to affect the immune system and suppress specific immune functions and affect the rate of physiological functions (Hayward et al. 2009; Wang et al. 2023).

The effects of temperature on specific immune responses, especially the humoral response, have been studied (Wang et al. 2023) and revealed that the primary antibody response is suppressed at low temperatures. In contrast, the secondary response can be elicited at low temperatures if immunological memory is established at high temperatures (Avtalion et al. 1972). Oppositely, some results showed that lower temperatures induce a delay in the peak of the primary response but have no effect on the magnitude of the response (Rijkers et al. 1981). Studies showed that tilapia acclimated to 27 °C and then exposed to cooler water (19 and 23 °C) experienced a drop in measurable lysozyme and an elevation when exposed to higher temperatures (31 and 35 °C) (Ndong et al. 2007). In addition, Atlantic halibut showed increased lysozyme levels when the culture temperature was elevated from 8 to 18 °C (Langston et al. 2002). Moreover, sea bass showed that high temperatures (23 °C) significantly increased the IgM concentrations (Varsamos et al. 2005). In different studies we performed recently, changes in temperature resulted in significant changes in immune-related parameters. For yellow perch, we tested both low (8 °C) and high (29 ± 0.50 °C) temperatures over the ambient temperature (17 ± 1.2 °C), and results showed a significant decrease in ALT, AST, SOD, and CAT activities (Elabd et al. 2017). Regarding bluegill sunfish (*Lepomis macrochirus*), we evaluated the effect of different temperatures beyond 10.2 ± 0.05, 20 ± 0.06, and above the optimum temperature (29.7 ± 0.09) 34.7 ± 0.05 °C; results revealed significantly higher ($P < 0.05$) erythrocyte lysate superoxide dismutase (SOD) in 10 and 35 °C groups (Elabd et al. 2016b).

6.5 Conclusion

The aquaculture industry is subjected to many challenges, among which are diseases. Diseases can be infectious or noninfectious according to the causative agent. Infectious diseases include bacterial, viral, mycotic, and parasitic diseases. On the contrary, noninfectious may include nutritional and environmental diseases, which predispose for the occurrence of other diseases plus its detrimental effect itself. Diseases conditions can cause serious negative effects on the aquatic animals and the aquaculture industries. Diseases challenge the immune system leaving it in an impaired status to perform its normal functions of protection. Reporting those conditions is necessary to evaluate their ecological effect and the economic impacts in fish farms. The correct identification of fish diseases is crucial to properly address the causative agents and decide the appropriate control approaches and measures.

References

Abarike ED, Jian J, Tang J, Cai J, Sakyi EM, Kuebutornye FKA (2020) A mixture of Chinese herbs and a commercial probiotic Bacillus species improves hemato-immunological, stress, and antioxidant parameters, and expression of HSP70 and HIF-1α mRNA to hypoxia, cold, and heat stress in Nile tilapia, Oreochromis niloticus. Aquac Rep 18:100438. https://doi.org/10.1016/j.aqrep.2020.100438

Abd El-Hamid MI, Ibrahim SM, Eldemery F, El-Mandrawy SAM, Metwally AS, Khalifa E, Elnahriry SS, Ibrahim D (2021) Dietary cinnamaldehyde nanoemulsion boosts growth and transcriptomes of antioxidant and immune related genes to fight Streptococcus agalactiae infection in Nile tilapia (Oreochromis niloticus). Fish Shellfish Immunol 113:96–105. https://doi.org/10.1016/j.fsi.2021.03.021

Abdel Rahman AN, Mansour DA, Abd El-Rahman GI, Elseddawy NM, Zaglool AW, Khamis T, Mahmoud SF, Mahboub HH (2022a) Imidacloprid toxicity in Clarias gariepinus: protective role of dietary Hyphaene thebaica against biochemical and histopathological disruption, oxidative stress, immune genes expressions, and Aeromonas sobria infection. Aquaculture 555:738170. https://doi.org/10.1016/j.aquaculture.2022.738170

Abdel Rahman AN, Shakweer MS, Algharib SA, Abdelaty AI, Kamel S, Ismail TA, Daoush WM, Ismail SH, Mahboub HH (2022b) Silica nanoparticles acute toxicity alters ethology, neurostress indices, and physiological status of African catfish (Clarias gariepinus). Aquac Rep 23:101034. https://doi.org/10.1016/j.aqrep.2022.101034

Abdel-Aziz M, Eissa AE, Hanna M, Okada MA (2013) Identifying some pathogenic Vibrio/Photobacterium species during mass mortalities of cultured Gilthead seabream (Sparus aurata) and European seabass (Dicentrarchus labrax) from some Egyptian coastal provinces. Int J Vet Sci Med 1(2):87–95. https://doi.org/10.1016/j.ijvsm.2013.10.004

Abdelsalam M, Elgendy M, Shaalan M, Moustafa M, Fujino M (2017) Rapid identification of pathogenic streptococci isolated from moribund red tilapia (Oreochromis spp.). Acta Vet Hung 65:50–59. https://doi.org/10.1556/004.2017.005

Abdel-Tawwab M, Adeshina I, Jenyo-Oni A, Ajani EK, Emikpe BO (2018a) Growth, physiological, antioxidants, and immune response of African catfish, Clarias gariepinus (B.), to dietary clove basil, Ocimum gratissimum, leaf extract and its susceptibility to Listeria monocytogenes infection. Fish Shellfish Immunol 78:346–354. https://doi.org/10.1016/j.fsi.2018.04.057

Abdel-Tawwab M, Samir F, Abd El-Naby AS, Monier MN (2018b) Antioxidative and immunostimulatory effect of dietary cinnamon nanoparticles on the performance of Nile tilapia, Oreochromis niloticus (L.) and its susceptibility to hypoxia stress and Aeromonas hydrophila infection. Fish Shellfish Immunol 74:19–25. https://doi.org/10.1016/j.fsi.2017.12.033

Abu-Elala NM, Shaalan M, Ali SE, Younis NA (2021) Immune responses and protective efficacy of diet supplementation with selenium nanoparticles against cadmium toxicity in Oreochromis niloticus. Aquac Res 52(8):3677–3686. https://doi.org/10.1111/are.15212

Adeshina I, Abdel-Tawwab M, Tijjani ZA, Tiamiyu LO, Jahanbakhshi A (2021) Dietary Tridax procumbens leaves extract stimulated growth, antioxidants, immunity, and resistance of Nile tilapia, Oreochromis niloticus, to monogenean parasitic infection. Aquaculture 532:736047. https://doi.org/10.1016/j.aquaculture.2020.736047

Aly SM, Eisa AE, ElBanna N (2020) Characterization of Vibrio parahaemolyticus infection in gilthead seabream (Sparus auratus) cultured in Egypt. Egypt J Aquat Biol Fish 24(1):553–571. https://doi.org/10.21608/ejabf.2020.76562

Alzahrani OM, Elumalai P, Nada HS, Ahmed SAA, Zaglool AW, Shawky SM, Alkafafy M, Mahboub HH (2023) Pseudomonas putida: sensitivity to various antibiotics, genetic diversity, virulence, and role of formic acid to modulate the immune-antioxidant status of the challenged Nile tilapia compared to carvacrol oil. Aust Fish 8(1):6. https://doi.org/10.3390/fishes8010006

As P, Sj P, Ar B, Ba P, Sn G, Patel D (2018) Fungal diseases of fish: a review. Open Access J Vet Sci Res 3:000164. https://doi.org/10.23880/oajvsr-16000164

Avila BW, Huyvaert KP, Winkelman DL, Fetherman ER (2022) Factors affecting post-challenge survival of Flavobacterium psychrophilum in susceptible rainbow trout from the literature. Pathogens 11(11):1318. https://doi.org/10.3390/pathogens11111318

Avtalion RR, Wojdani A, Ducsyminer M (1972) Antibody production in carp (Cyprinus carpio): temperature effect and mechanism. In: Etude Phylogénique et Ontogénique de la Réponse Immunitaire et son Apport à la Théorie Immunologique. INSERM, pp 75–87

Ben Hamed S, Tavares Ranzani-Paiva MJ, Tachibana L, de Carla Dias D, Ishikawa CM, Esteban MA (2018) Fish pathogen bacteria: adhesion, parameters influencing virulence and interaction with host cells. Fish Shellfish Immunol 80:550–562. https://doi.org/10.1016/j.fsi.2018.06.053

Bowden TJ (2008) Modulation of the immune system of fish by their environment. Fish & Shellfish Immunology 25(4):373–383. https://doi.org/10.1016/j.fsi.2008.03.017

Chen Z, Galli M, Gallavotti A (2022) Mechanisms of temperature-regulated growth and thermotolerance in crop species. Curr Opin Plant Biol 65:102134. https://doi.org/10.1016/j.pbi.2021.102134

Chi TTK, Murrell KD, Madsen H, Khue NV, Dalsgaard A (2009) Fishborne zoonotic trematodes in raw fish dishes served in restaurants in Nam Dinh Province and Hanoi, Vietnam. J Food Prot 72(11):2394–2399. https://doi.org/10.4315/0362-028X-72.11.2394

Chong RS-M (2022a) Chapter 51—Ichthyophoniasis. In: Kibenge FSB, Baldisserotto B, Chong RS-M (eds) Aquaculture pathophysiology. Academic Press, London, pp 629–633. https://doi.org/10.1016/B978-0-12-812211-2.00051-2

Chong RS-M (2022b) Chapter 65—Endocrine disruption in fish. In: Kibenge FSB, Baldisserotto B, Chong RS-M (eds) Aquaculture pathophysiology. Academic Press, London, pp 727–740. https://doi.org/10.1016/B978-0-12-812211-2.00065-2

de Bruijn I, Liu Y, Wiegertjes GF, Raaijmakers JM (2018) Exploring fish microbial communities to mitigate emerging diseases in aquaculture. FEMS Microbiol Ecol 94(1). https://doi.org/10.1093/femsec/fix161

Désiré Adéyèmi A, Polycarpe Kayodé AP, Morelle Atcheon I, Bienvenue Chabi I, Rob Nout MJ, Linnemann AR (2023) Performance of sustainable fish feeds in Benin for *Clarias gariepinus* (Burchell 1822). Egypt J Aquat Res 49(3):387–393. https://doi.org/10.1016/j.ejar.2022.09.002

Dominguez M, Takemura A, Tsuchiya M (2005) Effects of changes in environmental factors on the non-specific immune response of Nile tilapia. Oreochromis niloticus L. Aquaculture Research 36(4):391–397. https://doi.org/10.1111/j.1365-2109.2005.01220.x

El Latif AMA, Elabd H, Amin A, Eldeen AIN, Shaheen AA (2019) High mortalities caused by Aeromonas veronii: identification, pathogenicity, and histopathological studies in Oreochromis niloticus. Aquac Int 27(6):1725–1737. https://doi.org/10.1007/s10499-019-00429-8

Elabd H, El Asely A, Shaheen A, Abbass A (2012) Effects of Zingiber officinalis and Cyanodon dactylon on the growth performance and immune parameters of Macrobrachium rosenbergii. World J Fish Mar Sci 4(3):301–307

Elabd H, Wang H-P, Shaheen A, Yao H, Abbass A (2016a) Feeding Glycyrrhiza glabra (liquorice) and Astragalus membranaceus (AM) alters innate immune and physiological responses in yellow perch (Perca flavescens). Fish Shellfish Immunol 54:374–384. https://doi.org/10.1016/j.fsi.2016.04.024

Elabd H, Wang H-P, Shaheen A, Yao H, Abbass A (2016b) Astragalus membranaceus (AM) enhances growth performance and antioxidant stress profiles in bluegill sunfish (Lepomis macrochirus). Fish Physiology and Biochemistry 42(3):955–966. https://doi.org/10.1007/s10695-015-0188-9

Elabd H, Wang H-P, Shaheen A, Yao H, Abbass A (2017) Anti-oxidative effects of some dietary supplements on Yellow perch (Perca flavescens) exposed to different physical stressors. Aquac Rep 8:21–30. https://doi.org/10.1016/j.aqrep.2017.09.002

Elabd H, Wang H-P, Shaheen A, Matter A (2020a) Astragalus membranaceus nanoparticles markedly improve immune and anti-oxidative responses; and protection against Aeromonas veronii in Nile tilapia Oreochromis niloticus. Fish Shellfish Immunol 97:248–256. https://doi.org/10.1016/j.fsi.2019.12.025

Elabd H, Wang H-P, Shaheen A, Matter A (2020b) Nano spirulina dietary supplementation augments growth, antioxidative and immunological reactions, digestion, and protection of Nile tilapia, Oreochromis niloticus, against Aeromonas veronii and some physical stressors. Fish Physiol Biochem 46(6):2143–2155. https://doi.org/10.1007/s10695-020-00864-y

Fadel A, Mabrok M, Aly S (2018) Epizootics of Pseudomonas anguilliseptica among cultured seabream (Sparus aurata) populations: control and treatment strategies. Microb Pathog 121:1–8. https://doi.org/10.1016/j.micpath.2018.04.021

Farag MR, Alagawany M, Khalil SR, Abd El-Aziz RM, Zaglool AW, Moselhy AAA, Abou-Zeid SM (2022) Effect of parsley essential oil on digestive enzymes, intestinal morphometry, blood chemistry and stress-related genes in liver of Nile tilapia fish exposed to Bifenthrin. Aquaculture 546:737322. https://doi.org/10.1016/j.aquaculture.2021.737322

Ganesan M, Mani R, Sai S (2023) Isolation and identification of Aeromonas sp. from fishes. In: Thomas J, Amaresan N (eds) Aquaculture microbiology. Springer, New York, pp 3–10. https://doi.org/10.1007/978-1-0716-3032-7_1

Goh KW, Abdul Kari Z, Wee W, Zakaria NNA, Rahman MM, Kabir MA, Abdul Hamid NK, Tahiluddin AB, Kamarudin AS, Téllez-Isaías G, Wei LS (2023) Exploring the roles of phytobiotics in relieving the impacts of Edwardsiella tarda infection on fish: a mini-review. Front Vet Sci 10:1149514. https://doi.org/10.3389/fvets.2023.1149514

Haraz E, Khalil R, El-Hoshy S (2015) Some bacterial and fungal affections causing disease problems in cultured seabream Sparus aurata = بعض الإصابات البكتيرية والفطرية المسببة للأمراض في أسماك الدنيس المستزرع. 10, 60–47. https://doi.org/10.12816/0033710

Harder A, Ardren W, Evans A, Futia M, Kraft C, Marsden JE, Richter C, Rinchard J, Tillitt D, Christie M (2018) Thiamine deficiency in fishes: causes, consequences, and potential solutions. Rev Fish Biol Fish 28:865. https://doi.org/10.1007/s11160-018-9538-x

Hayward AD, Wilson AJ, Pilkington JG, Pemberton JM, Kruuk LEB (2009) Ageing in a variable habitat: environmental stress affects senescence in parasite resistance in. St Kilda Soay sheep Proceedings of the Royal Society B: Biological Sciences 276(1672):3477–3485. https://doi.org/10.1098/rspb.2009.0906

Hegde A, Kabra S, Basawa RM, Khile DA, Abbu RUF, Thomas NA, Manickam NB, Raval R (2023) Bacterial diseases in marine fish species: current trends and future prospects in disease management. World J Microbiol Biotechnol 39(11):317. https://doi.org/10.1007/s11274-023-03755-5

Jackson C, Furnes M, Rød-Eriksen L, Yap KN, Davey M, Fossøy F, Flagstad Ø, Eide NE, Mjøen T, Ulvund K (2024) Subclinical thiamine deficiency results in failed reproduction in Arctic foxes. Vet Med Sci 10(2):e1358. https://doi.org/10.1002/vms3.1358

Jordán M, Wojtys-Tekiel S, Merino S, Tomás JM, Kaszowska M (2023) Structural diversity among Edwardsiellaceae core oligosaccharides. Int J Mol Sci 24(5):4768. https://doi.org/10.3390/ijms24054768

Jørgensen LG (2020) Zebrafish as a model for fish diseases in aquaculture. Pathogens 9(8):609. https://doi.org/10.3390/pathogens9080609

Kamchoo K, Chai J-Y (2023) Trematode metacercariae and adults in cyprinoid fish from Khun Thale swamp in Surat Thani province, Thailand. Parasites Hosts Dis 61(2):163–171. https://doi.org/10.3347/PHD.23030

Khalil RH, Diab A, Abdelhamed H, Shakweer M, Gohary M, Rashed MA (2019) Molecular characterization of vibrio Harveyi and vibrio alginolyticus with the impact of stressful environment on some naturally infected marine fish. Alex J Vet Sci 60:71. https://doi.org/10.5455/ajvs.28497

Kuebutornye FKA, Abarike ED, Lu Y (2019) A review on the application of bacillus as probiotics in aquaculture. Fish Shellfish Immunol 87:820–828. https://doi.org/10.1016/j.fsi.2019.02.010

Kumar G, Shivam S, Sudhagar A, Saravanan K, Sathish Kumar T, El-Matbouli M (2023) Chapter 18—Proteomic approaches and their applications in salmonid health research. In: Lakra WS, Goswami M, Trudeau VL (eds) Frontiers in aquaculture biotechnology. Academic Press, London, pp 249–257. https://doi.org/10.1016/B978-0-323-91240-2.00005-1

LaFrentz BR, Králová S, Burbick CR, Alexander TL, Phillips CW, Griffin MJ, Waldbieser GC, García JC, de Alexandre Sebastião F, Soto E, Loch TP, Liles MR, Snekvik KR (2022) The fish pathogen Flavobacterium columnare represents four distinct species: Flavobacterium columnare, Flavobacterium covae sp. nov., Flavobacterium davisii sp. nov. and Flavobacterium oreochromis sp. nov., and emended description of Flavobacterium columnare. Syst Appl Microbiol 45(2):126293. https://doi.org/10.1016/j.syapm.2021.126293

Langston WJ, Chesman BS, Burt GR, Pope ND, McEvoy J (2002) Metallothionein in liver of eels Anguilla anguilla from the Thames Estuary: an indicator of environmental quality? Marine Environmental Research 53(3):263–293. https://doi.org/10.1016/S0141-1136(01)00113-1

Lin W, Guo H, Yang L, Kuang Y, Li D, Yang P, Li L (2022) Alleviation of microcystin-LR-induced hepatic lipidosis and apoptosis in zebrafish by use of rice straw-derived biochar. Ecotoxicol Environ Saf 229:113054. https://doi.org/10.1016/j.ecoenv.2021.113054

Magray AR, Ganai BA, Ahmad F (2020) Isolation, identification and pathogenicity patterns of Mucor hiemalis in cultured Cyprinis Carpio communis using challenged system. Aquaculture 518:734837. https://doi.org/10.1016/j.aquaculture.2019.734837

Mahboub HH, Shaheen A (2020) Prevalence, diagnosis and experimental challenge of Dermocystidium sp. infection in Nile tilapia (Oreochromis niloticus) in Egypt. Aquaculture 516:734556. https://doi.org/10.1016/j.aquaculture.2019.734556

Mahboub HH, Shaheen AA (2021) Mycological and histopathological identification of potential fish pathogens in Nile tilapia. Aquaculture 530:735849. https://doi.org/10.1016/j.aquaculture.2020.735849

Mahboub HH, Nada HS, Abdel-Ghany HM, Ghanem R, Ahmed Ismail T, Abdel Rahman AN (2022) Detection, diagnosis, Koch's postulate, hepatorenal and antioxidant indicators for some systemic pathogenic fungi invading the liver and kidneys of African catfish (Clarias gariepinus) in Egypt with a histopathological approach. Aquac Res 53(7):2670–2685. https://doi.org/10.1111/are.15783

Mahdy OA, Abdelsalam M, Salem MA (2023) Molecular characterization and immunological approaches associated with yellow grub trematode (Clinostomid) infecting Nile tilapia. Aquac Res 2023:e5579508. https://doi.org/10.1155/2023/5579508

Manchanayake T, Salleh A, Amal MNA, Yasin ISM, Zamri-Saad M (2023) Pathology and pathogenesis of vibrio infection in fish: a review. Aquac Rep 28:101459. https://doi.org/10.1016/j.aqrep.2022.101459

Marana MH, Dalsgaard I, Kania PW, Mohamed A, Hannibal J, Buchmann K (2022) Flavobacterium psychrophilum: response of vaccinated large rainbow trout to different strains. Biology 11(12):1701. https://doi.org/10.3390/biology11121701

Mishra B, Tiwari A, Mahmoud AED (2022) Microalgal potential for sustainable aquaculture applications: bioremediation, biocontrol, aquafeed. Clean Techn Environ Policy 25:1–13. https://doi.org/10.1007/s10098-021-02254-1

Mohd Ali NS, Saad MZ, Azmai MNA, Salleh A, Zulperi ZM, Manchanayake T, Zahaludin MAD, Basri L, Mohamad A, Md Yasin IS (2023) Immunogenicity and efficacy of a feed-based bivalent vaccine against Streptococcosis and motile aeromonad septicemia in red hybrid tilapia (Oreochromis sp.). Animals 13(8):1346. https://doi.org/10.3390/ani13081346

Moustafa M, Gaafar A, Elgendy M, Abumourad I (2015) Investigations into the potential causes of mass kills in mari-cultured Gilthead Sea bream (Sparus aurata) at northern Egypt. Res J Pharm, Biol Chem Sci 6:466–477

Muniesa A, Basurco B, Aguilera C, Furones D, Reverté C, Sanjuan-Vilaplana A, Jansen MD, Brun E, Tavornpanich S (2020) Mapping the knowledge of the main diseases affecting sea bass and sea bream in Mediterranean. Transbound Emerg Dis 67(3):1089–1100. https://doi.org/10.1111/tbed.13482

Muñoz G, Rebolledo M, Landaeta MF (2023) Effects of metacercariae of Prosorhynchoides sp. (Trematoda: Bucephalidae) on the swimming ability and blood parameters of the intertidal fish Girella laevifrons (Osteichthyes: Kyphosidae). Exp Parasitol 246:108473. https://doi.org/10.1016/j.exppara.2023.108473

Ndong M, Uehara M, Katsumata S, Suzuki K (2007) Effects of Oral Administration of Moringa oleifera Lam on Glucose Tolerance in Goto-Kakizaki and Wistar Rats Journal of Clinical Biochemistry and Nutrition 40(3) 229-233. https://doi.org/10.3164/jcbn.40.229

Neethu, S., & Jyothis, M. (2023). Chapter 6—Applications of microbially derived nanomaterials in disease prevention and treatment. In J. Mathew, M. S. Jose, Radhakrishnan E.K., & A. Kumar (Eds.), Recent advances in aquaculture microbial technology (pp. 103–125). Academic Press, London. https://doi.org/10.1016/B978-0-323-90261-8.00004-3

Nissa MU, Banerjee A, Pinto N, Srivastava S (2023) Chapter 19—Proteo-metabolomic technologies in context of fishery/aquacultural research and applications. In: Lakra WS, Goswami M, Trudeau VL (eds) Frontiers in aquaculture biotechnology. Academic Press, London, pp 259–274. https://doi.org/10.1016/B978-0-323-91240-2.00015-4

Ogbon AM, Afoakwah R, Mireku KK, Tossavi ND, MacKenzie K (2023) Parasites of Sardinella maderensis (Lowe, 1838) (Actinopterygii: Clupeidae) and their potential as biological tags for stock identification along the coast of West Africa. Biology 12(3):389. https://doi.org/10.3390/biology12030389

Oliveira e Silva R, Copatti CE, Pereira GA, Macedo JS, de Souza AM, Dutra LM, Almeida JRGS, Le Reste G, Melo JFB (2024) Promotion of growth and resistance against Aeromonas hydrophila in Nile tilapia juveniles supplemented with Citrus limon extract. Aquaculture 578:740115. https://doi.org/10.1016/j.aquaculture.2023.740115

Othman R, Wang H-P, Elabd H, Xie D-K, Yao H, O'Bryant P, Rapp D (2022) The effect of density on sex differentiation, sexual dimorphism, stress, and related gene expression in yellow perch. PLoS One 17(5):e0267904. https://doi.org/10.1371/journal.pone.0267904

Özcan F, Arserim NB (2022) Fungal Diseases in Fish Black Sea Journal of Agriculture 5(1) 48–52. https://doi.org/10.47115/bsagriculture.983345

Pandey V, Hussain Bhat RA, Chandra S, Tandel RS, Dubey MK, Sharma P, Gehlot B, Dash P, Joshi R (2021) Clinical signs, lethal dose and histopathological lesions in grass carp, Ctenopharyngodon idella experimentally infected with Edwardsiella tarda. Microb Pathog 161:105292. https://doi.org/10.1016/j.micpath.2021.105292

Pradhan SK, Devi R, Khan MIR, Kamilya D, Gon Choudhury T, Parhi J (2023) Isolation of Aeromonas salmonicida subspecies salmonicida from aquaculture environment in India: polyphasic identification, virulence characterization, and antibiotic susceptibility. Microb Pathog 179:106100. https://doi.org/10.1016/j.micpath.2023.106100

Qian A, Zheng L, He Z, Zhou J, Tang S, Xing W (2024) Predictive value of hyponatremia for short-term mortality in supratentorial spontaneous intracerebral hemorrhage: a single center study. Front Neurol 15:1301197. https://doi.org/10.3389/fneur.2024.1301197

Rijkers GT, Van Oosterom R, Van Muiswinkel WB (1981) The immune system of cyprinid fish. Oxytetracycline and the regulation of humoral immunity in carp (Cyprinus carpio). Veterinary Immunology and Immunopathology 2(3):281–290. https://doi.org/10.1016/0165-2427(81)90029-5

Rohani MF, Islam SM, Hossain MK, Ferdous Z, Siddik MA, Nuruzzaman M, Padeniya U, Brown C, Shahjahan M (2022) Probiotics, prebiotics and synbiotics improved the functionality of aquafeed: upgrading growth, reproduction, immunity and disease resistance in fish. Fish Shellfish Immunol 120:569–589. https://doi.org/10.1016/j.fsi.2021.12.037

Sarjito S, Haditomo A, Sabdaningsih A, Desrina, Prayitno S (2019) Aspergillus diversity associated with fungal diseases on fish with molecular based. IOP Conf Ser Earth Environ Sci 246:012035. https://doi.org/10.1088/1755-1315/246/1/012035

Sarkar P, Stefi Raju V, Kuppusamy G, Rahman MA, Elumalai P, Harikrishnan R, Arshad A, Arockiaraj J (2022) Pathogenic fungi affecting fishes through their virulence molecules. Aquaculture 548:737553. https://doi.org/10.1016/j.aquaculture.2021.737553

Shafiq A, Abbas F, Hafeez-ur-Rehman M, Khan BN, Aihetasham A, Amin I, Hmidullah, Mothana RA, Alharbi MS, Khan I, Khalil AAK, Ahmad B, Mubeen N, Akram M (2023) Parasite diversity in a freshwater ecosystem. Microorganisms 11(8):1940. https://doi.org/10.3390/microorganisms11081940

Syanya FJ, Litabas JA, Mathia WM, Ntakirutimana R (2023) Nutritional fish diseases in aquaculture: a human health hazard or mythical theory: an overview. Eur J Nutr Food Saf 15(8):41–58. https://doi.org/10.9734/ejnfs/2023/v15i81326

Tanekhy M (2013) Some study on bacterial infection in some cultured marine fish. J Arab Aquac Soc 8:163–178

Tartor Y, Taha M, Mahboub H, El Ghamery M (2018) Yeast species associated with diseased fish: occurrence, identification, experimental challenges and antifungal susceptibility testing. Aquaculture 488:134–144. https://doi.org/10.1016/j.aquaculture.2018.01.033

Taylor JF, Needham MP, North BP, Morgan A, Thompson K, Migaud H (2007) The influence of ploidy on saltwater adaptation acute stress response and immune function following seawater transfer in non-smolting rainbow trout. General and Comparative Endocrinology 152(2-3):314–325. https://doi.org/10.1016/j.ygcen.2007.02.029

Thirumalaikumar E, Lelin C, Sathishkumar R, Vimal S, Anand SB, Babu MM, Citarasu T (2021) Oral delivery of pVAX-OMP and pVAX-hly DNA vaccine using chitosan-tripolyphosphate (Cs-TPP) nanoparticles in Rohu, (Labeo rohita) for protection against Aeromonas hydrophila infection. Fish Shellfish Immunol 115:189–197. https://doi.org/10.1016/j.fsi.2021.06.004

Tort L, Balasch JC (2022) Stress and immunity in fish. In: Buchmann K, Secombes CJ (eds) Principles of fish immunology: from cells and molecules to host protection. Springer, pp 609–655. https://doi.org/10.1007/978-3-030-85420-1_20

Van Doan H, Soltani M, Leitão A, Shafiei S, Asadi S, Lymbery AJ, Ringø E (2022) Streptococcosis a re-emerging disease in aquaculture: significance and phytotherapy. Animals 12(18):2443. https://doi.org/10.3390/ani12182443

Varsamos S, Flik G, Pepin J, Bonga SW, Breuil G (2005) Husbandry stress during early life stages affects the stressresponse and health status of juvenile sea bass. Dicentrarchus labrax. Fish & Shellfish Immunology 20(1):83–96. https://doi.org/10.1016/j.fsi.2005.04.005

Wang Z, Pu D, Zheng J, Li P, Lü H, Wei X, Li M, Li D, Gao L (2023) Hypoxia-induced physiological responses in fish: from organism to tissue to molecular levels. Ecotoxicol Environ Saf 267:115609. https://doi.org/10.1016/j.ecoenv.2023.115609

Welker TL, Mcnulty ST, Klesius PH (2007) Effect of Sublethal Hypoxia on the Immune Response and Susceptibility of Channel Catfish Ictalurus punctatus to Enteric Septicemia Abstract Journal of the World Aquaculture Society 38(1) 12-23. 10.1111/jwas.2007.38.issue-1 https://doi.org/10.1111/j.1749-7345.2006.00069.x

Wolinski L, Souza MS, Modenutti B, Balseiro E (2020) Effect of chronic UVR exposure on zooplankton molting and growth. Environmental Pollution 267:115448. https://doi.org/10.1016/j.envpol.2020.115448

Yada T, Muto K, Azuma T, Fukamachi S, Kaneko T, Hirano T (2006) Effects of acid water exposure on plasma cortisol, ion balance, and immune functions in the "cobalt" variant of rainbow trout. Zool Sci 23:707–713. https://doi.org/10.2108/zsj.23.707

Younis G, Esawy A, Elkenany R, Deen M (2019) Conventional identification of pathogenic fungi isolated from fresh water aquarium fish (O. Niloticus and C. Gariepinus) combined with molecular identification of Saprolegnia parasitica in Egypt. Adv Anim Vet Sci 8. https://doi.org/10.17582/journal.aavs/2020/8.1.77.88

Zaheen Z, War AF, Ali S, Yatoo AM, Ali MN, Ahmad SB, Rehman MU, Paray BA (2022) Chapter 7—Common bacterial infections affecting freshwater fish fauna and impact of pollution and water quality characteristics on bacterial pathogenicity. In: Dar GH, Bhat RA, Qadri H, Al-Ghamdy KM, Hakeem KR (eds) Bacterial fish diseases. Academic Press, London, pp 133–154. https://doi.org/10.1016/B978-0-323-85624-9.00006-3

Zhi T, Huang C, Sun R, Zheng Y, Chen J, Xu X, Brown CL, Yang T (2020) Mucosal immune response of Nile tilapia Oreochromis niloticus during Gyrodactylus cichlidarum infection. Fish Shellfish Immunol 106:21–27. https://doi.org/10.1016/j.fsi.2020.07.025

Vibriosis in Aquaculture: Pathogenic Vibrio Species, Affected Animals, Diagnosis and Adverse Effects

7

Nur Syafiqah Shafiee, Nurzafirah Mazlan, Nur Nashrah Abd Rahman, Rossita Shapawi, and Muhammad Dawood Shah

Abstract

Various species of economically significant cultured marine animals have been infected by vibriosis for more than a decade. Although knowledge regarding the bacterial disease of many species of cultured fish in the global community is still lacking, it has shown some improvement in the prevention and control of infectious diseases in aquaculture. This review focuses on species of pathogens, animals affected and adverse effects while discovering the impacts of vibriosis in aquaculture that have driven serious economic losses to aquaculture farmers worldwide. Many *Vibrio* species foster severe human diseases with significant levels of antibiotic resistance that can be found in estuaries around the world. Between 1996 and 2005, the average annual incidence of all *Vibrio* infections increased by 41%. *Vibrio vulnificus* (Vv), a shellfish-associated species found in the Southeast of the United States, requires temperature (16–33 °C) and salinity (5–20 ppt) for optimal growth. In addition, a comprehensive study on vibriosis is one of the initiatives aimed at strengthening the capacity of countries to conduct assessments of the burden of food-borne disease and encouraging them to use the burden of food-borne disease estimates for cost-effectiveness analyses of prevention, intervention and control measures, particularly in many Asian countries with a tropical climate.

N. S. Shafiee
Faculty of Health and Life Sciences, Management and Science University, Shah Alam, Selangor, Malaysia

N. Mazlan (✉) · N. N. Abd Rahman · R. Shapawi · M. D. Shah
Higher Institution Centre of Excellence (HICoE), Borneo Marine Research Institute, Universiti Malaysia Sabah, Kota Kinabalu, Sabah, Malaysia
e-mail: nurzafirah@ums.edu.my

N. M. Faudzi et al. (eds.), *Essentials of Aquaculture Practices*,
https://doi.org/10.1007/978-981-97-6699-4_7

Keywords

Vibriosis · Diagnosis · Prevention · Antibiotic resistance · Control measures

7.1 Introduction

Aquaculture is the world's fastest-growing food sector, which is estimated to have reached about 47% of the total 171 million metric tons of fish supplies worldwide (FAO 2016). Various cultured marine fish, such as grouper (*Epinephelus* spp.), pompano (*Trachinotus blochii*) and Asian sea bass (*Lates calcarifer*), have been infected by pathogenic bacteria during disease outbreaks (Amalina et al. 2019). This caused mortalities and significant stock losses in terms of the efficiency of fish production which was obstructed by health problems (Yang et al. 2017).

Foods associated with illnesses due to raw food consumption with a higher prevalence of pathogenic *Vibrio parahaemolyticus* can lead to gastrointestinal infections. *V. parahaemolyticus* infection can cause varying degrees of illnesses including gastroenteritis, wound infection and septicaemia. Symptoms of gastroenteritis usually involve abdominal pain, diarrhea, nausea and fever. Based on the previous studies from the Foodborne Diseases Active Surveillance Network (FoodNet) data and Morbidity and Mortality Weekly Report (MMWR) published by the Centers for Disease Control and Prevention (CDC) in the United States during the year 2016, *V. parahaemolyticus* has become the leading cause of seafood-related disease compared to other *Vibrio* isolates. It is estimated that *V. parahaemolyticus* causes about 34,664 episodes of domestically acquired food-borne illness annually in the United States (Huang et al. 2016).

Vibriosis is one of the devastating bacterial diseases affecting diverse marine fish and shellfish (Amalina et al. 2019). As a result, numerous health management techniques, such as international codes, regional recommendations, new biosecurity measures, targeting research and novel diagnostics and therapy tactics, have been implemented to provide solutions to aquatic health concerns. Diseases in fish occur as a result of a complex interaction between the host, pathogen and external stressors such as environmental deterioration related to climate change, eutrophication in natural habitat and poor fish systems and management (Huicab-Pech et al. 2016). External factors are the most common causes of vibriosis. If untreated, the infection can spread throughout the body, increasing mortality. Vibriosis is caused by *Vibrio* spp., which are among the most recurrent bacterial diseases in fish. Clinical symptoms include skin ulcers and systemic illnesses (De Souza and Wan 2021).

This review aims to update the current findings on vibriosis in cultured marine fish, focusing on key aspects including the species of pathogens involved, affected marine species, diagnostic techniques and the resultant adverse effects. Additionally, it seeks to shed light on the challenges and opportunities inherent in future research on vibriosis in aquaculture.

7.2 Pathogenic Vibrio Species

7.2.1 *Vibrio anguillarum*

The species *V. anguillarum* alone has infected more than 50 species of fish in temperate countries. This species has been extensively studied and is the best-characterized species in the genus (Hickey and Lee 2018). *V. anguillarum* is an important fish pathogen since it has been commonly used as a model microorganism to study the pathogenesis of vibriosis in fish, leading to high mortalities and economic losses in aquaculture (Lages et al. 2019). Out of 23 identified O-serotypes, only three stereotypes, the O1, O2 and O3, have been linked to fish mortalities (Vasquez et al. 2020). The most frequent serotype that was isolated from Northern pike species is serotype O2A. In comparison, the pan-genome of the serotype O1 strains contained 18% more core genes than the serotype O2A strains. Meanwhile, the serotype O2A strains contained 46%, 35% and 21% more accessory, unique and total genes, respectively, than the serotype O1 strains. In this study, the serotype O1 strains are a very homogenous group compared to the serotype O2A strains, which demonstrate much more diversity. This has explained a lot about why *V. anguillarum* serotype O2A genomes are commercially available, and experimental vaccines based on only one serotype O2A strain, which has produced good protection from infection from some strains, but not for others (Hansen et al. 2020).

In Northern China, the development of half-smooth tongue sole (*Cynoglossus semilaevis*), has been hindered by serious infectious bacterial diseases. Specifically, *V. anguillarum* has resulted in a high mortality rate of infection leading to enormous economic losses (Zhang et al. 2015). The widespread of siderophores piscibactin genes in *V. anguillarum* strains have been found associated with a highly virulent phenotype for several fish species, such as turbot (*Psetta maxima*), cod (*Gadus morhua*), halibut (*Hippoglossus hippoglossus*), rainbow trout (*Oncorhynchus mykiss*) or sea bass (*Dicentrarchus labrax*) (Balado et al. 2018). Ayu (*Plecoglossus altivelis*), one of the most important species in Japanese fresh water fisheries, has also been infected by *V. anguillarum.* This pathogen was reported to be isolated from cultured sea bass in the Mediterranean Sea and gilthead seabream in Turkey (Türe and Alp 2016).

7.2.2 *Vibrio ordalii*

Vibrio ordalii was initially characterized as *V. anguillarum* (Steinum et al. 2016). This pathogen was first reported to cause bacterial haemorrhagic septicaemia in marine fishes in Japan, Australia and the Pacific Northwest of the United States (Amalina et al. 2019). *V. ordalii* is a γ-proteobacterium that is responsible for infections, namely, vibriosis and haemorrhagic septicaemia in several species of aquaculture fish, mainly salmonids (Ruiz et al. 2019). *V. ordalii* has also been reported in the population of Atlantic salmon (*Salmo salar*), Pacific salmon (*Oncorhynchus kisutch*) and rainbow trout (*Oncorhynchus mykiss*) in the south of Chile (Ruiz et al. 2015).

7.2.3 *Vibrio salmonicida*

The main symptoms of "cold water vibriosis" in farmed salmonid fish caused by *V. salmonicida* include a systemic disease with haemorrhages and anaemia. The outbreak of the disease appears in seawater and occurs only at low water temperatures usually below 10 °C, even though the species can grow until 20 °C (Guijarro et al. 2015). In 2018, however, cold water vibriosis that is caused by *V. salmonicida* was not identified in salmon (Hjeltnes et al. 2019). Cold water vibriosis has been widespread across Scotland, Iceland, Faroe Isles, Nova Scotia/New Brunswick, Canada and Maine, United States. It has been found that Atlantic salmon is more susceptible to this disease compared to rainbow trout and Atlantic cod (Foyn 2017).

7.2.4 *Vibrio parahaemolyticus*

The rising vibriosis and food-borne cases worldwide have put *Vibrio parahaemolyticus* in the limelight (Letchumanan et al. 2019). *V. parahaemolyticus* was identified as the causative agent of acute hepatopancreatic necrosis disease (AHPND). This disease is known to cause early mortality syndrome (EMS) in shrimp, which was first detected in China in 2009 and subsequently was reported in various countries such as a few Southeast Asian countries, Mexico, South America and the United States. This disease has caused a massive loss in shrimp farming in Vietnam since 2010, particularly in the Mekong Delta in the South region. This is because they provide approximately 95% of the total shrimp production in the country (Hong To et al. 2020). From shrimps, antimicrobial-resistant *V. parahaemolyticus* strains have been isolated and detected in Thailand, Malaysia and China (Letchumanan et al. 2015). Recently, *V. parahaemolyticus* strains have been found in seafood, such as oysters, clams and shellfish from numerous countries in Asia and several European countries (Yu et al. 2015; Kang et al. 2017). *V. parahaemolyticus* has been shown to inflict serious infections in ornamental fishes such as clownfish (*Amphiprion sebae*) and damsels (*Stegastes variabilis*) in India (Marudhupandi et al. 2017). It has also been reported in tropical fish species such as grouper (*Epinephelus* sp.) and yellow croaker (*Larimichthys polyactis*) in China (Liu et al. 2016).

7.2.5 *Vibrio alginolyticus*

Vibrio alginolyticus was preliminarily regarded as biotype 2 of *V. parahaemolyticus* (Fernández-Bravo et al. 2019). *V. alginolyticus*, although sometimes pathogenic, also has been considered a common marine flora. This pathogen has been shown in vivo to work well as a probiotic for Atlantic salmon that protects against *Aeromonas salmonicida*, *V. anguillarum* and *V. ordalii* (Egerton et al. 2018). *V. alginolyticus* has been reported as a bivalve pathogen (Dubert et al. 2017). Cases of high mortality rates among fish stock and severe economic losses in aquaculture throughout the Mediterranean area have been associated with pathogenic bacteria, *V.*

alginolyticus. This also gives rise to a lot of epizootic outbreaks among the gilthead seabream and European sea bass populations, which influence the high economic value of the Mediterranean communities (Rigos et al. 2020). *V. alginolyticus* strikes at several marine fishes, such as large yellow croaker (Liu et al. 2016), silver sea bream, cobia, grouper and Asian sea bass. Cobia aquaculture in Taiwan and India has been frequently affected by the said pathogenic bacteria (Rameshkumar et al. 2017). Additionally, the presence of *V. alginolyticus* has been reported among fresh water Nile tilapia *Oreochromis niloticus* in Saudi Arabia (Younes et al. 2016).

7.2.6 *Vibrio vulnificus*

Vibrio vulnificus can be found worldwide. This pathogen is known to cause gastrointestinal disease, wound infections and highly fatal septicaemia. Admittedly, it has the highest number of seafood-related deaths in the United States (95% of all cases) (Haftel and Sharman 2023). *V. vulnificus* is virulent in humans and fish. In terms of virulence, it has been classified as the biochemical and serological properties of the strain (Al-Dulaimi et al. 2014). *V. vulnificus* has been documented in numerous studies where it has induced primary septicaemia following ingestion of aquaculture marine fish such as raw molluscan shellfish (Oliver and Jones 2014). The second biotype of *V. vulnificus* has been found associated with fish epizootics; however, it is hardly found in humans (Amaro et al. 2015). Based on recent studies in Israel, about 62 cases of wound infections are associated with farmed tilapia and humans, which have been found related to biotype 3 of *V. vulnificus* (Kim et al. 2017). All of the first biotype 2 isolates were from Asia and Europe. The second biotype was reclassified from eel known as *Serovar E*. It was found to be highly virulent towards eels causing high mortality due to epizootics outbreaks in brackish-water eel farms (Amaro et al. 2015). The infection of *V. vulnificus* produces pathological changes of bacterial haemorrhagic septicaemia in Japanese sea bass from Japan (Li et al. 2019) and oysters from Delaware and Chesapeake Bays (Parveen et al. 2020), New England and the Northern Pacific (Aruta 2019).

7.2.7 *Vibrio mimicus*

Vibrio mimicus was found to initially reassemble *V. cholerae* in many characteristics (Guardiola-Avila et al. 2016). Severe diarrhoeal illness is potentially caused by this non-halophilic pathogen (Baker-Austin et al. 2018). Systemic infections with *V. mimicus* were found infecting cultured crustaceans such as red swamp crayfish (*Procambarus clarkii*) in the United States, red claw crayfish (*Cherax quadricarinatus*) and yabbie (*Cherax albidus*) in Australia and red swamp crayfish (*Procambarus clarkii*) in Assiut, Egypt (Sherry et al. 2016). In China, *V. mimicus* infection has led to heavy economic losses in catfish, including yellow catfish, southern catfish and Zhengchuan catfish which have approximately caused more than 70% cumulative mortality. The major clinical symptoms of infection of this pathogen are characterized by the presence of regularly shaped ulcers with clear boundaries (Yan-Wei et al. 2019). A few countries such as Singapore, China and Thailand have been associated with *V. mimicus* isolated from ornamental fishes (Zago et al. 2017).

7.2.8 *Vibrio harveyi*

Vibrio harveyi, formerly known as *Achromobacter harveyi*, is a Gram-negative bacterium that is widely distributed within the aquatic environment. Based on the genotyping data, two closely similar strains, which are *V. trachuri* and *V. carchariae*, are classified under the Harveyi clade (Montánchez and Kaberdin 2020). This pathogen is a major threat to diverse aquatic species specifically in Asia from invertebrates such as crustaceans and bivalves to vertebrates such as sharks and bony fishes (Zhang et al. 2020). In a few countries, namely, Malaysia, India, the Philippines and Thailand, *V. harveyi* poses a perilous impact on prawn *Penaeus* spp. farming. The high mortality rate in other marine invertebrates such as seahorses (*Hippocampus kuda* and *Hippocampus erectus*), sea cucumber (*Holothuria scabra*), Japanese abalone (*Haliotis discus hannai*) and Pacific oysters (*Crassostrea gigas*) due to infection (Zhang et al. 2020). There are many reported cases of infections in various fish species including sandbar (*Carcharhinus plumbeus*) and lemon sharks (*Negaprion brevirostris*), farmed sea perch (*Lateolabrax japonicus*), cage-reared grouper (*Epinephelus awoara*), silvery black porgy (*Acanthopagrus cuvieri*) and brown-spotted grouper (*Epinephelus tauvina*) in Kuwait, common dentex (*Dentex dentex*) and farmed sole (*Solea senegalensis*) in Spain and Japanese horse mackerel (*Trachurus japonicus*) (Mohamad et al. 2019a, b; Zhang et al. 2020). In addition to that, Montánchez and Kaberdin (2020) mentioned that eye disease in milkfish (*Chanos chanos*) from the Philippines, short sunfish (*Mola mola*) and common snook (*Centropomus undecimalis*) is caused by *V. harveyi*.

7.2.9 *Photobacterium damselae* subsp. *damselae*

The first name of *Photobacterium damselae* subsp. *damselae* was *V. damselae* which was named after a marine fish that was known as damselfish from where it was isolated (Osorio et al. 2018). This pathogen is a marine bacterium of the family Vibrionaceae which has been found in a wide variety of aquatic animals, including fish, molluscs and crustaceans (Terceti et al. 2016). It has been reported to trigger diseases in turbot (*Scophthalmus maximus*), sea bream (*Sparus aurata*) and other sparid fish species, sea bass (*Dicentrarchus labrax*), rainbow trout (*Oncorhynchus mykiss*) and shrimp (*Exopalaemon carinicauda*) (Terceti et al. 2018). Recently, outbreaks have been reported from newly cultured species of marine fish in the Mediterranean area such as gilthead seabream, white seabream (*Diplodus sargus*) and red-banded seabream (*Pagrus auriga*). In addition, *P. damselae* subsp. *damselae* also has been found in outbreaks of sea bass (*Dicentrarchus labrax*) cultures in Spain, Egypt and Tunisia (Terceti et al. 2016).

7.2.10 *Photobacterium damselae* subsp. *piscicida*

This pathogen remains as one of the main causative agents for an infectious disease known as Photobacteriosis (Valderrama et al. 2019). This pathogen is known as the main zoonotic pathogen acquired topically from fish and caused a high number of human cases particularly in coastal areas of the United States, Japan and Australia, during the warm season (Matanza and Osorio 2020). Table 7.1 summarizes the species of pathogens and species of animals affected by the vibriosis. *Photobacterium damselae* subsp. *piscicida* is harmful and able to infect a diverse marine fish species, including the yellowtail (*Seriola quinqueradiata*) in Japan; gilthead sea bream (*Sparus aurata*), sea bass (*Dicentrarchus labrax*) and sole (*Solea senegalensis* and *Solea solea*) in Europe; striped bass (*Morone saxatilis*), white perch (*Morone americana*) and hybrid striped bass (*Morone saxatilis* × *Morone chrysops*) in the United States; cobia (*Rachycentron canadum*) in Taiwan; and golden pompano (*Trachinotus ovatus*) in China (Andreoni and Magnani 2014).

Table 7.1 Species of pathogens and animals affected

Species	Animals	References
Vibrio anguillarum	Half-smooth tongue sole (*Cynoglossus semilaevis*) Turbot (*Psetta maxima*) Cod (*Gadus morhua*) Halibut (*Hippoglossus hippoglossus*) Rainbow trout (*Oncorhynchus mykiss*) Sea bass (*Dicentrarchus labrax*) Ayu (*Plecoglossus altivelis*)	Zhang et al. (2015), Türe and Alp (2016)
Vibrio ordalli	Atlantic salmon (*Salmo salar*) Pacific salmon (*Oncorhynchus kisutch*) Rainbow trout (*Oncorhynchus mykiss*) Atlantic cod (*Gadus morhua*)	Ruiz et al. (2015), Foyn (2017)
Vibrio parahaemolyticus	Clownfish (*Amphiprion sebae*) Damsels (*Stegastes variabilis*) Grouper (*Epinephelus* sp.) Yellow croaker (*Larimichthys polyactis*)	Liu et al. (2016), Marudhupandi et al. (2017)
Vibrio alginolyticus	Silver sea bream (*Chrysophrys auratus*) Cobia (*Rachycentron canadum*) Malabar grouper (*Epinephelus malabaricus*) Asian sea bass (*Lates calcarifer*) Large yellow croaker (*Larimichthys crocea*) Nile tilapia (*Oreochromis niloticus*)	Liu et al. (2016), Younes et al. (2016)

(continued)

Table 7.1 (continued)

Species	Animals	References
Vibrio vulnificus	Raw molluscan shellfish Nile tilapia (*Oreochromis niloticus*) Eel (*Monognathus ahlstromi*) Japanese sea bass (*Lateolabrax japonicus*) Oysters (*Magallana gigas*)	Oliver and Jones (2014), Amaro et al. (2015), Kim et al. (2017), Aruta (2019), Li et al. (2019), Parveen et al. (2020)
Vibrio mimicus	Red swamp crayfish (*Procambarus clarkii*) Yellow catfish (*Ameiurus natalis*) Southern catfish (*Pseudauchenipterus nodosus*) Zhengchuan catfish (*Ictalurus punctatus*)	Sherry et al. (2016), Yan-Wei et al. (2019)
Photobacterium damselae subsp. *damselae*	Turbot (*Scophthalmus maximus*) Sea bream (*Sparus aurata*) Sea bass (*Dicentrarchus labrax*) Rainbow trout (*Oncorhynchus mykiss*) Shrimp (*Exopalaemon carinicauda*) Gilthead seabream, white seabream (*Diplodus sargus*) Red-banded seabream (*Pagrus auriga*)	Terceti et al. (2016, 2018)
Photobacterium damselae subsp. *piscicida*	Yellowtail (*Seriola quinqueradiata*) Gilthead Sea bream (*Sparus aurata*) Sea bass (*Dicentrarchus labrax*) Sole (*Solea senegalensis* and *Solea solea* Striped bass (*Morone saxatilis*) White perch (*Morone americana*) Hybrid striped bass (*Morone saxatilis* × *Morone chrysops*) Cobia (*Rachycentron canadum*) Golden pompano (*Trachinotus ovatus*)	Andreoni and Magnani (2014)

7.3 Adverse Effect of Vibriosis

7.3.1 Economic Impacts

A variety of marine fauna such as fishes, crustaceans and bivalves have been affected by vibriosis that devastatingly causes high mortalities with severe economic losses worldwide. Mariculture represents an important investment for fishermen in Egypt; however, diseases and high feeding costs are the main barriers facing the sustainability and profitability of this sector (Abdelaziz et al. 2017). Most aquaculturists have focused on shrimps and prawns, as they promote the primary source of protein for consumption which generates major revenue in the food industries. Even though there is high production of global farmed shrimp, the largest exporting countries, especially Asia, have faced a decline in seafood export due to vibriosis (Santos et al. 2020a, b). In Asia, shrimp disease has resulted in huge losses of national income amounting to billions of dollars annually (Shinn et al. 2018).

For instance, cultured salmon in Norway and Iceland have been seriously affected by Hitra disease caused by *V. salmonicida* and winter ulcer disease caused by *V. viscosus*. Meanwhile, in Asia, vibriosis is the main cause of high mortality in cultured grass carp and turbot in China and cultured eels in Japan (Ji et al. 2020). However, vaccination is possible during vibriosis outbreaks in the salmon industry in Norway but does not appear to offer full protection from winter ulcer disease (Erkinharju and Dalmo 2021). Bacterial diseases are also a major problem that causes a disease known as acute hepatopancreatic necrosis disease (AHPND). The disease is associated with strains of bacterial *Vibrio* species carrying an insect toxin plasmid. The mortality increased up to 100% due to infections in post-larvae shrimp, and economic losses were estimated at more than USD 1 billion in Asia at the end of 2013 (Thornber et al. 2020). Based on the previous study in Malaysia, the average costs of vibriosis in Asian sea bass were USD 0.24 which comprised mortality costs (USD 0.19)/tail), treatment costs (USD 0.05)/tail) and diagnosis costs (USD 0.003)/tail).

Vibrio infection and the illnesses linked with these bacteria affect all aquatic crustaceans. Vibriosis outbreaks frequently result in significant production losses, and diseases such as acute hepatopancreatic necrosis disease (AHPND) have imposed harm to shrimp farm operations (e.g. *Penaeus monodon* and *P. vannamei*) (Prachumwat et al. 2020). The impact of AHPND has been implicated in a USD 44 billion loss in production in China, Malaysia, Mexico, Thailand and Vietnam between 2010 and 2016 (Tang and Bondad-Reantaso 2019). Hatchery and luminous vibriosis, shell disease syndrome, tail necrosis, limp lobster disease, red body disease, *Vibrio*-caused bacteremia and summer syndrome are some of the other vibriosis disorders that have been discovered in crustaceans (Radhakrishnan et al. 2019). Overall, *Vibriosis* has a negative impact on global aquaculture productivity, causing significant socio-economic losses and posing a threat to the global food security (Tang et al. 2020).

7.3.2 Mortality Impacts

Vibrio sp. such as *V. harveyi*, *V. parahaemolyticus*, *V. alginolyticus*, *V. anguillarum*, *V. vulnificus* and *V. splendidus* are usually associated with shrimp diseases. It was reported that *V. harveyi* is associated with luminescent vibriosis in shrimps, for example, *Litopenaeus vannamei* and *Penaeus monodon*, which are the most important etiological agents for mass mortality in *P. monodon*. The most frequently isolated marine *Vibrio* species that have been associated with large-scale losses of larval and juvenile penaeids and causing several opportunistic diseases in fishes are *V. harveyi* and *V. anguillarum* (Chatterjee and Haldar 2012; Mohamad 2019).

White spot syndrome virus (WSSV) has been determined as the major factor of mass mortalities in cultured shrimp in Asia and eventually the Americas (Klongklaew et al. 2020). *V. harveyi* was described as an emerging disease in the Asia region that is responsible for many cases of moderate mortality of farmed fishes. In Vietnam, outbreaks of diseases in commercial barramundi farms have occurred in all fish sizes with a cumulative mortality of up to 40% (Dong et al. 2017). Asian sea bass is in the highest demand in the aquaculture fish industry in Asia. However, intensive farming has led to frequent disease outbreaks among the said fish species. One of the common pathogens of vibriosis in Asian sea bass is *V. harveyi* which can spread infections widely and pose major impacts on high mortality and financial loss (Chong et al. 2016). Few other mass mortalities also have been reported where due to infections of *V. harveyi*, the cultured tiger shrimp are infected with black shell disease (*Penaeus monodon*) in India and bacterial white tail disease in white leg shrimp (*Litopenaeus vannamei*) in China (Zhang et al. 2020).

In other regions, such as Australia, *V. harveyi* is a persistent problem in barramundi culture reared in sea cages, resulting in fatalities due to heavy infections (Bøgwald and Dalmo 2019). There have been reported mortalities caused by *V. harveyi*, which has widely spread to silvery black porgy (*Acanthopagrus cuvieri*) and brown-spotted grouper (*Epinephelus tauvina*) in Kuwait, common dentex (*Dentex dentex*) and farmed sole (*Solea senegalensis*) in Spain and farmed sea perch (*Lateolabrax japonicus*) and cage-reared grouper (*Epinephelus awoara*) in China (Zhang et al. 2020). In 2016, *P. damselae* subsp. *damselae* was responsible for the high mortality of Australian native fresh water pygmy perch (*Nannoperca vittata*). The outbreak has caused mortalities in 50% of the population, which was lower at 28 °C (4.82×10^5 CFU mL^{-1}) than at 17 °C (2.81×10^6 CFU mL^{-1}) (Lawlor 2016). Mortality cases reported among caged-cultured cobia fishes in India are caused *by V. alginolyticus*, *V. harveyi*, *V. parahaemolyticus* and *V. vulnificus* (Sharma et al. 2017).

Another case has been reported in India for cobia cultured with 50% mortality caused by *V. alginolyticus*. A similar case has been reported to that of the *V. alginolyticus* infection in cobia cultured in Taiwan and in sea bass (Rameshkumar et al. 2014). The significant outbreaks of vibriosis in cultured cobia sub-adults have been documented in Mandapam, Tamil Nadu, India. In August 2013, a total of 99 numbers of cobia sub-adults died with a total cumulative mortality of approximately 33% in all the cages and a morbidity rate of 67% (Rameshkumar et al. 2017). Chile is the second-largest producer of salmon which has been reported to be affected by *V. ordalii* (Miranda et al. 2018). *V. anguillarum* has affected many brackish water areas and marine Mediterranean species, mainly European sea bass (*Dicentrarchus labrax*), gilthead seabream (*Sparus aurata*), sole (*Solea* spp.), sea mullet (*Mugil* spp.), turbot (*Scophthalmus maximus*) and European eel (*Anguilla anguilla*) which give rise to mortality rate, above all in fingerlings and juveniles (Zrncic 2020). The diseases of aquatic animals associated with vibriosis is summarized in Table 7.2.

Table 7.2 List of diseases in aquatic animals associated with vibriosis

Name of disease	Host	Etiological agent	References
Luminescent vibriosis	Pacific white shrimp (*Litopenaeus vannamei*)	*V. harveyi*	Chatterjee and Haldar (2012), Mohamad (2019)
White spot syndrome virus	Tiger shrimp (*Penaeus monodon*)		Klongklaew et al. (2020)
Black shell disease	Tiger shrimp (*Penaeus monodon*)		Zhang et al. (2020)
Scale drop and muscle necrosis disease	Silvery black porgy (*Acanthopagrus cuvieri*) Brown-spotted grouper (*Epinephelus tauvina*) Dentex (*Dentex dentex*) Farmed sole (*Solea senegalensis*) Farmed Sea perch (*Lateolabrax japonicus*) Cage-reared grouper (*Epinephelus awoara*)		Zhang et al. (2020)
Extreme eye haemorrhage and ventral lesions	Australian native fresh water pygmy perch (*Nannoperca vittate*)	*P. damselae* subsp. *damselae*	Lawlor (2016)
Pasteurellosis	Cage-farmed cobia (*Rachycentron canadum*)	*P. damselae* subsp. *damselae* *V. alginolyticus* *V. harveyi* *V. parahaemolyticus* *V. vulnificus*	Sharma et al. (2017)
Haemorrhagic septicaemia and ulcer on the body surface	Cobia (*Rachycentron canadum*) Asian sea bass (*Lates calcarifer, Bloch*)	*V. alginolyticus*	Rameshkumar et al. (2014)
Winter ulcer	Atlantic salmon (*Salmo salar*)	*V. ordalii*	Miranda et al. (2018)
Pasteurellosis Septicaemia Gill disease	European sea bass (*Dicentrarchus labrax*) Gilthead seabream (*Sparus aurata*)	*V. anguillarum*	Zrncic (2020)
Gill disease Red spot disease Gill disease	Sole (*Solea* spp.) Sea mullet (*Mugil* spp.)		
Red pest of eels/red boil disease	Turbot (*Scophthalmus maximus*) European eel (*Anguilla anguilla*)		

7.3.3 Indirect Consequences

Despite mortalities, there were few cases that have been reported from the infection of *Vibrio* sp., which can impair the growth of cultured marine fishes. From the previous studies, *Vibrio* sp. was believed to slow down the growth of grouper (*E. coioides*), which caused vibriosis that influences slow growth following stress and inappetence (Egerton et al. 2018). Productivity in the economy is important to continue the world's economic development to achieve the target market size in the aquaculture industry. These side effects have contributed to economic loss. The clinical signs of *Vibrio* sp. infections on the fish can be observed by gross visible external lesions, reducing their marketability. This condition can be seen in adult fish with chronic skin and haemorrhage.

Other than social and economic impact, fish diseases reduce local employment in fish processing, thereby affecting the large farm's income, food security, job security of the workers, processing food for animals and processing plants which will give a great impact on job loss in rural and regional area flows through the local community. Small-scale grouper and barramundi farmers in Thailand had accounted for losses of up to about USD 5000 due to fish diseases that have been reported. Besides, a small-scale grouper farmer in the Philippines has been reported to face an income reduction of 75% and their debt increased to 19% due to disease-related fish losses. A few studies have indicated that the disease has a significant impact on the production of groupers and subsequently on the economic stability of this small sector (Santos et al. 2020a, b). The disease has caused intensification in aquaculture activities and outbreaks of diseases, which have been increasingly reported as a significant challenge to the growth of aquaculture production and affect the industrial development of some countries (Novriadi 2016). The food-borne disease appears to be a problem that can be solved, based on the growth of food exports from developed countries and the progress of initiatives to improve food safety in informal markets. While the full health and economic costs of contaminated food are unknown, the global impact on health, commerce and growth is estimated to be immense. The food-borne disease affects hundreds of millions of people worldwide each year, costing billions of dollars (Grace 2015).

7.4 Disease Diagnosis

The range of illnesses caused by *Vibrio* spp. is wide, but two syndromes stand out: classical vibriosis and gastroenteritis. Common vibriosis is defined by the affected animal's slow motion. Various skin ulcerations, fin rotting and body pigmentation are all present (Hashima 2020). Figure 7.1 shows the ulceration on the skin of sea cucumber caused by vibriosis.

Mouth ulcers are potentially more dangerous than skin ulcers because they cause loss of appetite and, inevitably, death from starvation. Fin lesions, likewise, prevent movement to compete for food. Internal organs, especially the liver, kidney and spleen, appear swollen, haemorrhagic and congested in gastroenteritis (Ina-Salwany

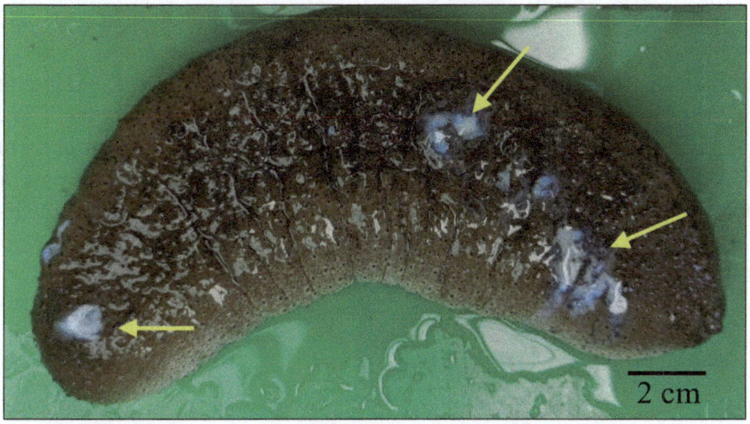

Fig. 7.1 Skin ulceration (arrow) on the skin of sea cucumber (*Holothuria scabra*) caused by *V. parahaemolyticus*

et al. 2019). Following infections with *V. harveyi*, *V. alginolyticus* and *P. damselae* subsp. *damselae*, the gastroenteritis symptom was observed. Ascites is the most common gross lesion associated with gastroenteritis. Eyesight alterations are rare cases, beginning with opacity and progressing to ulceration and exophthalmia, which is also known as "pop-eye" (Azmai et al. 2019).

Even though vibriosis is caused by the same species of *Vibrio*, clinical symptoms can differ greatly from case to case. Multiple factors influence this, including the type of host, the age of the host, the length of exposure and the virulence of the virus. In young fish larvae and fry, vibriosis is common. It spreads so quickly after infection that most of the infected young fish die without presenting any symptoms. As a result, sudden death is often observed in young fish that die with no external symptoms. Infected adult fish, on the other hand, typically develop a chronic infection with noticeable skin ulcerations and pigmentation. In previous research, skin ulcers caused by *V. alginolyticus* infection were observed in natural cases. In the current study, there were no skin ulcers among infected fingerling cobia in either the natural outbreak or the experimentally infected fish (Rameshkumar et al. 2017). When high-virulence *Vibrio* species are involved, the gross pathology is normally minimal, and death occurs before the gross lesions manifest. External and internal symptoms, on the other hand, are often unrelated.

Infection with various *Vibrio* sp. resulted in a variety of macro- and microscopically visible lesions. As a result, clinical signs and histopathological changes in diseased fish are insufficient to distinguish between *Vibrio* sp. infections. Necrosis with haemorrhage is a common lesion, particularly in the liver and kidney. Congestion, haemorrhage and vascular ectasia are typically shown in the liver (Dong et al. 2017). Fatty degeneration was found in the liver parenchyma from sea cage cultured cobia. The nucleus was pushed to the periphery, while all the tiny vacuoles merged to form a large fatty vacuole. Some of the regions had mild to moderate bile duct obstruction and hypertrophy was observed. In naturally infected sea bass juveniles, necrotic changes in hepatic tissues resulted in honeycomb vacuolation (Rameshkumar et al. 2017). In kidneys, tubular degeneration, haemorrhage and inflammatory cell infiltration can be seen (Sharma et al. 2017). Haemorrhage and vessel obstruction may occur in the brain, but they are uncommon. Typical histological variations can, however, be seen in certain cases. For example, collapsed tubules and sloughing of epithelial cells were observed in the kidney of infected barramundi in the case of *V. harveyi* infection (Dong et al. 2017). Damages to the spleen may involve haemorrhages and disaggregation of the melanomacrophage centres in certain *Vibrio* infections (Mancuso et al. 2015).

7.5 Pathogenesis of Vibriosis

Vibrio is commonly linked to food-borne illnesses and disorders in aquaculture. *Vibrio* infection has been linked to diseases and even mortality in aquatic creatures across the world, with an upward tendency (Jeamsripong et al. 2020). As a result, vibriosis has had a tremendous impact on human health and aquaculture development (Qin et al. 2020). Multiple virulence factors expressed by *Vibrio* virulence genes determine the pathogenicity of the bacteria (Ramamurthy et al. 2020). *Vibrio* infects and kills the host through a variety of mechanisms, including adhesion, invasion, immune evasion, in vivo proliferation and toxin generation (Huang et al. 2015). Adhesion factors, capsule and polysaccharides, cytotoxin and other virulence factors are mostly found in *Vibrio*. Hence, *Vibrio* infection and pathogenesis are the results of the combined action of many virulence factors, rather than a single virulence factor.

Previous research has identified four different kinds of pathogenesis for experimental vibriosis in bivalve larvae (Wang et al. 2021). The first descriptions of colonization routes for various bivalve species were solely based on histology findings (Dubert et al. 2017). Pathogenic *Vibrio* adheres to the larval shell and develops along the mantle, producing gradual rupture of the mantle from the periphery inward in type I pathogenesis. The germs then infiltrate the visceral cavity, finally killing the larvae. As the disease continues, type II pathogenesis shows velar distortion without *Vibrio* invasion into tissues, followed by total necrosis of digestive organs and subsequent bacterial colonization within the visceral cavity. Vibriosis with pathogenesis type III is characterized by gradual and widespread visceral atrophy with decreased visceral mass occupying only the dorsal region, followed by bacterial invasion with focal lesions in digestive tract organs. Wang et al. (2021) used GFP-tagging, histopathology and ultrastructural investigations in blue mussel (*M. edulis*) larvae to describe a new kind of pathogenesis (type IV) of bivalve vibriosis. It was comparable to pathogenesis type II, but it happened the other way around: the pathogenic *Vibrio* grows along the digestive system, while the velar abnormalities appear after the bacteria have colonized the soft tissues.

Wang et al. (2021) reported the same sort of vibriosis pathogenesis in Pacific oyster (*C. gigas*) larvae, as we found in our current study. The GFP-tagged ME9 and NB10 strains were filtered via the velum and then reached the stomach through the larval oesophagus, according to the green fluorescence. Following that, it was obvious from the integrating histopathological observations that the pathogenic *Vibrio* multiplied fast in the stomach by 6 h post challenge and that it took 42 h to produce necrosis from larval digestive organs in the dorsal area to the oesophagus and velum. From then onwards, the destroyed larval tissues were expelled from the body. By 96 h post challenge, all soft tissues had been impaired with only a few remnants left in the shell. These results are consistent with previous studies reporting that in the advanced stage of disease progression, bivalve larvae exhibited nearly complete necrosis of digestive organs (Dubert et al. 2016). Because the immune system is not well established in the early phases of ontogenesis, bivalve larvae are more susceptible to infections (Zannella et al. 2017). The immune system of *C. gigas* larvae was thought to be initiated during the trochophore stage, while the early umbo larvae stage had a fully formed immune system (Song et al. 2016). The immune response occurred sooner in the presence of pathogens as a protection against bacterial assault, and all 12 immune genes analysed were highly up-regulated in D-veliger larvae of *C. gigas* following infection (Song et al. 2016). However, the significant mortality observed in our investigation suggests that the immune system of *C. gigas* D-veliger larvae did not provide adequate defensive activity in response to pathogenic *Vibrio* infection. The failure of most of the immune genes studied may provide a plausible explanation for why pathogenic germs are able to damage tissues before the immune system responds. Table 7.3 summarizes the clinical signs of animals infected with vibriosis.

Table 7.3 Clinical signs exhibited by host infected with different *Vibrio* sp.

Etiological agent	Infected fish	Source of infected fish	Clinical signs	References
V. anguillarum	Lumpfish (*Cyclopterus lumpus*)	Natural outbreak	Haemorrhagic septicaemia, which is characterized by skin lesions, gill haemorrhages and bacterial aggregations in lymphoid organs	Vasquez et al. (2020)
V. ordalii	Farmed Atlantic cod (*Gadus morhua*)	Natural outbreak	Necrotic lesions and varying degrees of cutaneous haemorrhage on both body and fins, with or without exophthalmia	Lillehaug and Colquhoun (2020)
V. viscosus	Atlantic salmon (*Salmon salar*)	Experimentally infected	Greyish skin with necrotic skin and slight petechial haemorrhaging. Necrosis in underlying muscle	Benediktsdóttir et al. (2000)

(continued)

Table 7.3 (continued)

Etiological agent	Infected fish	Source of infected fish	Clinical signs	References
V. salmonicida	Atlantic salmon (*Salmon salar*)	Experimentally infected	Anaemia and extended internal and external haemorrhages	Bjelland et al. (2013)
V. parahaemolyticus	Indian prawn (*Penaeus indicus*)	Natural outbreak	Moribund shrimp with acute bacteraemia and septic haemocytic nodules on the lymphoid organ, heart and loose connective tissue	De Souza and Wan (2021)
V. harveyi	Whiteleg shrimp (*Litopenaeus vannamei*)	Natural outbreak	White or opaque lesions in the tail, which were attributed to muscle necrosis	Aguilera-Rivera et al. (2019)
	Tiger shrimp (*Penaeus monodon*)	Natural outbreak	Lethargic, displaying excessive mucus production; fin rot; congestion of the brain, liver and kidneys; and splenic enlargement Scale drop and muscle necrosis	Zhang et al. (2020)
	Juvenile hybrid grouper (*E. polyphekadion* × *E. fuscoguttatus*)	Natural outbreak	Severe fin necrosis, eye muddy and proptosis and subcutaneous haemorrhage	Zhang et al. (2020)
	Japanese flounder (*Paralichthys olivaceus*)	Natural outbreak	Scale drop and muscle necrosis	Zhu et al. (2020)
	Crimson snapper (*Lutjanus erythropterus*)	Natural outbreak		Zhu et al. (2020)
	Barramundi (*Lates calcarifer*)	Natural outbreak		Dong et al. (2017)

(continued)

Table 7.3 (continued)

Etiological agent	Infected fish	Source of infected fish	Clinical signs	References
P. damselae subsp. *damselae*	Hybrid grouper cantang (*E. lanceolatus* × *E. fuscoguttatus*)	Natural outbreak	Decreased appetite, weak and tilted swimming behaviour, pale kidneys and darker body colour	Istiqomah et al. (2020)
	Cobia (*Rachycentron canadum*)	Natural outbreak	Abdominal cavities were filled with cloudy gelatinous fluid	Sharma et al. (2017)
	Turbot (*Scophthalmus maximus*)	Natural outbreak	Haemorrhaged areas and ulcerative lesions on the body surface	Fajardo et al. (2023)
	Gilthead seabream (*Sparus aurata*)	Natural outbreak	Granulomatous-like lesions in the internal viscera, particularly in the spleen and kidney. External symptoms are weight loss, a darkening of the skin, localized necrosis in the gills and skin, distended stomach and external haemorrhages	Vallecillos et al. (2021)
V. mimicus	Pacific white shrimp (*Litopenaeus vannamei*)	Natural outbreak	Reddish carapace, melanosis of the skin, necrosis of the tail and redness of the swimming legs and walking legs	Istiqomah et al. (2020)
V. alginolyticus	Nile tilapia (*Oreochromis niloticus*)	Experimentally infected	Dark colouration of the skin with detached scales. Haemorrhages at the base of the fins and some fishes showed eroded fins	Abdelaziz et al. (2017)
	Cobia (*Rachycentron canadum*)	Natural outbreak	Haemorrhage, ulcers on the skin, dark skin and fluid accumulation in the peritoneal cavity Inappetence, exophthalmia and pale liver	Rameshkumar et al. (2017)

7.6 Management Strategies for Vibriosis Disease Prevention and Control

The aquaculture industry needs to be sustained through the implementation of various disease control and prevention alternative strategies including the use and application of non-chemical products, biosecurity practices and use of immunostimulants, probiotics and antimicrobials. Vibriosis has spread rapidly from adhesion to colonization within the host which could be the preliminary findings for therapeutic strategies for disease prevention and treatment (Huzmi et al. 2019). Some interventions have been done on target marine pathogens such as through effective direct elimination of pathogens or application of microbial-manipulation strategies to lower pathogen virulence. In another method, multi-trophic aquaculture has been created between mussels (*Mytilus edulis*) and salmon farms. The mussels have been placed near salmon farms to filter larval sea lice and pathogenic *Vibrio* bacterial species from the water column. This implementation will develop a sustainable management for various types of infections in aquaculture and continued development economically (Groner et al. 2016).

In control and prevention, some strategies have been introduced to reduce the risk of *Vibrio* infections, especially from *V. parahaemolyticus* associated with seafood consumption. The strategies that are based on physical and chemical methods have been developed. The thermal process is introduced to lessen the risk of infections of *V. parahaemolyticus*, by inactivating the residue of strains in seafood by storing the seafood at low temperature (18 °C or −24 °C) (Wang et al. 2018). In a recent study by the Food and Agriculture Organization (FAO) of the United Nations and the World Health Organization (WHO), high infection of seafood has been reported to be caused by *Vibrio* spp., leading to an increase in seawater temperature. Most manufacturers have used high-pressure processing (HPP) as a food control measure against *Vibrio* spp. without changing the quality of seafood products (FAO, WHO 2021). Gamma irradiation in combination with sodium hypochlorite (NaOCl) is another important method to reduce *V. parahaemolyticus* in shucked oysters. Pak and Ha have found that 2 kGy of gamma irradiation combined with 60 ppm of NaOCl reduced the concentration of *V. parahaemolyticus* by 3.6 orders of magnitude. Besides, the combination of NaOCl to 80 ppm with 2 kGy of gamma irradiation decreased the concentration of *V. parahaemolyticus* in shucked oysters by 2.2 orders of magnitude (Ndraha et al. 2020). Other than that, chemical reagents have been used to minimize bacterial contamination in seafood including chlorine, electrolyzed oxidizing water and iodophors (Marino 2016).

Similarly, the ongoing development of vaccines against vibriosis has provided a new revelation on disease management and prevention. According to Ma et al. (2019), generic fish vaccines produce antigens that stimulate innate and/or adaptive immunity as of 2019; there have been more than 26 commercial fish approved by the United States Department of Agriculture (USDA) that are available worldwide for various aquaculture fish species (Ma et al. 2019). There have been several successes recorded as a result of the administration of inactivated vaccines. For example, the formalin-inactivated vaccine is successful in protecting tilapia against

Table 7.4 Phylogenetic markers with their status of validation and limitation

Target genes	Status of validation	Limitation	References
toxR	High resolution of *Harveyi clade*: *V. parahaemolyticus*, *V. harveyi*, *V. campbellii*, *V. alginolyticus*, *V. rotiferianus* and *V. natriegens*	Lower successful identification—Lack of available *Vibrio* spp. sequences in the genomic banks	Turner et al. (2018)
pyrH	Differentiate *V. harveyi* from *V. campbellii*; differentiate *V. mimicus* from *V. cholera*	Low discriminatory level between *V. Campbellii* and *V. rotiferianus* and *V. Communis* and *V. Owensii*	Amalina et al. (2019)
rpoD	High resolution of *Harveyi clade*: *V. parahaemolyticus*, *V. harveyi*, *V. campbellii*, *V. alginolyticus*, *V. rotiferianus* and *V. natriegens*	–	Martín-Rodríguez et al. (2018)

genetically diverse *V. vulnificus* infections, while orange-spotted grouper against *V. harveyi* infections (Ji et al. 2020). However, the vaccine for vibriosis infections is limited to inactivated vaccines only where it is said to be less effective compared to live vaccines and exhibits several disadvantages that include a temporary immunization period, multiple rounds of vaccination and harmful allergic reactions due to recurrent stimulation (Ji et al. 2020). Extensive research on live vector and DNA vaccines for *Vibrio* spp.-induced infections must be done to encourage the development of the said vaccines for commercial use in aquaculture. Table 7.4 summarizes the phylogenetic markers used in identifying the target genes for vibriosis and their status of validation and limitation.

7.7 Conclusion

Vibrio is a genus of bacteria that can be found in both the aquatic environment and direct contact with fish, crustaceans and bivalves. It is virtually impossible to remove these bacteria in an aquaculture context, and not all species/strains cause a disease. Nonetheless, most *Vibrio* species are opportunistic pathogens. Pathogenic strains can be deadly to aquatic creatures, including fish, molluscs and corals, as well as crustaceans. It is imperative to control and regulate the *Vibrio* population. Preventing the initiation of infectious processes and, eventually, the appearance of diseases is the most important mitigating element. A complete disease management plan, including immune resistance strengthening through nutrition, using robust stock animals and preserving high-water quality, should be part of the prevention strategy.

While recent biotechnological developments such as quorum sensing disruption, immunological priming and phage therapy have the potential to reduce vibriosis outbreaks, further farm-scale validation of biosafety and efficacy is necessary. Aquaculture is a long-term solution for guaranteeing food security around the world. The risk of *Vibrio* illnesses in the industry could make it impossible to accomplish this goal. Science and industry must work together to address these concerns by improving *Vibrio* management strategies, particularly by gaining a better knowledge of how vibriosis spreads on farms and how to minimize the adverse effects. Although *Vibrio* in marine fishes and farmed shrimp has been extensively studied, there are still gaps in our understanding, specifically for crabs and lobsters. As aquaculture becomes a more important industry, future studies should focus on these topics.

Acknowledgement This research was supported by the HiCoE grant (Grant Code: HIC2404) from the Ministry of Higher Education Malaysia and the MSU Seed Research Grant (Grant Code: SG 465 0518 HLS) from Management and Science University Malaysia.

References

Abdelaziz M, Ibrahem MD, Ibrahim MA, Abu-Elala NM, Abdel-moneam DA (2017) Monitoring of different *Vibrio* species affecting marine fishes in Lake Qarun and Gulf of Suez: phenotypic and molecular characterization. Egypt J Aquat Res 43(2):141–146. https://doi.org/10.1016/j.ejar.2017.06.002

Aguilera-Rivera D, Prieto-Davó A, Rodríguez-Fuentes G, Escalante-Herrera KS, Gaxiola G (2019) A vibriosis outbreak in the Pacific white shrimp, Litopenaeus vannamei reared in biofloc and clear seawater. J Invertebr Pathol 167:107246. https://doi.org/10.1016/j.jip.2019.107246

Al-Dulaimi MMK, Mutalib SA, Ghani MA (2014) Isolation and identification among cockle isolates of *Vibrio vulnificus* isolated from Selangor, Malaysia. AIP Conf Proc 1614:208–212. https://doi.org/10.1063/1.4895197

Amalina NZ, Santha S, Zulperi D, Amal MNA, Yusof MT, Zamri-Saad M, Ina-Salwany MY (2019) Prevalence, antimicrobial susceptibility, and plasmid profiling of *Vibrio* spp. isolated from cultured groupers in Peninsular Malaysia. BMC Microbiol 19(1):1–15. https://doi.org/10.1186/s12866-019-1624-2

Amaro C, Sanjuán E, Fouz B, Pajuelo D, Lee C-T, Hor L-I, Barrera R (2015) The fish pathogen *Vibrio vulnificus* biotype 2: epidemiology, phylogeny, and virulence factors involved in warm-water vibriosis. Microbiol Spectr 3(3):1–23. https://doi.org/10.1128/microbiolspec.ve-0005-2014

Andreoni F, Magnani M (2014) Photobacteriosis: prevention and diagnosis. J Immunol Res 2014:793817. https://doi.org/10.1155/2014/793817

Aruta, D. (2019, January 24) Implications of Vibrio vulnificus Infections in Healthcare. Labroots. https://www.labroots.com/trending/health-and-medicine/13616/implications-vibrio-vulnificus-infections-healthcare

Azmai MN, Roseli M, Sutra J, Roslan S, Mazlan S, Basri L, Lee J, Mohamad N, Ismail A, Saad MZ, Yasin IS, Nasruddin NS (2019) Javanese medaka (Oryzias javanicus Bleeker, 1854) as potential model organism for aeromoniasis and vibriosis study in fish.

Baker-Austin C, Oliver JD, Alam M, Ali A, Waldor MK, Qadri F, Martinez-Urtaza J (2018) *Vibrio* spp. infections. Nat Rev Dis Prim 4(1):8. https://doi.org/10.1038/s41572-018-0005-8

Balado M, Lages MA, Fuentes-Monteverde JC, Martínez-Matamoros D, Rodríguez J, Jiménez C, Lemos ML (2018) The siderophore piscibactin is a relevant virulence factor for *Vibrio anguillarum* favored at low temperatures. Front Microbiol 9:1–16. https://doi.org/10.3389/fmicb.2018.01766

Benediktsdóttir E, Verdonck L, Spröer C, Helgason S, Swings J (2000) Characterization of *Vibrio viscosus* and *Vibrio wodanis* isolated at different geographical locations: a proposal for reclassification of *Vibrio viscosus* as *Moritella viscosa* comb. nov. Int J Syst Evol Microbiol 50(2):479–488. https://doi.org/10.1099/00207713-50-2-479

Bjelland AM, Fauske AK, Nguyen A, Orlien IE, Østgaard IM, Sørum H (2013) Expression of *Vibrio salmonicida* virulence genes and immune response parameters in experimentally challenged Atlantic salmon (*Salmo salar L.*). Front Microbiol 4:1–12. https://doi.org/10.3389/fmicb.2013.00401

Bøgwald J, Dalmo RA (2019) Review on immersion vaccines for fish: an update 2019. Microorganisms 7(12):1–28. https://doi.org/10.3390/microorganisms7120627

Chatterjee S, Haldar S (2012) *Vibrio* related diseases in aquaculture and development of rapid and accurate identification methods. https://doi.org/10.4172/2155-9910.S1-002

Chong V, Al-Azad S, Shapawi R (2016) Comparison of two edible mushroom extract as aquaculture feed additive to enhance immune response of Asian Seabass. Trans Sci Technol 3:427–432

Hong To, T. T., Yanagawa, H., Khanh Thuan, N., Hiep, D. M., Cuong, D. V., Khai, L. T. L., Taniguchi, T., Kubo, R., & Hayashidani, H. (2020). Prevalence of Vibrio parahaemolyticus Causing Acute Hepatopancreatic Necrosis Disease of Shrimp in Shrimp, Molluscan Shellfish and Water Samples in the Mekong Delta, Vietnam. Biology, 9(10), 312. https://doi.org/10.3390/biology9100312

De Souza C, Wan AHL (2021) Vibrio and major commercially important vibriosis diseases in decapod crustaceans. J Invertebr Pathol 181:107527. https://doi.org/10.1016/j.jip.2020.107527

Dong HT, Taengphu S, Sangsuriya P, Charoensapsri W, Phiwsaiya K, Sornwatana T, Khunrae P (2017) Recovery of *Vibrio harveyi* from scale drop and muscle necrosis disease in farmed barramundi, *Lates calcarifer* in Vietnam. Aquaculture 473:89–96. https://doi.org/10.1016/j.aquaculture.2017.02.005

Dubert J, Nelson DR, Spinard EJ, Kessner L, Gomez-Chiarri M, da Costa F, Prado S, Barja JL (2016) Following the infection process of vibriosis in Manila clam (*Ruditapes philippinarum*) larvae through GFP-tagged pathogenic *Vibrio* species. J Invertebr Pathol 133:27–33. https://doi.org/10.1016/j.jip.2015.11.008

Dubert J, Barja JL, Romalde JL (2017) New insights into pathogenic vibrios affecting bivalves in hatcheries: present and future prospects. Front Microbiol 8:1–16. https://doi.org/10.3389/fmicb.2017.00762

Egerton S, Culloty S, Whooley J, Stanton C, Ross RP, Ross RP (2018) The gut microbiota of marine fish. Front Microbiol 9:1–17. https://doi.org/10.3389/fmicb.2018.00873

Erkinharju T, Dalmo RA (2021) Cleaner fish in aquaculture: review on diseases and vaccination. Rev Aquac 13:189–237. https://doi.org/10.1111/raq.12470

FAO (2016) The State of World Fisheries and Aquaculture 2016. In The state of world fisheries and aquaculture. https://doi.org/10.18356/8e4e0ebf-en

FAO, WHO (2021) Advances in science and risk assessment tools for Vibrio parahaemolyticus and V. vulnificus associated with seafood. Meeting report. Microbiological Risk Assessment Series no. 35. FAO, Rome, Italy. https://doi.org/10.4060/cb5834en

Fajardo C, Santos P, Passos R, Vaz M, Azeredo R, Machado M, Fernández-Boo S, Baptista T, & Costas B (2023) Early Molecular Immune Responses of Turbot (Scophthalmus maximus L.) Following Infection with Aeromonas salmonicida subsp. salmonicida. International Journal of Molecular Sciences, 24(16), 12944. https://doi.org/10.3390/ijms241612944

Fernández-Bravo A et al (2019) *Vibrio alginolyticus* infections: report of two cases from Spain with literature review. Med Microbiol Diagn 8(1):8–13. https://doi.org/10.4172/2161-0703.1000298

Foyn S (2017) Microbe-associated molecular patterns of *Aliivibrio salmonicida*: roles in the pathogenesis of cold water vibriosis. (Doctoral thesis. Norwegian University of Life Sciences). http://hdl.handle.net/11250/2497987

Grace D (2015) Food safety in developing countries: an overview. A learning resource for DFID livelihoods advisers about evidence on demand and professional development. pp 1–83. https://doi.org/10.12774/eod_er.oct2015.graced

Groner ML, Maynard J, Breyta R, Carnegie RB, Dobson A, Friedman CS, Froelich B, Garren M, Gulland FMD, Heron SF, Noble RT, Revie CW, Shields JD, Vanderstichel R, Weil E, Wyllie-Echeverria S, Harvell CD (2016) Managing marine disease emergencies in an era of rapid change. Philos Trans R Soc B Biol Sci 371(1689):20150364. https://doi.org/10.1098/rstb.2015.0364

Guardiola-Avila I, Acedo-Felix E, Sifuentes-Romero I, Yepiz-Plascencia G, Gomez-Gil B, Noriega-Orozco L (2016) Molecular and genomic characterization of *Vibrio mimicus* isolated from a frozen shrimp processing facility in Mexico. PLoS One 11(1):1–15. https://doi.org/10.1371/journal.pone.0144885

Guijarro JA, Cascales D, García-torrico AI, García-domínguez M (2015) Temperature-dependent expression of virulence genes in fish-pathogenic bacteria. Front Microbiol 6:1–11. https://doi.org/10.3389/fmicb.2015.00700

Haftel A, Sharman T (2023) Vibrio vulnificus Infection. In: StatPearls. StatPearls Publishing, Treasure Island (FL)

Hansen MJ, Kudirkiene E, Dalsgaard I (2020) Analysis of 44 *Vibrio anguillarum* genomes reveals high genetic diversity. PeerJ 8:1–19. https://doi.org/10.7717/peerj.10451

Hashima AA (2020) Differential diagnosis, prevention, and control Argulosis and Vibriosis that spread in the summer season. July, 0–11. https://doi.org/10.13140/RG.2.2.30675.40489

Hickey ME, Lee JL (2018) A comprehensive review of *Vibrio* (*Listonella*) *anguillarum*: ecology, pathology, and prevention. Rev Aquac 10(3):585–610. https://doi.org/10.1111/raq.12188

Hjeltnes B, Britt B-J, Geir B, Asle H, Cecilie SW (2019) The health situation in Norwegian aquaculture. Veterinaerinstituttet, pp 37–41. https://www.vetinst.no/rapporter-og-publikasjoner/rapporter/2019/fish-health-report-2018

Huang L, Hu J, Su Y, Qin Y, Kong W, Ma Y, Xu X, Lin M, Yan Q (2015) Identification and characterization of three *Vibrio alginolyticus* non-coding RNAs involved in adhesion, chemotaxis, and motility processes. Front Cell Infect Microbiol 5:56. https://doi.org/10.3389/fcimb.2015.00056

Huang JY, Henao OL, Griffin PM, Vugia DJ, Cronquist AB, Hurd S, Tobin-D'Angelo M, Ryan P, Smith K, Lathrop S, Zansky S, Cieslak PR, Dunn J, Holt KG, Wolpert BJ, Patrick ME (2016) Infection with pathogens transmitted commonly through food and the effect of increasing use of culture-independent diagnostic tests on surveillance—Foodborne Diseases Active Surveillance Network, 10 U.S. sites, 2012–2015. MMWR. Morb Mortal Wkly Rep 65(14):368–371. https://doi.org/10.15585/mmwr.mm6514a2

Huicab-Pech ZG, Eros-Sánchez C, Castañeda-Chávez MD, Lango-Reynoso F, López-Collado CJ, De PR (2016) Current state of bacteria pathogenicity and their relationship with host and environment in tilapia *Oreochromis niloticus*. J Aquac Res Dev 7:1–10

Huzmi H, Yasin I-SM, Natrah FMI, Syukri F, Karim M (2019) Strategies of controlling vibriosis in fish. Asian J Appl Sci 7(5):513–521. https://doi.org/10.24203/ajas.v7i5.5933

Ina-Salwany MY, Al-saari N, Mohamad A, Mursidi FA, Mohd-Aris A, Amal MNA, Kasai H, Mino S, Sawabe T, Zamri-Saad M (2019) Vibriosis in fish: a review on disease development and prevention. J Aquat Anim Health 31(1):3–22. https://doi.org/10.1002/aah.10045

Istiqomah I, Sukardi, Murwantoko, Isnansetyo A (2020) Review vibriosis management in indonesian marine fish farming. E3S Web Conf 147. https://doi.org/10.1051/e3sconf/202014701001

Jeamsripong S, Khant W, Chuanchuen R (2020) Distribution of phenotypic and genotypic antimicrobial resistance and virulence genes in *Vibrio parahaemolyticus* isolated from cultivated oysters and estuarine water. FEMS Microbiol Ecol 96(8):fiaa081. https://doi.org/10.1093/femsec/fiaa081

Ji Q, Wang S, Ma J, Liu Q (2020) A review: progress in the development of fish *Vibrio* spp. vaccines. Immunol Lett 226:46–54. https://doi.org/10.1016/j.imlet.2020.07.002

Kang C, Shin Y, Jang S, Yu H, Kim S, An S, Park K, So J (2017) Characterization of *Vibrio parahaemolyticus* isolated from oysters in Korea: resistance to various antibiotics and prevalence of virulence genes. Mar Pollut Bull 118:261. https://doi.org/10.1016/j.marpolbul.2017.02.070

Kim BS, Gavin HE, Satchell KJF (2017) Variable virulence of biotype 3 *Vibrio vulnificus* due to MARTX toxin effector domain composition. MSphere 2(4):1–7. https://doi.org/10.1128/mspheredirect.00272-17

Klongklaew, N., Praiboon, J., Tamtin, M., & Srisapoome, P. (2020). Antibacterial and Antiviral Activities of Local Thai Green Macroalgae Crude Extracts in Pacific White Shrimp (Litopenaeus vannamei). Marine Drugs, 18(3), 140. https://doi.org/10.3390/md18030140

Lages MA, Balado M, Lemos ML (2019) The expression of virulence factors in *Vibrio anguillarum* is dually regulated by iron levels and temperature. Front Microbiol 10:1–11. https://doi.org/10.3389/fmicb.2019.02335

Lawlor E (2016) Interaction of disease and temperature on the aerobic scope of freshwater fish and implications for changing climates. (Thesis. Murdoch University). https://researchportal.murdoch.edu.au/esploro/outputs/991005544101307891

Letchumanan V, Yin W, Lee L, Chan K (2015) Prevalence and antimicrobial susceptibility of *Vibrio parahaemolyticus* isolated from retail shrimps in Malaysia. Front Microbiol 6:1–11. https://doi.org/10.3389/fmicb.2015.00033

Letchumanan V, Ab Mutalib NS, Wong SH, Chan K,Lee, LH, (2019) Determination of antibiotic resistance patterns of Vibrio parahaemolyticus from shrimp and shellfish in Selangor, Malaysia. Prog Microbes Mol Biol 2019; 2(1): a0000019

Li L, Wang L, Zhang C, Chen P, Luo X (2019) A case of *Vibrio vulnificus* related wound infection diagnosed by next-generation sequencing. IDCases 15:e00497. https://doi.org/10.1016/j.idcr.2019.e00497

Lillehaug, Atle; Colquhoun, Duncan (2020). Vibriosis caused by Vibrio anguillarum and V. ordalii. ICES Identification Leaflets for Diseases and Parasites in Fish and Shellfish. Report. https://doi.org/10.17895/ices.pub.7622

Liu L, Ge M, Zheng X, Tao Z, Zhou S, Wang G (2016) Investigation of *Vibrio alginolyticus*, *V. harveyi*, and *V. parahaemolyticus* in large yellow croaker, *Pseudosciaena crocea* (Richardson) reared in Xiangshan Bay, China. Aquac Rep 3:220–224. https://doi.org/10.1016/j.aqrep.2016.04.004

Ma J, Bruce TJ, Jones EM, Cain KD (2019) A review of fish vaccine development strategies: conventional methods and modern biotechnological approaches. Microorganisms 7(11):569. https://doi.org/10.3390/MICROORGANISMS7110569

Mancuso M, Genovese M, Guerrera MC, Casella G, Genovese L, Piccolo G, Maricchiolo G (2015) First episode of vibriosis in wild specimens of *Pagellus bogaraveo* (Brünnich, 1768) in the Mediterranean Sea. Cah Biol Mar 56(4):355–361. https://doi.org/10.21411/cbm.a.2724367d

Marino D (2016) Evaluation of the properties of electrochemically activated water on different salt compositions and its capabilities as an antimicrobial, cleaning, and allergen control solution. (Master thesis. Univesity of Minnesota). https://hdl.handle.net/11299/185079

Martín-Rodríguez AJ, Suárez-Mesa A, Artiles-Campelo F, Römling U, Hernández M (2018) Multilocus sequence typing of Shewanella algae isolates identifies disease-causing *Shewanella chilikensis* strain 6I4. FEMS Microbiol Ecol 95(1):1–10. https://doi.org/10.1093/femsec/fiy210

Marudhupandi T, Thankappan T, Kumar A, Prakash S, Balamurugan J, Balachandran N (2017) *Vibrio parahaemolyticus* a causative bacterium for tail rot disease in ornamental fish, Amphiprion sebae. Aquac Rep 8:39–44. https://doi.org/10.1016/j.aqrep.2017.09.004

Matanza XM, Osorio CR (2020) Exposure of the opportunistic marine pathogen *Photobacterium damselae* subsp. *damselae* to human body temperature is a stressful condition that shapes the transcriptome, viability, cell morphology, and virulence. Front Microbiol 11:1–17. https://doi.org/10.3389/fmicb.2020.01771

Miranda CD, Godoy FA, Lee MR (2018) Current status of the use of antibiotics and the antimicrobial resistance in the Chilean salmon farms. Front Microbiol 9:1–14. https://doi.org/10.3389/fmicb.2018.01284

Mohamad, N., Amal, M. N. A., Yasin, I. S. M., Saad, M. Z., Nasruddin, N. S., Al-Saari, N., Mino, S., & Sawabe, T. (2019). Vibriosis in cultured marine fishes: a review. Aquaculture, 512, 734289. https://doi.org/10.1016/j.aquaculture.2019.734289

Mohamad N, Amal MNA, Yasin ISM, Zamri Saad M, Nasruddin NS, Al-saari N, Mino S, Sawabe T (2019a) Vibriosis in cultured marine fishes: a review. Aquaculture 512:734289. https://doi.org/10.1016/j.aquaculture.2019.734289

Mohamad N, Mohd Roseli FA, Azmai MNA, Saad MZ, Md Yasin IS, Zulkiply NA, Nasruddin NS (2019b) Natural concurrent infection of *Vibrio harveyi* and *V. alginolyticus* in cultured hybrid groupers in Malaysia. J Aquat Anim Health 31(1):88–96. https://doi.org/10.1002/aah.10055

Montánchez I, Kaberdin VR (2020) *Vibrio harveyi*: a brief survey of general characteristics and recent epidemiological traits associated with climate change. Mar Environ Res 154:104850. https://doi.org/10.1016/J.MARENVRES.2019.104850

Ndraha N, Wong H, chung, & Hsiao, H. I. (2020) Managing the risk of Vibrio parahaemolyticus infections associated with oyster consumption: a review. Compr Rev Food Sci Food Saf 19(3):1187–1217. https://doi.org/10.1111/1541-4337.12557

Novriadi R (2016) Vibriosis in aquaculture. Omni-Akuatika 12(1):1–12. https://doi.org/10.20884/1.oa.2016.12.1.24

Oliver JD, Jones JL (2014) Vibrio parahaemolyticus and *Vibrio vulnificus*. In: Molecular medical microbiology, vol 2–3, 2nd edn. Elsevier Ltd. https://doi.org/10.1016/B978-0-12-397169-2.00066-4

Osorio, C. R., Vences, A., Matanza, X. M., & Terceti, M. S. (2018). Photobacterium damselae subsp. damselae, a generalist pathogen with unique virulence factors and high genetic diversity. Journal of bacteriology, 200(15), e00002–e18. Advance online publication. https://doi.org/10.1128/JB.00002-18

Parveen S, Jacobs J, Ozbay G, Chintapenta LK, Almuhaideb E, Meredith J, Ossai S, Abbott A, Grant A, Brohawn K, Chigbu P, Richards GP (2020) Seasonal and Geographical Differences in Total and Pathogenic Vibrio parahaemolyticus and Vibrio vulnificus Levels in Seawater and Oysters from the Delaware and Chesapeake Bays Determined Using Several Methods. Applied and Environmental Microbiology 86(23). https://doi.org/10.1128/aem.01581-20

Prachumwat A, Wechprasit P, Srisala J, Kriangsaksri R, Flegel TW, Thitamadee S, Sritunyalucksana K (2020) *Shewanella khirikhana* sp. nov.—a shrimp pathogen isolated from a cultivation pond exhibiting early mortality syndrome. Microb Biotechnol 13(3):781–795. https://doi.org/10.1111/1751-7915.13538

Qin Z, Yang X, Chen G, Park C, Liu Z (2020) Crosstalks between gut microbiota and *Vibrio cholerae*. Front Cell Infect Microbiol 10:1–11. https://doi.org/10.3389/fcimb.2020.582554

Radhakrishnan EV, Phillips BF, Lakshmi Pillai S, Padua S (2019) Ecology and global distribution pattern of lobsters. In: Lobsters: biology, fisheries and aquaculture. https://doi.org/10.1007/978-981-32-9094-5_5

Ramamurthy T, Nandy RK, Mukhopadhyay AK, Dutta S, Mutreja A, Okamoto K, Miyoshi SI, Nair GB, Ghosh A (2020) Virulence regulation and innate host response in the pathogenicity of *Vibrio cholerae*. Front Cell Infect Microbiol 10:1–22. https://doi.org/10.3389/fcimb.2020.572096

Rameshkumar P, Kalidas C, Tamilmani G, Sakthivel M, Nazar AKA, Maharshi VA, Rao KS, Gopakumar G (2014) Microbiological and histopathological investigations of *Vibrio alginolyticus* infection in cobia Rachycentron canadum (Linnaeus, 1766) cultured in sea cage. Indian J Fish 61(1):124–127. https://epubs.icar.org.in/ejournal/index.php/IJF/article/view/16405/; http://epubs.icar.org.in/ejournal/index.php/IJF/article/download/16405/17827

Rameshkumar P, Nazar AKA, Pradeep MA, Kalidas C, Jayakumar R, Tamilmani G, Sakthivel M (2017) Isolation and characterization of pathogenic *Vibrio alginolyticus* from sea cage cultured cobia (*Rachycentron canadum* (Linnaeus 1766)) in India. Lett Appl Microbiol 65(5):423–430. https://doi.org/10.1111/lam.12800

Rigos G, Kogiannou D, Fioravanti M, Zarza C, Florio D (2020) Best therapeutic practices for the use of antibacterial agents in finfish aquaculture: a particular view on European seabass (*Dicentrarchus labrax*) *and* gilthead seabream (*Sparus aurata*) in Mediterranean aquaculture. Rev Aquac 13(3):1285–1323. https://doi.org/10.1111/raq.12523

Ruiz P, Poblete M, Yáñez AJ, Irgang R, Toranzo AE, Avendaño-Herrera R (2015) Cell-surface properties of *Vibrio ordalii* strains isolated from Atlantic salmon Salmo salar in Chilean farms. Dis Aquat Org 113(1):9–23. https://doi.org/10.3354/dao02820

Ruiz P, Balado M, Fuentes-Monteverde JC, Toranzo AE, Rodríguez J, Jiménez C, Avendaño-Herrera R, Lemos ML (2019) The fish pathogen *Vibrio ordalii* under iron deprivation produces the *Siderophore piscibactin*. Microorganisms 7(9):313. https://doi.org/10.3390/microorganisms7090313

Santos HM, Tsai CY, Maquiling KRA, Tayo LL, Mariatulqabtiah AR, Lee CW, Chuang KP (2020a) Diagnosis and potential treatments for acute hepatopancreatic necrosis disease (AHPND): a review. Aquac Int 28(1):169–185. https://doi.org/10.1007/s10499-019-00451-w

Santos R, Pabon A, Silva W, Silva H, Pinho M (2020b) Population structure and movement patterns of blackbelly rosefish in the NE Atlantic Ocean (Azores archipelago). Fish Oceanogr 29(3):227–237. https://doi.org/10.1111/fog.12466

Sharma SRK, Pradeep MA, Sadu N, Dube PN, Vijayan KK (2017) First report of isolation and characterization of *Photobacterium damselae* subsp. *damselae* from cage-farmed cobia (Rachycentron canadum). J Fish Dis 40(7):953–958. https://doi.org/10.1111/jfd.12557

Sherry YE, Mahmoud MM, Hassan ES, Sabah I, Elkamel AA (2016) Subclinical vibriosis in red swamp crayfish, *Procambarus clarkii*. Int J Fish Aquat Stud 4(3):119–123

Shinn AP, Pratoomyot J, Griffiths D, Trong TQ, Vu NT, Jiravanichpaisal P, Briggs M (2018) Asian shrimp production and the economic costs of disease. Asian Fish Sci 31:29–58. https://doi.org/10.33997/j.afs.2018.31.s1.003

Song X, Wang H, Xin L, Xu J, Jia Z, Wang L, Song L (2016) The immunological capacity in the larvae of Pacific oyster Crassostrea gigas. Fish Shellfish Immunol 49:461–469. https://doi.org/10.1016/j.fsi.2016.01.009

Steinum TM, Karatas S, Colquhoun DJ (2016) Multilocus sequence analysis of close relatives *Vibrio anguillarum* and *Vibrio ordalii*. Appl Environ Microbiol 82(18):5496–5504. https://doi.org/10.1128/AEM.00620-16.Editor

Tang KFJ, Bondad-Reantaso MG (2019) Impacts of acute hepatopancreatic necrosis disease on commercial shrimp aquaculture. Rev Sci Tech 38(2):477–490. https://doi.org/10.20506/rst.38.2.2999

Tang KFJ, Bondad-Reantaso MG, Arthur JR, MacKinnon B, Hao B, Alday-Sanz V (2020) Shrimp acute hepatopancreatic necrosis disease strategy manual. In FAO eBooks. https://doi.org/10.4060/cb2119en

Terceti MS, Ogut H, Osorio R (2016) *Photobacterium damselae* subsp. *damselae*, an emerging fish pathogen in the Black Sea: evidence of a multiclonal origin. Appl Environ Microbiol 82(13):3736–3745. https://doi.org/10.1128/AEM.00781-16.Editor

Terceti MS, Vences A, Matanza XM, Dalsgaard I, Pedersen K, Osorio CR (2018) Molecular epidemiology of *Photobacterium damselae* subsp. *damselae* outbreaks in marine rainbow trout farms reveals extensive horizontal gene transfer and high genetic diversity. Front Microbiol 9:2155. https://doi.org/10.3389/fmicb.2018.02155

Thornber K, Verner-jeffreys D, Hinchliffe S, Rahman MM, Bass D, Tyler CR (2020) Evaluating antimicrobial resistance in the global shrimp industry. Rev Aquac 12:966–986. https://doi.org/10.1111/raq.12367

Türe M, Alp H (2016) Identification of bacterial pathogens and determination of their antibacterial resistance profiles in some cultured fish in Turkey. J Vet Res (Poland) 60(2):141–146. https://doi.org/10.1515/jvetres-2016-0020

Turner JW, Tallman JJ, Macias A, Pinnell LJ, Elledge NC, Azadani DN, Nilsson WB, Paranjpye RN, Armbrust EV, Strom MS (2018) Comparative genomic analysis of *Vibrio diabolicus* and six taxonomic synonyms: a first look at the distribution and diversity of the expanded species. Front Microbiol 9:1–14. https://doi.org/10.3389/fmicb.2018.01893

Valderrama K, Balado M, Rey-Varela D, Rodríguez J, Vila-Sanjurjo A, Jiménez C, Lemos ML (2019) Outer membrane protein FrpA, the siderophore piscibactin receptor of *Photobacterium damselae* subsp. *piscicida*, as a subunit vaccine against photobacteriosis in sole (*Solea senegalensis*). Fish Shellfish Immunol 94:723–729. https://doi.org/10.1016/j.fsi.2019.09.066

Vallecillos A, Chaves-Pozo E, Arizcun M, Perez R, Afonso JM, Berbel C, Pérez-Sánchez J, María-Dolores E, Armero E (2021) Genetic parameters for *Photobacterium damselae* subsp. *piscicida* resistance, immunological markers and body weight in gilthead seabream (*Sparus aurata*). Aquaculture 543:736892. https://doi.org/10.1016/j.aquaculture.2021.736892

Vasquez I, Cao T, Chakraborty S, Gnanagobal H, O'Brien N, Monk J, Boyce D, Westcott JD, Santander J (2020) Comparative genomics analysis of *Vibrio anguillarum* isolated from lumpfish (*Cyclopterus lumpus*) in newfoundland reveal novel chromosomal organizations. Microorganisms 8(11):1–25. https://doi.org/10.3390/microorganisms8111666

Wang Y, Zhang HY, Fodjo EK, Kong C, Gu RR, Han F, Shen XS (2018) Temperature effect study on growth and survival of pathogenic *Vibrio parahaemolyticus* in Jinjiang oyster (*Crassostrea rivularis*) with rapid count method. J Food Qual 2018:1. https://doi.org/10.1155/2018/2060915

Wang D, Loor A, De Bels L, Van Stappen G, Van den Broeck W, Nevejan N (2021) Dynamic immune response to vibriosis in pacific oyster Crassostrea gigas larvae during the infection process as supported by accurate positioning of GFP-tagged vibrio strains. Microorganisms 9(7):1523. https://doi.org/10.3390/microorganisms9071523

Yang Y, Xie J, Li H, Tan S, Chen Y, Yu H (2017) Prevalence, antibiotic susceptibility, and diversity of *Vibrio parahaemolyticus* isolates in seafood from South China. Front Microbiol 8:1–9. https://doi.org/10.3389/fmicb.2017.02566

Yan-Wei L, Xiang Z, Yi-Jie C, Shu-Yan C, Hong-Yan S, Xue-Ming D (2019) Effective method to control *Vibrio mimicus* infection in channel catfish *Ictalurus punctatus*. Afr J Biotechnol 18(27):603–611. https://doi.org/10.5897/ajb2019.16859

Younes AM, Fares MO, Gaafar AY, Mohamed LA (2016) Isolation of *Vibrio alginolyticus* and *Vibrio vulnificus* strains from cultured *Oreochromis niloticus* around Qarun lake, Egypt. Glob Vet 16(1):1–5. https://doi.org/10.5829/idosi.gv.2016.16.01.10214

Yu Q, Niu M, Yu M, Liu Y, Wang D, Shi PX (2015) MOST-USDA joint research center for food safety, school of agriculture and SC. In: Food Control. Elsevier Ltd. https://doi.org/10.1016/j.foodcont.2015.08.005

Zago V, Zambon M, Civettini M, Zaltum O, Manfrin A (2017) Virulence-associated factors in *Vibrio cholerae* non-O1/non-O139 and *V. mimicus* strains isolated in ornamental fish species. J Fish Dis 40(12):1857–1868. https://doi.org/10.1111/jfd.12659

Zannella C, Mosca F, Mariani F, Franci G, Folliero V, Galdiero M, Tiscar PG, Galdiero M (2017) Microbial diseases of bivalve mollusks: infections, immunology, and antimicrobial defense. Mar Drugs 15(6):182. https://doi.org/10.3390/md15060182

Zhang X, Wang S, Chen S, Chen Y, Liu Y, Shao C, Wang Q, Lu Y, Gong G, Ding S, Sha Z (2015) Transcriptome analysis revealed changes of multiple genes involved in immunity in *Cynoglossus semilaevis* during *Vibrio anguillarum* infection. Fish Shellfish Immunol 43(1):209–218. https://doi.org/10.1016/j.fsi.2014.11.018

Zhang X-H, He X, Austin B (2020) *Vibrio harveyi*: a serious pathogen of fish and invertebrates in mariculture. Mar Life Sci Technol 2(3):231–245. https://doi.org/10.1007/s42995-020-00037-z

Zhu Z, Duan C, Dong C, Weng S, He J (2020) Epidemiological situation and phylogenetic relationship of *Vibrio harveyi* in marine-cultured fishes in China and Southeast Asia. Aquaculture 529(135):735652. https://doi.org/10.1016/j.aquaculture.2020.735652

Zrncic S (2020) Diagnostic manual for the main pathogens in European seabass and gilthead seabream aquaculture. In: Diagnostic manual for the main pathogens in European seabass and gilthead seabream aquaculture, vol 75. http://om.ciheam.org/om/pdf/b75/b75.pdf

Biosecurity in Aquaculture: Nurturing Health and Ensuring Sustainability

8

Rafidah Othman, Hiam Elabd,
Fittrie Meyllianawaty Pratiwy, Noora Barzkar,
Leong-Seng Lim, and Kenneth Francis Rodrigues

Abstract

"*Biosecurity Management in Aquaculture: Nurturing Health and Ensuring Sustainability*" explores the transformative potential of biosecurity in shaping the future of aquaculture. Aquaculture is emerging as an important contributor to global food security, accounting for a considerable share of the world's seafood production. Nevertheless, the industry's sustainable development is dependent on a critical pillar: biosecurity. This topic will highlight the critical importance of biosecurity in aquaculture, emphasizing its role in protecting aquatic populations' health, guaranteeing sustainability, and contributing to global food security. Biosecurity in aquaculture is a holistic disease prevention, control, and management approach. It includes a wide range of strategies, from strictly enforced quarantine and advanced monitoring systems to selective breeding for disease resistance. The industry faces a range of global challenges, particularly disease outbreaks, environmental pressures, and the pressing need to meet rising

R. Othman (✉) · N. Barzkar · L.-S. Lim
Higher Institution Centre of Excellence (HICoE), Borneo Marine Research Institute, Universiti Malaysia Sabah, Kota Kinabalu, Sabah, Malaysia
e-mail: rafidaho@ums.edu.my

H. Elabd
Department of Fish Diseases and Management, Aquatic Animals' Medicine Department, Faculty of Veterinary Medicine, Benha University, Moshtohor, Toukh, Egypt

F. M. Pratiwy
Department of Fisheries, Faculty of Fisheries and Marine Science, Universitas Padjadjaran, Bandung, Indonesia

K. F. Rodrigues
Biotechnology Research Institute, Universiti Malaysia Sabah, Kota Kinabalu, Sabah, Malaysia

N. M. Faudzi et al. (eds.), *Essentials of Aquaculture Practices*,
https://doi.org/10.1007/978-981-97-6699-4_8

seafood demand. Hence, effective biosecurity management is crucial to navigating these challenges. As technology and research advances, the same goes for the tools and strategies for biosecurity management in aquaculture. The most recent advances in disease diagnostics, surveillance systems, and immunization methods reveal how innovation contributes to progress in aquatic health management and disease prevention. Meanwhile, the environmental stewardship inherent in biosecurity management demonstrates how ethical practices benefit not only the aquaculture industry but also aquatic ecosystems and biodiversity. In conclusion, "Biosecurity Management in Aquaculture" is of indisputable importance as the foundation of success in the aquaculture sector. The aquaculture industry could not just flourish but also play a critical role in increasing global food security by effectively addressing global concerns, embracing innovation, and upholding environmental responsibility.

Keywords

Disease management · Enforced quarantine · Advanced monitoring · Food safety

8.1 Introduction

Aquaculture is the practice of cultivating aquatic organisms in controlled environments such as ponds, tanks, or cages. It includes a diverse range of species, such as fish, mollusks, crustaceans, and aquatic plants, and is a key source of seafood production to fulfill expanding worldwide demand. A total of 3.3 billion people globally relied on fish for 20% of their animal protein in 2020 (FAO 2020). According to the FAO, global fish consumption is projected to reach 21.9 kg per capita by 2030, up from 20.5 kg per capita in 2018 (FAO 2020). The Food and Agriculture Organization (FAO) has highlighted the importance of increasing aquaculture production to meet the challenge of feeding a world population that is estimated to reach 9.6 billion people by 2050 (FAO 2014). Due to this constraint, the global food demand for protein has increased drastically to fulfill the food requirements.

Aquaculture is an important economic activity and is the fastest-growing food-producing industry, surpassing that of capture fisheries (Balcázar et al. 2006; FAO 2020). Biosecurity in aquaculture is a maturing activity, still in need of improved information on diagnostics, disease transmission, clean-up, and eradication (Pruder 2004). In Malaysia, the fisheries sector plays a significant role in its economic development and is also a source of employment. The changing aquaculture practices, such as the use of various aquaculture chemicals that supposedly control the spread of diseases, could lead to increased concerns about the safety of food fish. Thus, the Department of Fisheries (DOF) Malaysia established its Fisheries Biosecurity Division (FBD) for the control and management of fish health and food safety, particularly for fish and fishery products (Sekaran 2014).

Biosecurity in aquaculture consists of practices that minimize the risk of introducing an infectious disease and spreading it to the animals at a facility. It also involves managing the risk that diseased animals or infectious agents will leave a facility and spread to other sites and other susceptible species. Biosecurity and health management are currently top concerns for proper aquaculture management as they go beyond farmers' control and need significant resources, coordinated action, and cooperation (FAO 2018). Environmental diseases and degradation are common in large-scale production facilities where aquatic animals are subjected to stressful conditions, resulting in significant economic losses (Balcázar et al. 2006). One of the most significant obstacles to the expansion and advancement of sustainable aquaculture is aquatic animal disease (FAO 2020).

Biosecurity practices also reduce stress on the animals, thus making them less susceptible to disease. Infectious disease is by far the leading cause of death in farmed fish; an outbreak can wipe out entire stocks, causing expensive decontamination of the facilities and equipment affected by it. Aquatic animal diseases are often caused by viruses and occasionally by bacteria and parasites as the causative agents (FAO 2020). Meanwhile, some bacterial fish pathogens have been linked to human diseases, making aquaculture goods a possible health concern for customers with either zoonotic or foodborne diseases (Castro et al. 2008). Therefore, aquaculture can guarantee food safety and maximum compliance with hygiene-sanitary criteria through constant inspection and examination of the fish and their feed, which is expressed in their complete traceability (FAO 2020). Aquaculture goods have several characteristics, such as freshness, homogenous quality, year-round supply, and steady prices, which give them significant market advantages over their wild counterparts.

Disease management is a tough prospect because of the pathogen's peculiar ecology, in which the pathogen is continually seeking an opportunity when the host's health is compromised. Disease outbreaks are linked to environmental degradation, which causes stress in cultured animals (Raman et al. 2015). Every 3–5 years, a previously undiscovered virus that causes a new and unknown disease develops; spreads fast, even across national lines; and causes significant productivity losses in aquaculture (FAO 2016). Examining specific points in farming or production activities when pathogens (parasites, viruses, bacteria, and fungi) may be introduced or disease may develop will help farmers develop precautionary measures. Disease outbreaks are one of the most critical challenges in aquaculture, producing mass, often explosive, eruptions of fish species, resulting in production losses in commercial aquaculture and threatening valuable wild aquatic animal stocks (Murwantoko et al. 2018). Hence, good biosecurity minimizes the fish's exposure and susceptibility to pathogens, reduces economic losses from mortalities, and helps prevent mandatory depopulation. This is because pathogens are transmitted through various routes, including water, vectors, and the movement of infected individuals (Johansen et al. 2011). Therefore, management is crucial to either fully avoid or significantly reduce the loss (Otta and Hoque 2020). The greatest strategy to avoid disease is to prevent pathogens from entering and spreading within the system. Pathogens can be kept out of aquaculture practices by using biosecurity, which is a component of good management (Phu et al. 2016). Figure 8.1 shows an example of biosecurity signage at the main entrance of a farm.

Implementing effective biosecurity measures is necessary to prevent the entry of disease into aquaculture systems. As an example, pathogen and pest introduction is facilitated by the open setting of aquaculture operations and the addition of freshwater or seawater to land-based systems (Johansen et al. 2011; Fitridge et al. 2012). Aquaculture farmers may protect the environmental quality and integrity of their farming ecosystems by mitigating risk factors and dealing with potential pathways for disease introduction. Hence, one of the vital components of disease control and biosecurity is the identification of entry points and routes through which diseases and pests might get into aquaculture facilities. These entry points and pathways can change based on the aquaculture system and the environment around it, but a few common ones include the following:

1. Water sources
2. Broodstock and stock introductions
3. Fish stock or stocking material movements
4. Shared infrastructures, type of equipment, and vessels/transportation
5. Staff, contractors, and visitors
6. Wildlife and vector species
7. Feed and feed ingredients

Fig. 8.1 Biosecurity and safety notice signage displayed at the main entrance of the farm

8.2 The Importance of Aquaculture Biosecurity

The introduction and discussion of biosecurity in aquaculture occurred in 1997 in a World Aquaculture Society (WAS) Special Session titled "*Sustainable Shrimp Farming: Emerging Technologies and Products for Biosecurity and Zero Discharge*." The term biosecurity was first used in the agricultural sector in the 1980s as "the sum of risk management practices in the defense against biological threats." A biosecurity program identifies the potential routes for pathogen introduction and transmission, between and within farms, and describes the measures required to reduce the risk of these events (Subasinghe et al. 2023). Two of the most significant components of an efficient biosecurity program are the use of basic epidemiological concepts and a logical and science-based approach to identifying and managing hazards (Subasinghe 2005).

Infectious disease has been identified as a potential limiting factor to aquaculture production (Jansen et al. 2012) and causes damage to the livelihood of farmers, loss of jobs, reduced incomes, and food insecurity. As a growing sector, current aquaculture practices have resulted in disease outbreaks, which have been identified as one of the key concerns affecting aquaculture fields. Aquaculture practices and the environment interact in ways that are unique to each combination of production and the physical and biological characteristics of the site (Iwama 2009). Numerous pathogens are prevalent in the surrounding environment and could be borne by the animals used for production. Nevertheless, according to Murray and Peeler (2005), clinical disease outbreaks are frequently influenced by the environment and farming practices. The main goals of aquaculture biosecurity are (i) animal management, (ii) pathogen management, and (iii) people management (Fig. 8.2).

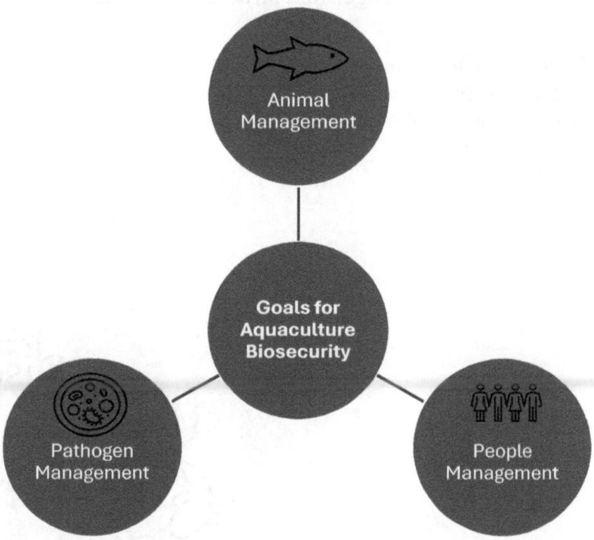

Fig. 8.2 Three main goals for the implementation of aquaculture biosecurity

Diseases in aquaculture are the result of a chain of events that include interactions between the host, the environment, and the presence of a pathogen (Ananda Raja and Jithendran 2015). The epidemiological triad is a conceptual model for the multi-factorial context of disease occurrence that reflects the interaction between host, pathogen, and environment (Fig. 8.3). These three overlapping circles are commonly used to represent how the host, environment, and pathogen interact, where the shared core region symbolizes the conditions that will lead to the onset of the disease, while a decrease in the impact of any one of the three components corresponds to an equivalent reduction in the threat of disease (Warren and Marshall 1983). Many current aquaculture diseases are the result of increased production techniques that lack a fundamental knowledge of the fine balance between host, pathogen, and environment. Aquatic animals need more care since they live in a more complex environment compared to terrestrial animals.

Although the initial expenses associated with infrastructure, equipment, and management procedures are significant, biosecurity measures can result in long-term cost benefits by reducing the likelihood of disease outbreaks and the need for costly interventions and treatments. Aquaculture operations can become more financially viable by optimizing production efficiency, lowering feed conversion ratios, and reducing overall production costs through the implementation of effective biosecurity procedures. Diseases emerged as a result of anthropogenic stock transfer, pathogen interaction with natural populations, and evolution from non-pathogenic microorganisms (Murray and Peeler 2005). Biofouling's effects on finfish aquaculture are usually attached to infrastructure, which leads to the following:

- Restriction on water exchange
- Increased disease risk
- Deformation of cages and structures (Fitridge et al. 2012)

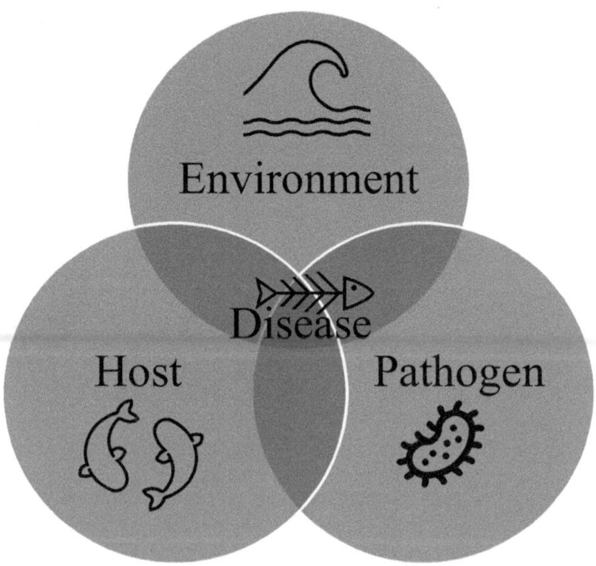

Fig. 8.3 Fish health or disease concept is a set of interactions among the host, pathogen, and environment

8.3 Vital Steps for the Implementation of Aquaculture Biosecurity

8.3.1 Facilities Design and Structures

Site selection, water supply features, facility design, and farming practices all have an impact on the conditions that stock experiences (Westers 1983; Munro and Waddell 1984). Therefore, the location, design, and construction of the aquaculture facility should include the interaction with wildlife, including the minimization of attraction, and the exclusion of predators (e.g., birds, seals) and scavengers (e.g., birds, cats, dogs, and rodents) throughout the planning phase. The design of aquaculture facilities and how they are handled can have significant implications on biosecurity (Westers 1983; Donovan 2006). Hence, biosecurity concerns during the design and construction stages of aquaculture facilities are critical for affecting the rearing environment, stock health planning, and the implementation of optimum biosecurity measures. An example, proper positioning of farms and cages can reduce parasite population growth, extending the time between stressful and costly parasite treatments (Chambers and Ernst 2005). This generally leads to long-term cost savings for the facility.

8.3.2 Disease Prevention and Control

The farming environment increases the risk of disease outbreaks since stressed and weakened individuals are more susceptible to pathogens and parasites and may transmit them to healthy individuals (Robertsen 2011). Infectious diseases caused by a varied array of bacterial pathogens are prominent in fish reared in confined aquaculture facilities. The disease is an inevitable part of existence for all multicellular creatures on the entire globe, including plants, fish, and terrestrial animals. Infectious diseases continue to be one of the most significant barriers to the successful multiplication of aquatic animals. Intensive farming with high fish-rearing density in aquaculture increases the introduction and proliferation of infectious diseases and has become the root cause of such devastating outbreaks (Rohani et al. 2022).

Fish diseases caused by parasites, bacteria, or viruses can spread from pond to pond or farm to farm through the transfer of diseased fish, as well as through animals, people, equipment, and water contaminated by contact with sick fish or fish pathogens. These pathogenic microorganisms and parasites are opportunistic and can be found in habitats or as asymptomatic carriers on some fish, making aquaculture facilities vulnerable to disease outbreaks and impeding the growth of an efficient, cost-effective, and reliable aquaculture process (Jørgensen 2020). Hence, there should be no contact between the fish on the farm and any potential disease carriers to prevent the introduction of new diseases. Therefore, effective control approaches primarily based on pathogen reduction, alteration of the environment, strengthening host resistance, prophylactic and preventive medicines, and immunization should be implemented accordingly.

8.3.3 Considerations for the Economic Impacts

The economic impact of a disease is determined by various factors such as the incidence and severity of infection, the impact the disease has on animal productivity and mortality, the impact the disease has on human health, and the efforts made to combat the disease (Rushton et al. 2018).

Financial Impact of Disease Outbreaks

Globally, 56% of fish that are directly used for human consumption come from aquaculture (FAO 2020). Over the years, the aquaculture industry has experienced enormous challenges with disease outbreaks and epidemics, which have resulted in significant economic losses. The aquaculture industry loses about USD 6 billion, or 2% of total production, each year, with diseases playing a major role (Austin 2023). Hence, economic effects on biosecurity in aquaculture play a major role in the industry's viability and profitability. According to Mohd Yazid et al. (2021), the estimated economic loss due to vibriosis is 7.06% of the total production cost of Asian seabass per kilogram. Diseases could harm aquaculture production by damaging essential resources, diminishing the physical output or unit value of the process of production, decreasing efficiency, and directly affecting the well-being of people (Mohamad et al. 2019).

Disease control has been viewed as a primary driving force for the establishment of several thriving aquaculture industries. Managing aquatic animal disease is quite difficult, as even minor losses in aquaculture production could have a detrimental impact on profitability (Subasinghe 2005). The world's largest aquaculture producer, the People's Republic of China (PR China), reported 205,000 metric tons of production losses at an estimated cost of USD 401 million in 2018 due to aquatic diseases (FAO 2020). Demand for aquaculture is anticipated to reach 109 million tons in 2030, a 32% (26 million tons) increase from 2018. Asia will continue to lead the aquaculture business, accounting for more than 89% of demand growth by 2030 and accounting for 89% of global aquaculture production in 2030 (Amorim et al. 2019). Therefore, effective biosecurity policies must be implemented to help stop diseases from entering and spreading within aquaculture facilities.

The industry will benefit from the implementation of appropriate biosecurity practices, as the costs involved are relatively minimal when considering the anticipated increases in productivity and product quality. Outbreaks of infectious diseases pose a significant barrier to the sustainable growth of aquaculture (FAO 2020). For instance, significant losses were documented in oyster farming in 2017, amounting to a total of USD 540 million, attributed to various pathogens (Bureau of Fisheries 2018). Similarly, in carp farming between 2002 and 2004, losses ranging from USD 0.2 to 25 million were incurred due to the koi herpesvirus (Bondad-Rentaso et al. 2001).

The Cost-Effectiveness of Biosecurity Measure Implementation

The aquaculture sector has experienced a range of fish diseases, which have led to huge financial losses and put the long-term viability of the industry at risk (Ridzuan et al. 2022). Therefore, the main aim of establishing biosecurity practices is to limit the likelihood of disease introduction and subsequent spread. Several factors must be considered when creating and executing an economical, biosecure aquaculture protocol. These factors include aggressively developing and implementing techniques for disease exclusion from culture systems (pathogen exclusion from water and carriers) as well as an efficient screening of broodstock and post-larvae.

Findings in a recent study stated that one of the main reasons many farmers, particularly in traditional small-scale aquaculture, do not implement biosecurity measures on their farms is a lack of information about these costs and benefits (Zulbainarni and Megawati 2019). Farmers can prevent costs related to mortality, slower growth rates, and lower-quality products by reducing the frequency of disease outbreaks. This will raise harvests and increase profitability. According to Jahangiri et al. (2023), having a thorough biosecurity plan on a typical marine fish farm in Hong Kong was only financially advantageous when a high percentage of mortality (46%) was expected on the farm and when the farm was stocked with a high number of fish (100 K). However, adding a strategy for the timely removal of dead fish and early treatment of disease on farms could reduce the risk of loss (Jahangiri et al. 2023). Effective biosecurity programs can be developed to prevent and manage disease on aquaculture farms (Perera et al. 2005).

8.3.4 Environmental Implications

Aquaculture has a detrimental impact on the environment in terms of genetics, water quality, ecology, health, and resource consumption (Salin and Arome Ataguba 2018). Nevertheless, aquaculture growth is still necessary to meet rising fish demand. Environmental impacts are determined by species, systems, production methods (whether extensive, semi-intensive, or intensive), location, and management quality, as well as the nature of the receiving ecosystem (Beveridge and Brummett 2015). Meanwhile, health management in aquaculture plays a vital role in food security, safety, and environmental sustainability (Su et al. 2021). The adoption of aquaculture management strategies is influenced by factors such as awareness, access to credit facilities, and interaction with extension agents (Ijah et al. 2020).

Potential Impacts of Disease Spread on Wild Populations

Each year, infectious diseases cost the shellfish and finfish aquaculture sectors billions of dollars and put wild fish in danger of spreading diseases (Lafferty et al. 2015). The dynamics of disease transmission between wild and farmed fish can lead to the creation of new pathways for transmission or bioamplification of pathogen abundance, which may hurt or threaten the survival of wild fish (Krkošek 2017). Few quantitative studies have been conducted to show that wild animals living close to farms are more susceptible to infectious diseases than animals in other locations (Lafferty et al. 2015). Diseases that emerge in aquaculture operations may spread to wild populations, which may have several impacts on public health, ecosystems, economies, and conservation initiatives. The rise of pathogenic pathogens in aquaculture poses a global transmission risk to wild species, highlighting the necessity for strong biosecurity measures (Mordecai et al. 2021). The increase in parasite transmission between humans, domesticated animals, and wildlife is frequently associated with the emergence of infectious diseases (Daszak et al. 2000). The spread of diseases from farm fish to wild populations can have significant consequences for both aquatic ecosystems and human activities.

Key epizootiological criteria of emerging infectious diseases (EIDs) in free-living wild animals were classified by Daszak et al. (2000) into three main groups:

1. "Spill-over" from domestic animals to the wildlife population
2. Human intervention through host or parasite translocations
3. No obvious involvement of domestic animals or humans

The occurrences of these EIDs have two significant biological consequences: first, numerous wildlife serve as reservoirs for pathogens that endanger domestic animal and human health; second, wildlife endangered by EIDs creates a serious challenge to the preservation of biodiversity worldwide. The dynamics of infectious diseases are affected by the frequent migrations of marine fish species as well as the migratory culling, escape, halting, and allopatry of host populations (Altizer et al. 2011). Salmon and sea lice are two examples of macroparasite-host systems with thresholds separating endemic from epidemic circumstances based on density-dependent transmission (Jansen et al. 2012). Thus, effective biosecurity measures are crucial to minimize the risks and impacts of disease transmission on both aquaculture and wild fish populations.

Minimizing Reliance on Chemical Treatments and Antibiotics

The potential environmental and human health risks associated with the use of veterinary medicines and feed additives in aquaculture highlight the importance of comprehensive evaluation and monitoring (Rico et al. 2013; Senarathna et al. 2021). Conventional disease treatment, which involves the use of chemicals, harms consumers, the environment, and organisms under culture. The indiscriminate use of antibiotics in aquaculture raises concerns about the emergence of antimicrobial resistance in fish pathogens (Kennedy et al. 2016). It was reported that the most common antibiotic resistance is reported to oxytetracycline, ampicillin, and florfenicol, while few studies indicate antibiotic resistance of *Aeromonas* species to oxytetracycline, sulfadiazine, fluoroquinolones, quinolones, etc. (Preena et al. 2020). Antibiotic resistance (AMR) in aquaculture has been suggested to be primarily caused by the overuse and misuse of antibiotics (Watts et al. 2017; Santos and Ramos 2018; Reverter et al. 2020).

Antibiotics are chemical substances that are produced naturally, semi-synthetically, or synthetically (Manyi-Loh et al. 2018), and they are often used to inhibit the growth of bacteria and kill them. Inappropriate and overuse of antibiotics will result in a high level of antibiotics ending up in the environment (Gillings 2013). Thus, only five antibiotics are approved to be used in aquaculture:

oxytetracycline, florfenicol premix, sarafloxacin, erythromycin, and sulphonamides potentiated with trimethoprim or ormethoprim. Furunculosis, enteric red-mouth disease, and vibriosis are all treated with these antibiotics (Serrano 2005). Nevertheless, many therapeutic plant chemicals, such as curcumin, quercetin, silymarin, ajoene, and caffeic acid, have been utilized as an alternative to conventional approaches to disease prevention since they strengthen non-specific immunity (Jeyavani et al. 2022).

Regulations regarding aquaculture include those about the environment, property, labor, animal welfare, and food safety (Hishamunda et al. 2014). According to Abate et al. (2016), the first empirical model of aquaculture expansion as a function of environmental policies showed that the tighter environmental controls, the less aquaculture can grow. Nevertheless, aquaculture policy lessons could be learned from Norway in identifying efficient strategies for aquaculture policy, which has maintained its competitiveness in international markets and presently provides more than half of the world's salmon supply (Naylor et al. 2023). Aquaculture policy has developed over time, taking a flexible and balanced approach to both environmental preservation and the growth of the industry (Hersoug 2021). Long-term growth and profitability in aquaculture depend on environmental laws since over-intensification, parasite and pathogen issues, and disputes with other stakeholders, including capture fisheries, can harm the sector (Naylor et al. 2023). In many cases, aquaculture provides a healthier and more environmentally friendly substitute for the consumption of red meat (Gephart et al. 2021; Golden et al. 2021). This increasing production in aquaculture must be more environmentally friendly to protect natural resources for future generations.

8.3.5 Sustainable Aquaculture

Sustainable aquaculture aims to address environmental, social, and economic issues to meet the demands of current and future generations. It emphasizes responsible stewardship of aquatic resources, ethical business practices, and holistic management approaches to improve the health and resilience of aquaculture-dependent ecosystems and communities.

The Linkage of Biosecurity with Long-Term Sustainability

The approach to infectious disease control has shifted because of the deregulation of the global trade in animals and animal products (Thiermann 2005; Oidtmann et al. 2013). Aquaculture encounters increasing challenges to its expansion from emerging aquatic diseases and recurring outbreaks. According to World Bank projections, world fish supply might increase to 187 million metric tons by 2030, with aquaculture surpassing global capture production (The World Bank 2013). Sustainable aquaculture is crucial to meeting the increasing global demand for seafood products while protecting natural resources, strengthening resilient ecosystems, and improving community well-being all around the world.

Numerous contributing factors put farmed stocks at risk of infection and subsequent stock losses, making adequate biosecurity a complex and multifaceted task. Hence, the focus has shifted from the recovery of diseases to prevention through enhanced biosecurity measures (Chiew et al. 2019). To support the Sustainable Development Goals (SDGs), the policy environment could not be stronger, or it is now ideal to shift toward aquatic food systems that are more resilient, inclusive, efficient, and sustainable (FAO 2022). One of the key factors influencing sustainable production in the aquaculture sector is the effective implementation of biosecurity and disease control strategies (Kyule-Muendo et al. 2022).

Mitigating the Risk of Introducing Non-native Species

Aquaculture is anticipated to continue growing and intensifying as the world's food demand is predicted to rise significantly (FAO 2022). Aquaculture's rapid growth is mostly due to the introduction and utilization of non-native species (Lin et al. 2015). Compared to other regions of the world, Asia has made the most use of the potential of species that are available and are being cultivated in other countries (Gozlan 2017). Nonetheless, there may be risks to native ecosystems, biodiversity, and economic activity when these non-native species are introduced into aquaculture systems. Aquatic non-native species can spread throughout new areas in several ways, even though people are the ones who move them around the most. As shown in Table 8.1, several factors can lead to non-native fish escaping from aquaculture facilities into new environments.

The original fish biogeography has been altered by escaping non-native fish brought in through aquaculture, and this has led to a loss of assemblage distinctiveness and homogeneity of fauna (Kang et al. 2023). Therefore, it is necessary to incorporate responsible management strategies into sustainable aquaculture to mitigate potential threats associated with the introduction and use of non-native species (Lin et al. 2015). Countries must have strict, well-coordinated regulations and policies for transporting live aquatic animals and goods to minimize the risk of introducing and spreading diseases that affect aquatic animals and their effects on biodiversity. These regulations and policies need to be properly implemented and maintained at the national, regional, and worldwide levels. The two regions with the highest production of non-native species worldwide, accounting for 62% of all taxa, are Europe and the former USSR (Gozlan 2017).

Table 8.1 Risk of the introduction of aquaculture non-native species into a new environment

Aquaculture practice	Factors contributing to fish escape
Land-based aquaculture	1. Inadequate screening over pond outflow pipes 2. Pond overflow during flood events 3. Predatory birds transporting and dropping non-native species into adjacent bodies of water
Marine aquaculture	1. Inadequate facility maintenance 2. Powerful storms and unfavorable weather 3. Other marine species destroying nets or cages

Table 8.2 Several strategies to mitigate the risk of non-native species escape from aquaculture

Mitigation strategies	Measures	Aim
Strengthening infrastructure	Regular maintenance, repair, and reinforcement of aquaculture infrastructure	Reduce the likelihood of physical damage and escape events
Training and education	Providing comprehensive training and education to aquaculture personnel on • Biosecurity protocols • Equipment operation • Emergency response procedures	Minimize human error and improve risk management
Enhanced biosecurity measures	Implementing robust biosecurity measures: • Strict quarantine procedures • Thorough screening of incoming stock • Controlled access to facilities	Prevent non-native species introductions and escapes
Risk assessment and monitoring	Conducting risk assessments and monitoring programs for effective management and response	• Identif]y potential escape risks • Evaluate environmental impacts • Detect early signs of non-native species introductions

According to earlier research conducted in China, 25% of non-native fish raised for aquaculture have effectively formed a wild population in natural waters, and 15% have successfully invaded (Kang et al. 2023). Evaluating non-native species' effects on ecosystems and the cost of using them to produce fish for food for nearly 95.3% of the world's annual production of non-native aquatic species is very challenging (Gozlan 2017). Thus, to properly analyze risks and implement effective management for the sustainable development of aquaculture, researchers from academia, government, and the aquaculture sector must work together (Lin et al. 2015), which is essential to addressing this complex challenge and ensuring the long-term integrity and resilience of aquatic ecosystems. As an example, aquaculture operators can minimize these risks and support the sustainable management of aquatic resources by understanding the reasons that lead to escape incidents and putting suitable mitigation strategies into place, as shown in Table 8.2.

8.4 Successful Biosecurity Components

Several factors and practices have been linked to the introduction and spread of infectious diseases. These include the presence of wild aquatic animals, the sharing of personnel and equipment between farms, the interconnection of farms through waterways, shorter following periods, and the use of non-disease-resistant or disease-free seed (Corsin et al. 2001; Tang and Bondad-Reantaso 2019). As for shrimp farms, since shrimp lack mechanisms for adaptive immunity, biosecurity

Table 8.3 The key elements in implementing biosecurity plans

Key elements	Approaches
Risk assessment	Identify the aquaculture system's potential risks and limitations, including the entry points and routes for the spread of disease
Biosecurity measures	Use a combination of chemical, physical, and operational controls to mitigate the risks that have been identified • Controlling the entry points for access to facilities • Implementing hygiene protocols for personnel and equipment • Implementing optimum health management practices • Regular aquatic health and disease monitoring program • Employing appropriate diagnostic and detection methods for infectious diseases • Obtaining high-quality animal stock from reputable suppliers • Protocols for quarantine of newly acquired stocks • Water quality management procedures • Use of effective disinfection and pathogen eradication techniques
Training and education	Provide training and educational programs for aquaculture personnel to ensure they understand and follow biosecurity protocols • Identifying signs of diseases • Sanitation practices • Importance of biosecurity measures
Record keeping	Maintain records of biosecurity activities • Health monitoring • Stock movements • Disease incidences This information is important for monitoring compliance, identifying patterns, and guiding decision-making
Communication and collaboration	Promote interaction and collaboration between the stakeholders • Aquaculture farmers • Government agencies • Researchers, academics • Industry/trade associations
Regular review and improvement	Review and evaluate biosecurity strategies regularly to identify areas of improvement or adaptation

measures are crucial for preventing the introduction and spread of infections both within and between shrimp farms (Ali et al. 2018). Hence, biosecurity and health management strategies at the farm level are crucial for mitigating the effects of diseases (Hasan et al. 2013) and are determined to be the highest priority for sustainable aquaculture management (FAO 2018). Nevertheless, designs for biosecurity that perform successfully must be adjusted to the specific features of the species being raised and their production environments. Nevertheless, farm-level biosecurity procedures include stringent quarantine, egg disinfection, traffic control, water treatment, clean feed, and the disposal of mortalities (Assefa and Abunna 2018). The key elements in implementing biosecurity plans are tabulated (Table 8.3).

8.4.1 Physical Barriers

The prevention of disease in both farms and hatcheries is crucial and involves various measures such as proper system preparation, equipment and worker sanitation, pathogen barricades, quarantine, regular disease monitoring, and water treatment (Assefa and Abunna 2018). In general, biosecurity operations should have a defined structure and barriers, such as fences and gates, in place (Pruder 2004). Physical barriers, such as nets or partitions, can be used to separate different batches or species of stock in aquaculture systems, preventing direct contact and reducing pathogen spread. Separating stocks based on size, species, or status of health reduces disease transmission risks, while personnel, tools, and personal protection equipment across all sectors should be separate and facility-specific. The following mitigation strategies should be considered when evaluating and handling the risk of pathogen transmission by vectors:

- Enclose aquaculture production systems with a wall or fence with a gate to prevent entry of animals and people, except for authorized staff and guests
- Installing nets over open or outdoor aquaculture production systems to keep birds away
- Establishing a clear boundary between the inner biosecure area and the outer risk area to control access for both staff and visitors to the aquaculture farm
- To ensure that the biosecurity plan is followed, all visitors should be briefed and supervised

Many predatory species, including mammals, birds, and invertebrates, can be found in coastal and inland water bodies (Bevan et al. 2002). However, most of these animals are rarely seen or have limited contact with aquaculture facilities. There are only a few animals that pose a substantial threat to farmed aquatic animals or the facilities in which they are confined. As an example, bird predation can have a significant economic impact on aquaculture production by spreading diseases, weed seeds, and parasites from one facility to another. It is common to observe that a broad range of birds have been frequent in aquaculture operations. Therefore, various methods have been proposed and deployed to mitigate farmed fish losses due to predation by piscivorous birds; however, establishing physical barriers to limit bird access is largely recognized as the most effective alternative (Farrell and Leonard 2001; Nwadukwe and Arimoro 2012). The effectiveness of the chosen combination of measures will be determined by the target bird species' feeding habits (Otieno and Shidavi 2022).

Utilizing Fencing, Netting, and Physical Barriers for Security

Effective systems should be in place to prevent aquatic microorganisms and predators from entering facilities. Aquaculture farms should be aware of locally important, invasive, rare, and endangered species and record organisms encountered at their location (e.g., date, identification, abundance). Figure 8.4 shows the perimeter fencing around the prawn farm.

Fig. 8.4 Protected prawn farm enclosed with bird netting and perimeter fencing

Implementation of Controlled Access Points for Facility Security

On the other hand, limiting access, movement restrictions, and routine monitoring are essential to reducing disease transmission within and between livestock facilities (Vora et al. 2023). The inevitable movement of fish and equipment contributes to the spread of infectious diseases (Subasinghe et al. 2001). Roughly 85% of the employees frequently switched between locations (Muniesa et al. 2022). Therefore, every facility should have limited access, require visitors to follow biosecurity protocols, and always be accompanied by company staff (Fig. 8.5).

Fig. 8.5 Ensuring biosecurity compliance through visitor briefing and supervision

8.4.2 Hygiene Protocols

The pathogenic agent's stability in the environment, the kind and quantity of organic matter present on the fomite surface, the surface's water-holding capacity, and other factors will all affect the possible spread of pathogenic agents. Disinfection and cleaning are key components of pathogen risk management and disease prevention in the aquaculture industry (Hill et al. 2013).

The Significance of Equipment Sanitation

The primary methods for preventing bacterial diseases in fish farms include the use of sanitary certificates and quarantine procedures, the introduction of brood stocks free of specific pathogens, feed optimization, water quality improvement, and proper sanitation, together with health management (FAO 2007). Transporting aquaculture stock and equipment around is a major way for biological contamination and related "hitchhiker" species to spread (Inglis et al. 2013). Hence, all trucks, hauling tanks, pumps, nets, buckets, waders, or anything that may meet the delivery site water should be inspected, cleaned, and disinfected before coming back to the facility. Vehicles used to carry fish, eggs, or gametes should be cleaned and sanitized both before and after each shipment, whether by land, air, or sea.

However, according to Muniesa et al. (2022), worker disinfection stations were available in just 63% of the production units; in the on-growing phase, this percentage dropped to 50%. Regular maintenance must be performed to eradicate biofouling and the accumulation of organic materials in culture units, intake and discharge pipes, tanks, and ponds. Cleaning and drying can effectively control many fish infections in aquaculture (Assefa and Abunna 2018). All nets and equipment that are used for managing or harvesting fish should be rinsed, cleaned, and air-dried between uses.

Implementing Personal Hygiene Practices for Workers

Managing tanks or ponds, stocking fish, and harvesting fish are examples of aquaculture operations that might contaminate hands and clothing with harmful chemicals or microbes. Therefore, personal movement controls within and between facilities, along with other preventive measures for hygiene, could be improved by introducing disinfecting stations and company attire assigned to on-site usage (Muniesa et al. 2022). Aquaculture companies should consider implementing continuous internal training programs to provide all employees with education on biosecurity and hygiene practices (Muniesa et al. 2022).

Establishing Disinfection Procedures

Disinfection plays a significant role, with farmers acknowledging the benefits of disinfecting harvest nets and avoiding equipment sharing with contaminated farms (Chen et al. 2018). Water disinfection methods, such as ozonation and ultraviolet radiation, are effective in aquaculture systems (Semenov and Semenova 2022). The use of disinfectants like povidone-iodine has shown promise in reducing disease occurrence and improving survival rates (Sanawar et al. 2017). However, it is important to consider the environmental impact of disinfectants and the need for neutralization before discharge or reuse (Ivanitskaya et al. 2022). The evaluation of optimal disinfectant strategies and the assessment of the impact of probiotic supplementation on the microbiome in aquaculture systems are also crucial aspects to consider (Sanawar et al. 2017; Menanteau-Ledouble et al. 2020). Various chemical disinfectants such as potassium permanganate, phenol, 1% sodium hypochlorite, iodine solutions, glutaraldehyde, and formaldehyde are all regarded as effective, as are malachite green and copper sulfate, although excessive use of these substances in aquaculture may result in toxicity (Bornø and Colquhoun 2009). However, the use of disinfectants in fish health management should be carefully evaluated to address potential ecological and human health concerns (Faruk et al. 2021).

8.4.3 Disease Surveillance

Disease surveillance is necessary for ensuring the health and sustainability of aquaculture industries. Aquaculture producers can identify infections early, reduce economic losses, and maintain their efficiency by implementing effective surveillance programs.

Routine Health Monitoring of Aquatic Organisms

Surveillance is a systematic approach to collecting data on the incidence of important diseases and infections to create meaningful reports on the disease situation in each area. Disease surveillance is essential for reducing risks and strengthening aquaculture systems' resilience to changing disease pressures because it keeps a lookout for potential dangers to the system and acts quickly to address any emerging problems. Effective monitoring and regulation of fish health problems are essential for improving disease management in aquaculture (Opiyo et al. 2018). The most important defense against preventable disease losses is by monitoring as regularly as possible and taking appropriate action at the first signs of suspicious behavior, lesions, or mortalities (Bondad-Reantaso 2001). Disease surveillance is also crucial in ensuring food safety, maintaining the health and welfare of aquaculture species, facilitating regulatory compliance, and supporting the aquaculture industry's sustainable growth.

In aquaculture, disease surveillance is crucial for several purposes:

- Early detection
- Disease management
- Prevention of disease spread
- Protection of public health
- Regulatory compliance
- Sustainable aquaculture development

Farmers are the most suitable individuals to work with when it comes to disease surveillance, particularly when using farmer-based syndromic aquatic disease surveillance, which may be indicative of a probability of change in the health of a population that merits further research or enables a timely impact assessment and action requirement (Brugere et al. 2017). Disease reporting accuracy is just as crucial as disease detection, and a misdiagnosis could result in ineffective or poor control measures (Bondad-Reantaso 2001).

The Significance of Early Detection Methods in Disease Management

Significant productivity losses are being incurred because of silent mortality from diseases that remain undiagnosed and untreated. Cruz-lacierda et al. (2000) reported that leech infestations may be caused by fish carriers, trash feed, and the movement of currents. Young leeches hatch from the eggs, feed on the yolk for a few days, and then mature into adult bloodsuckers (Aloto and Eticha 2018). Early disease detection increases the likelihood of successful control and treatment, highlighting the importance of accurate and quick diagnostic methods (Hansen and Abd El Wahed 2020). The strategy of controlling infectious diseases has been shifted by epidemiological surveillance from depending on biosecurity to contain the disease at the

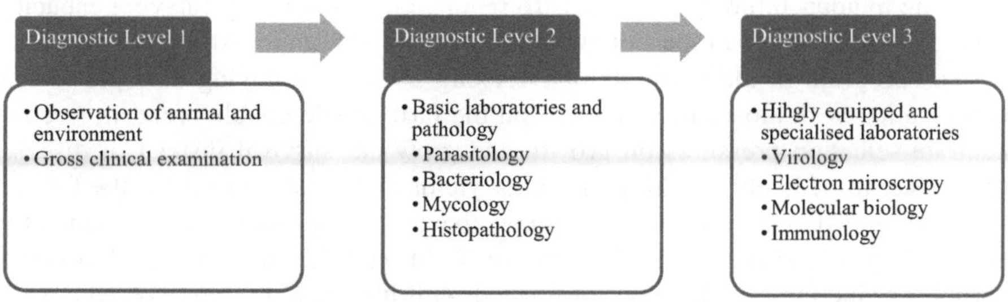

Fig. 8.6 Outlines the diagnostic activities at each level of disease diagnostic. (Modified from Bondad-Reantaso 2001)

point of invasion into a new population or territory to focus on changes in the distribution, frequency, and causes of disease within the population of origin (Brugere et al. 2017). Diagnostics for diseases should be based on an ongoing process of observations that begin on the farm and, in fact, even before the onset of the disease. Therefore, when evaluating a disease scenario, three different levels of disease diagnostics can be conducted, as shown in Fig. 8.6 (Bondad-Reantaso 2001).

8.4.4 Stocking Practices

The aquaculture community and industry have now been convinced through years of experience that using clean and healthy seeds should be a top priority in biosecurity to stop disease outbreaks and the losses that follow (Subasinghe et al. 2023). Stocking practices in aquaculture are crucial for promoting biosecurity measures, reducing the risk of disease introduction and transmission, and improving the health and sustainability of the industry.

The Importance of Selecting Disease-Resistant Stocks for Aquaculture

Throughout several generations, selective breeding of terrestrial animals has led to notable gains in growth, feed conversion efficiency, and reproductive performance (Boyle 2001). The most critical phase within aquaculture production is larval fish development, when the mortality of fish larvae can amount to 99% in the first 2 months following hatch (Bartl and Keckeis 2004). Disease resistance in farmed animals, including fish, can be enhanced through genetic selection (Gjedrem et al. 1991). Fish have been genetically modified to have increased resistance to bacterial (Silverstein et al. 2009) and viral (Wetten et al. 2007) infections. The majority of shrimp breeders' efforts have gone toward creating families of shrimp resistant to TSV and WSSV (Argue et al. 2002; Kong et al. 2003; Jiang et al. 2004; Gitterle et al. 2005; Cuéllar-Anjel et al. 2011).

High stocking density farming practices and declining environmental conditions combine to create stressful environments that exacerbate the incidence and severity of diseases (Bowden 2008). Disease-resistant stock can lead to improved animal welfare outcomes by reducing the incidence of disease-related stress and suffering in aquaculture populations. To regulate fish density by carrying capacity standards, Norway's salmon policy introduced the Maximum Allowable Biomass (MAB) program in the early 2000s (Hersoug 2021). Employing SPF prawns in combination with biosecurity controls on the farm would enable breeders to concentrate selection pressures on growth and grow-out survival instead of disease resistance, most likely resulting in higher yields and greater profit for the farmer (Moss et al. 2012). Nevertheless, an innovative technology that offers a financially and ecologically sustainable substitute for traditional shrimp culture is biosecure shrimp production systems that are loaded with high-health seeds (Wyban et al. 1993; Pruder 1995; Browdy and Bratvold 1998; Moss 2002). Selecting disease-resistant stock is essential for promoting disease prevention, economic viability, environmental protection, sustainable practices, public health, food safety, and animal welfare in aquaculture.

Enforcing Quarantine Protocols for Incoming Stock

It is critical to include quarantine facilities where required to prevent the introduction and spread of pathogens (Meyers 2010). Quarantine is a crucial practice in animal husbandry to isolate newly introduced or potentially exposed animals for observation and testing to prevent the spread of contagious diseases (Wei et al. 2023). Fish disease development is attributed to an intricate interaction involving the host, pathogen(s), and environment. When stress-inducing factors disturb this balance, fish tend to become sick. Cost-effective on-farm biosecurity practices may be a more realistic and economical approach to reducing losses due to disease than selective breeding programs targeted to increase disease resistance (Moss et al. 2012).

Understanding the survival and persistence of pathogens in hosts and the mode of disease transmission among individual hosts is crucial for developing effective measures for managing pathogens in aquaculture (Nasr-Eldahan et al. 2021). Additionally, aquaculture practices have been associated with repeated introductions of infectious diseases, particularly as high stocking densities can encourage the spread of parasitic infections (Pradeep et al. 2017). These measures include quarantining new animals, vaccination, and physical enclosures to separate animals and wildlife (Vora et al. 2023). When examining live fish and cleaning structures, equipment, and vessels, facility staff should be taught how to inspect for and detect biofouling or pest organisms.

Table 8.4 Optimum aquaculture water quality parameter (Mohanty et al. 2014)

Water parameter	Optimum level
Temperature	26–32 °C
Salinity	15–25 ppt (brackish water shrimp culture)
Dissolved oxygen	>4.0 ppm
pH	7.8–8.5
Total ammonia nitrogen	<1.0 ppm
Total nitrate nitrogen	<5.0 ppm
Nitrite nitrogen	<0.01 ppm
Sulfide	<0.03 ppm
Biological oxygen demand (BOD)	<10 ppm
Chemical oxygen demand (COD)	<70 ppm
Sacchi disk visibility (transparency)	25–45 cm
Ratio of phyto- and zooplankton	10:1

8.4.5 Water Quality Management

Effective water quality management is critical to the performance and sustainability of aquaculture operations. Water quality in aquaculture systems can rapidly degrade as fish use the water to live, feed, reproduce, grow, and excrete waste. Aquaculture producers can increase fish health and production efficiency and reduce negative environmental impacts by monitoring, maintaining, and optimizing water quality conditions, all of which contribute to the industry's long-term viability. Proper rearing environment management provides optimal circumstances for the growth and health of the cultivated fish and prawn species (Table 8.4).

Ensuring and Sustaining Optimal Water Conditions Through Monitoring

Water quality monitoring is crucial to make sure that the parameters governing the quality of the water are kept within the established acceptable ranges for the species and its developmental stage. Water quality parameters such as temperature, dissolved oxygen, hardness, pH, alkalinity, ammonia, and nitrites are crucial in aquaculture fish farms to ensure the growth and survival of different fish species (Carbonara et al. 2020). Environmental stress is a significant psychosomatic component in the spread of fish diseases, and many efforts have been made to investigate its impact on fish health (Raman et al. 2013). Hence, pre-treatment in water treatment for aquaculture farms is important for optimal water quality, supporting healthy aquatic ecosystems, and sustaining profitable aquaculture operations (Fig. 8.7). Paladini et al. (2017) have also shown that parasite species are responsible for slowing growth rates and feed conversion ratios of farmed animals, as well as direct mortality, product rejection during processing, and high treatment expenses. Feed availability and water quality are regarded as critical factors in fish growth. A rapid growth rate may indicate an abundance of food and optimum living conditions (Nuraini and Lutfiyah 2020).

Fig. 8.7 Water intake pre-treatment process for prawn culture ponds in a prawn farm

The Importance of Biofiltration Systems for Water Quality Maintenance

Aquaculture effluents contain various compounds that can be hazardous to the cultivated organism, the environment, and people, where it mostly contains lots of nitrogen (N) and phosphorus (P) in organic and inorganic forms (Hargreaves 1998). Hence, the demand for increasing production in aquaculture has driven the industry toward intensive operations, which has increased concern over the implications of aquaculture activities for the environment. Options for managing N biogeochemistry are limited, and goals are mostly determined by the intensity of fish production (Hargreaves 1998). The density of fish stocking in a containment area is a critical component affecting fish behavior, feed intake, immunity, growth, and water quality (Costa et al. 2017a, b). Poor water quality can compromise fish immunity and lead to the proliferation of harmful bacteria and fungi (Sidoruk and Cymes 2018). Therefore, the amount of organic waste released into the environment should be minimized.

Biofilters have become essential in aquaculture as they improve water quality while also improving the health and growth of cultivated organisms. These systems use microorganisms to eliminate harmful substances from the water, creating a healthy and favorable habitat for aquatic life. Nitrifying bacteria in the nitrogen cycle use oxygen and alkalinity to convert ammonia and nitrite into nitrate, which is less toxic to fish (Francis-Floyd 2020). The performance of biofilters is determined by a variety of elements, including the type and surface area of media used for bacterial enhancement, dissolved oxygen concentrations in the system, the amount of organic matter, temperature, pH, alkalinity, and salinity (Chen et al. 2018). Meanwhile, the nitrogen biogeochemistry of aquaculture ponds is influenced by feed and feeding strategies, water exchange and circulation, aeration, pond depth, and other management practices (Hargreaves 1998). Malone and Beecher (2000) proposed a volumetric oxygen conversion rate of 2.5–3.0 g O_2/m^3 media/day in recirculating aquaculture biofilters used in grow-out systems.

A biofilter with a higher surface area per unit volume is more efficient and cost-effective than a biofilter with a lower surface area. Nitrifying bacteria are very sensitive to their surroundings, so it is important to use non-corrosive materials like fiberglass, plastic, rock, or ceramic for biological filters (DeLong and Losordo 2012). These materials have large surfaces where the bacteria can grow. To reduce the use of plastic in filtering systems, biofilters made from bamboo, charcoal, coconut shells, and horns demonstrated the highest nitrification capacity (Mnyoro et al. 2022). In addition to that, they recommend using coconut shells instead of plastic components in simple recirculating aquaculture system (RAS) biofiltration, as it was found to be excellent in removing ammonia and nitrite.

8.5 Advanced Technologies and Innovations

Global aquaculture is becoming increasingly important in the quest against future food supply challenges. Thus, to ensure the successful scaling up of the aquaculture industry, biosecurity measures are a key aspect of aquaculture management. The labor-intensive nature of conventional fish farming systems pushes higher production costs since personnel are needed to oversee the farms, where sensitive and reliable determination of various parameters in aquaculture systems has largely relied on manual in situ and laboratory-based techniques. Therefore, innovations and technological advances are essential in strengthening aquaculture's biosecurity measures, preventing disease outbreaks, increasing productivity, and safeguarding the industry's sustainability. Significant advancements in fish genetics, culture systems, nutrition and feeding, and management have cumulatively improved fish performance and boosted global productivity. New technologies and innovative concepts are transforming aquaculture biosecurity, providing better ways to protect aquatic organisms while promoting sustainable production.

8.5.1 Vaccine and Vaccination

Vaccination is the most efficient method to prevent and control diseases caused by viruses and bacteria. It plays a crucial role in enhancing immunity in fish, particularly in aquaculture. The prevention of diseases through vaccination is a key strategy in fish health management aimed at reducing fish mortality rates in aquaculture farms (Mondal and Thomas 2022). Aquaculture faces a significant challenge in disease management due to the limited availability of effective vaccines. Vaccines are essential for preventing disease in animals with developed immune systems, while immunostimulants are used for those with less developed immune systems (Mohd-Aris et al. 2019). The promotion of aquatic vaccinations has the potential to minimize drug consumption, disease transmission, and environmental impact and improve the quality and safety of aquatic products. Hence, vaccination stands as a key practice among effective prevention and control strategies.

Aquatic vaccines can be classified according to the antipathogen group (bacteria, viruses, or parasites), composition (monovalent, multivalent, or mixed), or method of preparation (live, inactivated, or genetically engineered), and each type of this vaccine has advantages and disadvantages. The number and variety of aquatic vaccines (Table 8.5) that currently have production approvals worldwide are limited, and they are unable to completely cover the major diseases of the main breeding species (Du et al. 2022). Vaccination against diseases is commonly practiced in several economically significant fish species, including Atlantic salmon, rainbow trout, Nile tilapia, amberjack, and striped catfish (Nayak 2020).

Developing vaccines for aquatic organisms is complex due to their intricate immune systems. However, recent advancements, such as the successful development of a DNA vaccine providing protection against viral pathogens in farmed fish, offer hope (Sepúlveda et al. 2019). Various types of vaccines, including killed, attenuated, DNA, recombinant technology, and synthetic peptide vaccines, are being explored for use in fish (Assefa and Abunna 2018). The development of multivalent vaccines has also significantly advanced the aquatic animal vaccination program (Mohd-Aris et al. 2019).

Table 8.5 Various types of aquatic vaccines

Disease	Pathogen	Vaccine type	References
Vibriosis	*Vibrio anguillarum*	Genetic engineering attenuation	Liu et al. (2018)
Ulcer syndrome	*Aeromonas veronii*	Genetic engineering attenuation	Zhang et al. (2020a, b)
Carp disease	Carp nephritis and gill necrosis virus (CNGN)	Live attenuated	Eide et al. (2011), Perelberg et al. (2005), Ronen et al. (2003)
Hemorrhagic septicemia	Viral hemorrhagic septicemia virus (VHSV)	Genetic engineering attenuation	Kim et al. (2020), Kim and Kim (2019), Kim et al. (2016)
Enteric septicemia	*Edwardsiella ictaluri*	Genetic engineering attenuation	Abdelhamed et al. (2013)
Streptococcosis	*Streptococcus agalactiae*	Attenuated culture	Liu et al. (2019)
White spot disease	White spot syndrome virus (WSSV	Inactivated antigen, double-stranded DNA	Amar et al. (2021)
Vibriosis	*Vibrio harveyi, Vibrio mimicus, Vibrio alginolyticus*	Inactivated antigen	Nguyen et al. (2017), Wei et al. (2020), Cao et al. (2018a, b)
Furunculosis	*Aeromonas salmonicida*	Inactivated antigen	Liu et al. (2020)
Edwardsiellosis	*Edwardsiella tarda*	Inactivated antigen	Tang et al. (2019)
Infectious hematopoietic necrosis	Infectious hematopoietic necrosis virus (IHNV)	Inactivated antigen, a single strand of RNS	Kai and Chi (2008)
Viral hemorrhagic septicemia	Viral hemorrhagic septicemia virus (VHSV)	Inactivated antigen, ssRNA	Kim et al. (2016), Zhang et al. (2020a, b)
Viral nervous necrosis disease, grouper iridoviral disease	Viral nervous necrosis virus (VNNV); grouper iridovirus (GIV)	Inactivated antigen, single strand of RNA, DNA	Zeng et al. (2016)
Tilapia virus disease	Tilapia lake virus (TiLV)	Inactivated antigen, a single strand of RNA	Mai et al. (2021), Aich et al. (2022)

Live-attenuated oral vaccines have been developed for various fish diseases, contributing to the sustainability of aquaculture (Ma et al. 2019). Additionally, the use of genetically modified microalgae as a vaccine carrier has been explored, highlighting the potential for innovative vaccine delivery systems in aquaculture (Ma et al. 2020). Furthermore, the development of feed-based whole-cell polyvalent vaccines has demonstrated efficacy against multiple bacterial diseases in fish, emphasizing the continuous advancements in fish vaccine development (Mohamad et al. 2021) while suggesting that a combination of vaccination and immunostimulants such as probiotics (Mohd-Aris et al. 2019; Du et al. 2022) and herbal additives (Du et al. 2022) improves immune effect as part of important strategies for fish health management in aquaculture farms.

Viral infections pose a significant challenge in aquaculture, leading to substantial economic losses. The emergence of technologies such as RNA interference (RNAi) and the development of virus-like particles (VLPs) offer promising solutions, where the creation of VLPs holds great promise for the development of next-generation vaccines in aquaculture (Marsian et al. 2019). RNAi has shown efficacy in laboratory settings (Abo El-Ela et al. 2021), and the potential field application of this technology could help mitigate financial losses for farmers.

8.5.2 Diagnostic Tools

Disease diagnostics refers to the identification of the pathogen that causes the disease. Diagnostics are vital biosecurity tools in aquaculture, providing crucial information for disease prevention, surveillance, management, and research. Reliable diagnostic tools are necessary for a cost-effective health management and biosecurity approach to make timely and appropriate decisions in disease control practices (Fig. 8.8). Therefore, identifying pathogens is vital for rapid disease containment strategies. Conventional diagnostic procedures such as visual diagnosis, cell

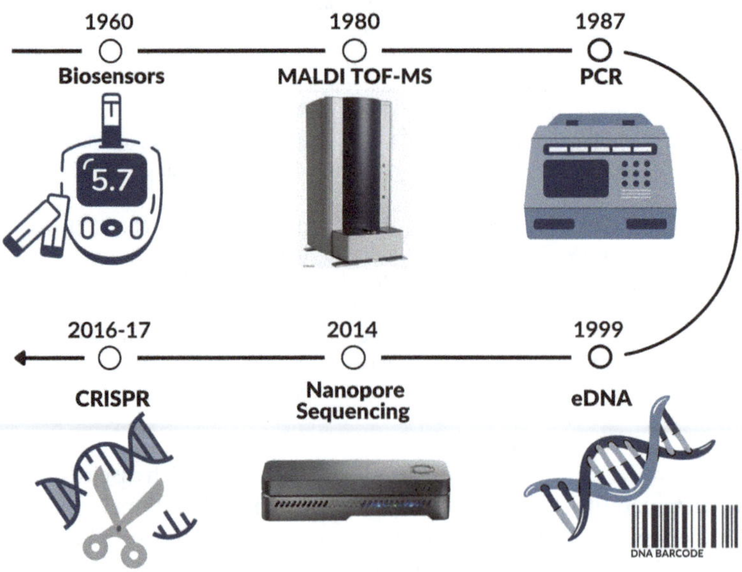

Fig. 8.8 A chronological overview of disease diagnostic tool development (Bohara et al. 2023)

culture, media culture, histopathology, and serology (Bohara et al. 2023) can take days for bacterial diseases and months for viral infections, resulting in significant losses. However, current technological advancements in aquaculture have led to more efficient and sustainable production (Rowan 2023) and have enabled rapid disease identification within minutes (Bohara et al. 2023). Aquaculture producers may improve biosecurity, safeguard their operations from disease risks, and contribute to the industry's sustainable development by investing in modern diagnostic tools and conducting rigorous testing procedures.

Modern molecular techniques have significantly improved the detection of diseases in aquatic animals, leading to early identification and control of diseases (Li et al. 2022). These techniques have been instrumental in enhancing the sensitivity, speed, and cost-effectiveness of diagnostic methods, ultimately improving the probability of successful disease control (Huang et al. 2022). Modern molecular techniques, such as clustered regularly interspaced short palindromic repeats (CRISPR)-based systems, offer rapid, sensitive, and cost-effective diagnostic capabilities (Li et al. 2022). The use of CRISPR-based diagnostics has shown great potential in providing convenient, efficient, low-cost, and practical tools for disease detection caused by pathogenic microorganisms (Fang et al. 2023). In aquaculture, CRISPR-Cas systems like cas12a (Cpf1) and Cas13 are extensively used for diagnostic purposes due to their unique characteristics (Chaijarasphong et al. 2019; Li et al. 2022).

The use of environmental DNA (eDNA) has emerged as a potential tool for monitoring aquatic animal diseases in aquaculture, offering a promising approach for early disease detection and surveillance (Kawato et al. 2021). Additionally, field-level diagnostic tools empower farmers to swiftly identify diseases and determine the most effective course of action for treatment or crop harvest (Hansen and Abd El Wahed 2020).

Next-generation sequencing (NGS) technology is a valuable diagnostic tool used to sequence infections and understand their evolution (Steinberg et al. 2008; Swain et al. 2022). Nanopore sequencing is a promising method for aquaculture that sequences the genomes of numerous fish species (Ge et al. 2019; Bi et al. 2023). It can readily detect pathogens without the need for samples to be taken to a laboratory. Hence, this allows for the early diagnosis of fish infections in remote regions, facilitating outbreaks to be contained quickly, and preventing future spread and economic losses (Bohara et al. 2023).

Matrix-assisted laser desorption ionization-time of flight mass spectrometry (MALDI-TOF MS) is a quick approach for diagnosing microbial species using mass spectrometry, making it a promising candidate for application in aquaculture diagnostics (Bohara et al. 2023). MALDI-TOF-MS is a faster, more precise, and less expensive identification method than traditional phenotypic techniques or molecular biology. A previous study by Moussa et al. (2021) created a database using several spectrum projections in MALDI-TOF MS to rapidly identify 25 different *Vibrio* species in mollusks.

8.5.3 Alternative Therapeutics

The massive use of antibiotics in aquaculture leads to a wide range of antibiotic resistance phenomena. Bacteria can acquire antibiotic resistance either through mutation or horizontal gene transfer in the environment via transformation, transduction, or conjugation. To combat microbial resistance, it is critical to set standards for the sensible use of antibiotics as well as the development of new medications and alternative therapies to treat bacterial infections (Castro et al. 2008). Plants are one of the richest sources of new medications due to their ability to produce a wide range of substances. Plants have long been thought to have therapeutic properties (Cowan 1999). According to Madhuri et al. (2012), plants have been used as traditional ways to treat several infections caused by bacteria, fungi, and viruses.

In aquaculture, medicinal plants and plant extracts have already been used long ago by rural fish farmers (Austin and Newaj-Fyzul 2017), either they process the plant or directly introduce the fresh plant into their rearing water, which has many beneficial effects on the rearing water and fish health, especially. The use of plant-based medicine in aquaculture is safe for the environment, farmers, and consumers. The previous study discovered a different type of plant as a remedy for various diseases in fish. Examples are garlic (*Allium sativum*), green tea (*Camellia sinensis*), ginger (*Zingiber officinale*), peppermint (*Mentha piperita*), black seed (*Nigella sativa*), and onion (*Allium cepa*) (Awad and Awaad 2017). Recent breakthroughs in nanotechnology open the door to using diverse medicinal plant-based nanoformulations to deal with aquatic microbial diseases (Jeyavani et al. 2022).

Utilizing marine organism products to prevent and treat infectious diseases appears to be a viable alternative. Marine species include a large diversity of pharmacologically active components and potential sources of novel therapeutic medications (Pérez et al. 2016). Blue, red, and green algae have been proven to have antioxidant, antiviral, antifungal, and microbial components (Barot et al. 2016). Brown and green seaweed extracts promote immunological protection and have a positive antibacterial effect on bacterial infections in fish (Thanigaivel et al. 2023). Terpenes, alkaloids, lectins, and halogenated chemicals are among the secondary metabolites that seaweed can create (Pérez et al. 2016). They also have antifungal, antiviral, antibacterial, antimacrofouling, and antiprotozoal properties. It is a cost-effective and environmentally friendly disease management technique.

Bacteriophages are viruses that infect bacteria. Bacteriophage therapy has been applied in a variety of biotechnology and medical sciences applications, including bacterial disease prevention, therapy, rapid disease diagnosis, and biological control (Haq et al. 2012). In the aquaculture industry, phage therapy offers significant promise for controlling harmful microorganisms (Liu et al. 2022). The first stage in aquatic phage therapy is to identify the fish disease agent, followed by discovering and isolating the phage capable of infecting the host bacteria. Phages that infect *Vibrio* pathogens, including *V. parahaemolyticus*, *V. harveyi*, *V. anguillarum*, and *V. splendidus*, have been isolated in aquaculture (Chen et al. 2019; Li et al. 2016), and the isolated phage vB_AsM_ZHF can effectively treat *A. salmonicida* subsp.

masoucida infections in turbot (Xu et al. 2021). Therefore, phage could be effective for long-term safety and harm the development of multidrug-resistant bacterial infections (Ninawe et al. 2020).

Quorum sensing (QS) is a type of cell-to-cell communication in bacteria that can also regulate the virulence potential of pathogenic strains. Bacteria create a variety of signal molecules, the most well studied of which are acylated homoserine lactones (AHLs) (Fuqua et al. 2001; Jayaraman and Wood 2008). Therefore, interfering with QS is being investigated as a novel biocontrol technique for combating bacterial infections. QS systems are used by a wide range of aquatic diseases, including *Edwardsiella*, *Vibrio*, and *Aeromonas* spp., to regulate gene expression, including virulent genes (Gupta and Kumar 2022). The capacity to interfere with QS offers a new biocontrol approach, as there is a relationship between QS and the pathogenesis of several key bacterial diseases (Donabedian 2003; Finch et al. 1998). Meanwhile, quorum sensing inhibitors (QSIs) not only prevent the expression of virulence-associated genes but also reduce the pathogenicity of aquaculture diseases (Zhao et al. 2015). As an example, marine macroalgae were also discovered to be able to disrupt QS (Jha et al. 2013). Hence, the use of QSIs and para probiotics is a potential approach to controlling pathogenic bacteria in aquaculture systems (Zhao et al. 2015). QSIs can be used as an innovative and practical method to address antibiotic resistance in aquaculture.

Fig. 8.9 Probiotics are important in aquaculture as they have many benefits by providing natural and sustainable solutions for enhancing the health, growth, and resilience of cultured aquatic organisms

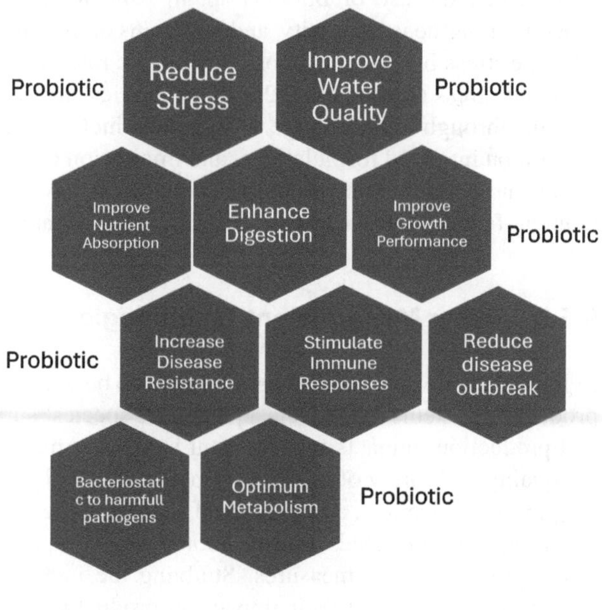

8.5.4 Probiotics

Probiotics, prebiotics, and synbiotics are a few examples of the beneficial feed additives that are being used in aquaculture these days to replace antibiotics and enhance immune responses, growth performance, and disease resistance (Fig. 8.9). The use of antibiotics increased the number of antimicrobial-resistant bacteria as well as antibiotic residues in the environment, which became one of the most serious issues, leading to the search for other alternative treatments (Pal 2015). Probiotics were first characterized by Parker (1974) as chemicals and organisms that influence the microbiota in the intestine. Alternatives to antibiotic use in aquaculture emphasize the necessity of antimicrobial agents for the treatment and prevention of infectious diseases in aquatic animals (Bondad-Reantaso et al. 2023). Research has shown that probiotics, such as lactic acid bacteria like *Lactobacillus* sp. and *Bacillus* sp., can modify gut flora, release antibiotic substances, and prevent pathogen adhesions (Aboleila et al. 2022). The use of probiotics in aquaculture health management is being investigated to prevent disease outbreaks and enhance the immune systems of aquaculture species (Abarike et al. 2018; Wei et al. 2022). The advantageous benefits of probiotics contribute to increased feed value and animal growth, as well as improved spawning and hatching rates in aquaculture systems (Hasan and Banerjee 2020).

Probiotic bacteria generally convert detrimental nitrogenous waste into food because they are unable to produce their own food in their natural environment (Porubcan 1991). Organic matter was transformed into CO_2 by gram-positive bacteria of this probiotic (Verschuere et al. 2000). Hence, probiotic action could regulate water quality parameters effectively. A previous study by Dalmin et al. (2001) showed that the use of *Bacillus* sp. in *Penaeus monodon* cultivation speeds up growth, strengthens immunity, and maintains optimum water quality. Probiotics can alleviate stress by slowing a host's metabolic rate (Varela et al. 2010) and lowering oxidative stress (Castex et al. 2009). Probiotics can be introduced into aquaculture systems through a wide range of methods, including direct addition to the water, integration into feed formulations, and immersion or oral administration to cultured organisms. Commercial probiotic products are available in liquid, powder, and granular forms, with every form suitable for application and aquatic organisms.

8.5.5 Water Management Technologies

In aquaculture, water quality parameters can be defined as any element of water in production systems that affects aquaculture species' survival, reproduction, growth, and production, impacts management strategies, hurts the environment, or lowers the quality and safety of products (Boyd et al. 1998). This "water quality" refers to the physical, chemical, and biological aspects of the water. Maintaining water quality in aquaculture needs routine monitoring, proper management practices, and effective preventative measures. Studying the interrelationship among key water quality parameters and their impact on productivity is crucial for effective water

Table 8.6 Key elements of innovative technologies and features in water quality management

Innovative technologies	Features
Real-time monitoring systems	• Provide real-time data that allows for immediate intervention in cases of fluctuations or abnormalities • Advanced sensors and monitoring devices to continuously measure key water quality parameters: – Temperature – pH – Dissolved oxygen – Ammonia – Nitrite – Nitrate
Automated control systems	• Manage water quality parameters based on real-time monitoring data • Regulate oxygenation, aeration, pH levels, temperature, and nutrient levels to maintain optimal conditions for aquatic organisms
Recirculating aquaculture systems (RASs)	• RAS technology recycles and treats water within the aquaculture system, minimizing the need for large water exchange volumes • Advanced RASs integrate the use of ozone and UV sterilization with mechanical, biological, and chemical filtering techniques to eliminate pollutants and pathogens from the water
Biofiltration	• Utilize beneficial microorganisms to convert harmful nitrite and ammonia into safer compounds
Nutrient management	• Nutrient management techniques to prevent the accumulation of nutrients and organic matter in the water, which can cause eutrophication and deterioration of the water's quality • Examples: – Biofloc technology – Algae-based water treatment – Integrated multi-trophic aquaculture (IMTA)
Remote monitoring and control	• Technology for remote monitoring and control that enables aquaculture operators to supervise water quality parameters from a distance • Facilitates proactive management and timely responses to fluctuations in water quality, minimizing the risk of adverse effects on aquatic organisms
Environmental modeling and predictive analytics	• Predictive analytics and environmental modeling to predict trends in water quality and identify potential challenges before they occur • Predict future water quality conditions and optimize management strategies

quality management (Onada et al. 2015). Therefore, fish farmers need to consider water management technologies to optimize fish health, improve production, and minimize environmental impact. Table 8.6 highlights a few important features of advanced water quality management, which aims to preserve optimum water conditions for aquatic organisms.

Fish farms have traditionally used handheld sensors to evaluate the water quality on regular schedules on-site. The development of low-cost small sensors and

Table 8.7 Few examples of research on innovative technologies for water quality management

Innovation technologies	References
Narrowband internet of things (NB-IoT) technology on water quality monitoring system for aquaculture	Huan et al. (2020)
Autonomous buoy system for coastal aquaculture and water quality monitoring	Schmidt et al. (2018)
Wireless sensor network (WSN) for water quality monitoring in fish farms	Cario et al. (2017)
ZigBee WSN for monitoring an RAS	Espinosa-Faller and Rendón-Rodríguez (2012)
WSN automatic monitoring system for fishery environment	Chen et al. (2016)
Wireless sensor networks (WSNs)	Nam et al. (2014)
Real-time remote monitoring system for aquaculture water quality	Luo et al. (2015)
Ocean remote monitoring system	Wang et al. (2012)

advances in various information and communication technologies (ICT) have made it more feasible to monitor multiple parameters simultaneously through wireless sensor networks (WSNs). Technologies make it possible to collect data in real time, enabling easy remote monitoring, adjustment, and assessment of system parameters and water quality in aquaculture farms. This real-time remote monitoring solution increases the amount and quality of data on both temporal and spatial scales by simplifying the information and reducing human error and time delays (Glasgow et al. 2004). Artificial intelligence–based aquaculture (AIOT) could help fish farmers remotely control and manage various fishpond equipment intelligently (Chiu et al. 2022). It can also help aquaculture operators perform professional aquaculture, minimize the barrier to entry, and promote aquaculture. A smart aquaculture monitoring and control system based on IoT devices can decrease feed residues, monitor fish growth, and boost fish survival rates, resulting in a higher feed conversion rate (Chiu et al. 2022). Research on innovation technologies for aquaculture has seen significant growth in recent years. Table 8.7 shows a few research trends and advancements in aquaculture.

Challenges to expanding to new areas, limited water availability, and concerns over pollution are considered key obstacles to the further expansion of conventional aquaculture systems. Recirculating aquaculture system (RAS) technology has been designed to ensure the lowest possible environmental cost while optimizing rearing conditions and controlling water quality parameters (Badiola et al. 2012). RAS can offer superior environmental conditions year-round for species that are sensitive to water quality, especially in subtropical settings (Zhang et al. 2011). RASs have been incorporated into an extensive range of species, ranging from hatchery/fingerling to grow-out production and from freshwater to seafood products (Schulz et al. 2003; Lymbery et al. 2006; Zachritz II et al. 2008). Despite its many benefits, this RAS technology is not yet widely used by farmers. This is mostly due to a lack of experience, a higher initial cost, and the high ongoing costs associated with the biofilter.

Nevertheless, the advantages of RAS will vary depending on the kind and location of their operation (Badiola et al. 2012).

8.6 Conclusion

Aquaculture is a rapidly expanding sector within global agriculture. The expansion, diversity, and trade of aquaculture pose additional obstacles to the industry's sustainable growth, with biosecurity concerns being the primary concern. Among the challenges that limit aquaculture production, infectious diseases take the largest portion, resulting in annual losses amounting to billions of dollars. Disease outbreaks emerged from the complexities of the farming environment, with most infections being opportunistic, pathogenic, and multi-pathogenic. Hence, maintaining effective biosecurity in aquaculture is becoming increasingly important. The increasing number of infectious and contagious diseases affecting worldwide aquaculture growth emphasizes the significance of biosecurity programs that effectively prevent, control, and eradicate infectious diseases. In addition to lowering the risk of diseases and pests for one premise, good operating procedures and effective biosecurity also lower the risk of diseases and pests for nearby sites and the industry. Biosecurity strategies in aquaculture can protect a facility from disease-causing pathogens that are not present in that system. Instead of treating each disease case reactively, it is advisable to adopt a preventive approach to mitigate the risk of disease outbreaks before they occur. As such, healthy seeds; appropriate stocking density; effective post-stocking management; good aquaculture practices, such as controlling stress and disease; and the implementation of preventive measures are necessary for producing healthy fish and sustainable aquaculture.

Acknowledgement The funding has been provided by the Ministry of Higher Education Malaysia through the Higher Institutions Centre of Excellence (HiCoE).

References

Abarike ED, Cai J, Lu Y, Yu H, Chen L, Jian J et al (2018) Effects of a commercial probiotic BS containing Bacillus subtilis and Bacillus licheniformis on growth, immune response and disease resistance in Nile tilapia, Oreochromis niloticus. Fish Shellfish Immunol 82:229–238

Abate TG, Nielsen R, Tveterås R (2016) Stringency of environmental regulation and aquaculture growth: a cross-country analysis. Aquac Econ Manag 20(2):201–221

Abdelhamed H, Lu J, Shaheen A, Abbass A, Lawrence ML, Karsi A (2013) Construction and evaluation of an Edwardsiella ictaluri fhuC mutant. Vet Microbiol 162(2–4):858–865

Abo El-Ela AA, Allam SM, Shaheen AM, Nagem NA (2021) Optimal allocation of biomass distributed generation in distribution systems using equilibrium algorithm. Int Trans Electr Energy Syst 31(2):e12727

Aboleila SM, Salah A, Mohamed RA, Diab AM (2022) Combined effect of a mixture of Bacillus species and vaccination on haemato-biochemical parameters, immune response and transcriptomic changes of Nile tilapia challenged with Streptococcus iniae. Aquac Res 53(14):5029–5044

Aich N, Paul A, Choudhury TG, Saha H (2022) Tilapia lake virus (TiLV) disease: current status of understanding. Aquac Fish 7(1):7–17

Ali H, Rahman MM, Rico A, Jaman A, Basak SK, Islam MM et al (2018) An assessment of health management practices and occupational health hazards in tiger shrimp (Penaeus monodon) and freshwater prawn (Macrobrachium rosenbergii) aquaculture in Bangladesh. Vet Anim Sci 5:10–19

Aloto D, Eticha E (2018) Leeches: a review on their pathogenic and beneficial effects. J Vet Sci Technol 9(01):511

Altizer S, Bartel R, Han BA (2011) Animal migration and infectious disease risk. Science 331(6015):296–302

Amar EC, Faisan JP Jr, Gapasin RS (2021) Field efficacy evaluation of a formalin-inactivated white spot syndrome virus (WSSV) vaccine for the preventive management of WSSV infection in shrimp grow-out ponds. Aquaculture 531:735907

Amorim P, Sousa P, Jardim E, Menezes GM (2019) Sustainability status of data-limited fisheries: global challenges for snapper and grouper. Front Mar Sci 6:654

Ananda Raja R, Jithendran K (2015) Aquaculture disease diagnosis and health management. In: Advances in marine and brackishwater aquaculture, pp 247–255

Argue BJ, Arce SM, Lotz JM, Moss SM (2002) Selective breeding of Pacific white shrimp (Litopenaeus vannamei) for growth and resistance to Taura syndrome virus. Aquaculture 204(3–4):447–460

Assefa A, Abunna F (2018) Maintenance of fish health in aquaculture: review of epidemiological approaches for prevention and control of infectious disease of fish. Vet Med Int 2018:5432497

Austin B (2023) The impact of disease on the sustainability of aquaculture. Sustain Aquat Res 2(1):74–91

Austin B, Newaj-Fyzul A (2017) Diagnosis and control of diseases of fish and shellfish. Wiley Online Library, Oxford

Awad E, Awaad A (2017) Role of medicinal plants on growth performance and immune status in fish. Fish Shellfish Immunol 67:40–54

Badiola M, Mendiola D, Bostock J (2012) Recirculating aquaculture systems (RAS) analysis: main issues on management and future challenges. Aquac Eng 51:26–35

Balcázar JL, De Blas I, Ruiz-Zarzuela I, Cunningham D, Vendrell D, Múzquiz JL (2006) The role of probiotics in aquaculture. Vet Microbiol 114(3–4):173–186

Barot M, Kumar N, Kumar RN (2016) Bioactive compounds and antifungal activity of three different seaweed species Ulva lactuca, Sargassum tenerrimum and Laurencia obtusa collected from Okha coast, Western India. J Coast Life Med 4(4):284–289

Bartl E, Keckeis H (2004) Growth and mortality of introduced fish larvae in a newly restored urban river. J Fish Biol 64(6):1577–1592

Bevan DJ, Chandroo KP, Moccia RD (2002) Predator control in commercial Aquaculture in Canada. Aec Order(02-001), pp 1–4

Beveridge MC, Brummett RE (2015) Aquaculture and the environment. In: Freshwater fisheries ecology, pp 794–803

Bi B, Yuan Y, Jia D, Jiang W, Yan H, Yuan G, Gao Y (2023) Identification and pathogenicity of emerging fish pathogen Acinetobacter johnsonii from a disease outbreak in Rainbow Trout (Oncorhynchus mykiss). Aquac Res 2023:1

Bohara K, Joshi P, Acharya KP, Ramena G (2023) Emerging technologies revolutionising disease diagnosis and monitoring in aquatic animal health. Rev Aquac 16:836

Bondad-Reantaso MG, McGladdery SE, East I, Subasinghe RP (eds) (2001) Asia Diagnostic Guide to Aquatic Animal Diseases. FAO Fisheries Technical Publication Paper No, 402, Supplement, vol 2. FAO, Rome, 240 pp

Bondad-Reantaso MG, MacKinnon B, Karunasagar I, Fridman S, Alday-Sanz V, Brun E et al (2023) Review of alternatives to antibiotic use in aquaculture. Rev Aquac 15:1421

Bornø G, Colquhoun D (2009) Classical furunculosis (in Norwegian). Fact sheet. Norwegian Veterinary Institute

Bowden TJ (2008) Modulation of the immune system of fish by their environment. Fish Shellfish Immunol 25(4):373–383

Boyd CE, Tucker CS, Boyd CE, Tucker CS (1998) Water quality and aquaculture: preliminary considerations. In: Pond aquaculture water quality management, pp 1–7

Boyle M (2001) Chicken breeding and genetics. Controlled and biosecure production systems: evolution and integration of chicken and shrimp models. In: Proceedings of a special session on integration of shrimp and chicken models, Sydney, Australia, World Aquaculture Society, pp 23–28

Browdy C, Bratvold D (1998) Preliminary development of a biosecure shrimp production system. Paper presented at the Proceedings of the US Marine Shrimp Farming Program Biosecurity Workshop

Brugere C, Onuigbo DM, Morgan KL (2017) People matter in animal disease surveillance: challenges and opportunities for the aquaculture sector. Aquaculture 467:158–169

Bureau of Fisheries (2018) Ministry of Agriculture and Rural Affairs of the People's Republic of China (BoF ARAAH). Retrieved from Beijing: China

Cao J, Huang A-L, Zhu X-C, Li L, Li J-N (2018a) Construction of Vibrio mimicus ghosts as a novel inactivated vaccine candidate and its protective efficacy against ascites disease in grass carps (Ctenopharyngodon idella). Aquaculture 485:147–153

Cao J, Zhang J, Ma L, Li L, Zhang W, Li J (2018b) Identification of fish source Vibrio alginolyticus and evaluation of its bacterial ghosts vaccine immune effects. Microbiology 7(3):e00576

Carbonara P, Alfonso S, Gai F, Gasco L, Palmegiano G, Spedicato MT et al (2020) Moderate stocking density does not influence the behavioural and physiological responses of rainbow trout (Oncorhynchus mykiss) in organic aquaculture. Aquac Res 51(7):3007–3016

Cario G, Casavola A, Gjanci P, Lupia M, Petrioli C, Spaccini D (2017) Long lasting underwater wireless sensors network for water quality monitoring in fish farms. Paper presented at the Oceans 2017-Aberdeen

Castex M, Lemaire P, Wabete N, Chim L (2009) Effect of dietary probiotic Pediococcus acidilactici on antioxidant defences and oxidative stress status of shrimp Litopenaeus stylirostris. Aquaculture 294(3–4):306–313

Castro S, Leal C, Freire F, Carvalho D, Oliveira D, Figueiredo H (2008) Antibacterial activity of plant extracts from Brazil against fish pathogenic bacteria. Braz J Microbiol 39:756–760

Chaijarasphong T, Thammachai T, Itsathitphaisarn O, Sritunyalucksana K, Suebsing R (2019) Potential application of CRISPR-Cas12a fluorescence assay coupled with rapid nucleic acid amplification for detection of white spot syndrome virus in shrimp. Aquaculture 512:734340

Chambers CB, Ernst I (2005) Dispersal of the skin fluke Benedenia seriolae (Monogenea: Capsalidae) by tidal currents and implications for sea-cage farming of Seriola spp. Aquaculture 250(1–2):60–69

Chen J, Yang N, Nishihara H (2016) Special issue in honor of Koichi Aoki Preface. J Electroanal Chem 779:1–6

Chen M, Chen Y, Dong S, Lan S, Zhou H, Tan Z, Li X (2018) Mixed nitrifying bacteria culture under different temperature dropping strategies: nitrification performance, activity, and community. Chemosphere 195:800–809

Chen L, Fan J, Yan T, Liu Q, Yuan S, Zhang H et al (2019) Isolation and characterization of specific phages to prepare a cocktail preventing Vibrio sp. Va-F3 infections in shrimp (Litopenaeus vannamei). Front Microbiol 10:2337

Chiew IKM, Salter AM, Lim YS (2019) The significance of major viral and bacterial diseases in Malaysian aquaculture industry. Pertanika J Trop Agric Sci 42(3):1023–1047

Chiu M-C, Yan W-M, Bhat SA, Huang N-F (2022) Development of smart aquaculture farm management system using IoT and AI-based surrogate models. J Agric Food Res 9:100357

Corsin F, Turnbull J, Hao N, Mohan C, Phi T, Phuoc L et al (2001) Risk factors associated with white spot syndrome virus infection in a Vietnamese rice-shrimp farming system. Dis Aquat Org 47(1):1–12

Costa A, Roubach R, Dallago BSL, Bueno GW, McManus C, Bernal FEM (2017a) Influence of stocking density on growth performance and welfare of juvenile tilapia (Oreochromis niloticus) in cages. Arq Bras Med Vet Zoot 69:243–251

Costa CR, da Costa MF, Barletta M, Alves LHB (2017b) Interannual water quality changes at the head of a tropical estuary. Environ Monit Assess 189:1–13

Cowan MM (1999) Plant products as antimicrobial agents. Clin Microbiol Rev 12(4):564–582

Cruz-Lacierda ER, Toledo JD, Tan-Fermin JD, Burreson EM (2000) Marine leech (Zeylanicobdella arugamensis) infestation in cultured orange-spotted grouper, Epinephelus coioides. Aquaculture 185(3–4):191–196

Cuéllar-Anjel J, Chamoro R, White-Noble B, Schofield P, Lightner D (2011) Testing finds resistance to WSSV in shrimp from Panamanian breeding program. In: Global aquaculture advocate, pp 65–66

Dalmin, G, Kathiresan, K, Purushothaman (2001). A. Effect of probiotics on bacterial population and health status of shrimp in culture pond ecosystem. Indian J Exp Biol. Sep;39(9):939-942.. PMID: 11831382

Daszak P, Cunningham AA, Hyatt AD (2000) Emerging infectious diseases of wildlife—threats to biodiversity and human health. Science 287(5452):443–449

DeLong DP, Losordo T (2012) How to Start a Biofilter. Southern Regional Aquaculture Center SRAC Publication No 2012:4502

Donabedian H (2003) Quorum sensing and its relevance to infectious diseases. J Infect 46(4):207–214

Donovan DJ (2006) Industry environmental codes of best practice for freshwater finfish aquaculture. Prepared by Kuruma Australia Pty Ltd. for the Department of Primary Industries and Fisheries and the Queensland Finfish Aquaculture Industry, 31 pp

Du Y, Hu X, Miao L, Chen J (2022) Current status and development prospects of aquatic vaccines. Front Immunol 13:1040336

Eide KE, Miller-Morgan T, Heidel JR, Kent ML, Bildfell RJ, LaPatra S et al (2011) Investigation of koi herpesvirus latency in koi. J Virol 85(10):4954–4962

Espinosa-Faller FJ, Rendón-Rodríguez GE (2012) A ZigBee wireless sensor network for monitoring an aquaculture recirculating system. J Appl Res Technol 10(3):380–387

Fang L, Yang L, Han M, Xu H, Ding W, Dong X (2023) CRISPR-Cas technology: a key approach for SARS-CoV-2 detection. Front Bioeng Biotechnol 11:1158672

FAO (2007) Biosecurity toolkit. Food and Agriculture Organization of United Nations, Rome, Italy. 128 pp

FAO (2014) Policy and governance in aquaculture lessons learned and way forward. FAO Fisheries and Aquaculture Technical Paper 577. Food And Agriculture Organization of The United Nations Rome, 2014. https://www.fao.org/3/i3156e/i3156e.pdf

FAO (2016). The State of World Fisheries and Aquaculture 2016: Contributing to Food Security and Nutrition for All. Italy: Food and Agriculture Organization (FAO) of the United Nations Rome 200 pp

FAO (2018) The state of world fisheries and aquaculture: meeting the sustainable development goals. FAO, Rome

FAO (2020) The state of world fisheries and aquaculture 2020. Sustainability in action. FAO, Rome

FAO (2022) Towards blue transformation. FAO, Rome

Farrell P, Leonard B (2001) Observations on the survival of the yabby, Cherax destructor, in ponds where access by piscivorous birds is inhibited. J Appl Aquac 11(3):75–80

Faruk MAR, Shorna HK, Anka IZ (2021) Use and impact of veterinary drugs, antimicrobials, and supplements in fish health management. J Adv Vet Anim Res 8(1):36

Finch R, Pritchard D, Bycroft B, Williams P, Stewart G (1998) Quorum sensing: a novel target for anti-infective therapy. J Antimicrob Chemother 42(5):569–571

Fitridge I, Dempster T, Guenther J, De Nys R (2012) The impact and control of biofouling in marine aquaculture: a review. Biofouling 28(7):649–669

Francis-Floyd R (2020) Diagnostic methods for the comprehensive health assessment of the long-spined sea urchin, Diadema antillarum: VM244/VM244, 05/2020. EDIS, 2020(3)

Fuqua C, Parsek MR, Greenberg EP (2001) Regulation of gene expression by cell-to-cell communication: acyl-homoserine lactone quorum sensing. Annu Rev Genet 35(1):439–468

Ge H, Lin K, Shen M, Wu S, Wang Y, Zhang Z et al (2019) De novo assembly of a chromosome-level reference genome of red-spotted grouper (Epinephelus akaara) using nanopore sequencing and Hi-C. Mol Ecol Resour 19(6):1461–1469

Gephart JA, Henriksson PJ, Parker RW, Shepon A, Gorospe KD, Bergman K et al (2021) Environmental performance of blue foods. Nature 597(7876):360–365

Gillings MR (2013) Evolutionary consequences of antibiotic use for the resistome, mobilome and microbial pangenome. Front Microbiol 4:4

Gitterle T, Salte R, Gjerde B, Cock J, Johansen H, Salazar M et al (2005) Genetic (co) variation in resistance to white spot syndrome virus (WSSV) and harvest weight in Penaeus (Litopenaeus) vannamei. Aquaculture 246(1–4):139–149

Gjedrem T, Salte R, Gjøen HM (1991) Genetic variation in susceptibility of Atlantic salmon to furunculosis. Aquaculture 97(1):1–6

Glasgow HB, Burkholder JM, Reed RE, Lewitus AJ, Kleinman JE (2004) Real-time remote monitoring of water quality: a review of current applications, and advancements in sensor, telemetry, and computing technologies. J Exp Mar Biol Ecol 300(1–2):409–448

Golden CD, Koehn JZ, Shepon A, Passarelli S, Free CM, Viana DF et al (2021) Aquatic foods to nourish nations. Nature 598(7880):315–320

Gozlan RE (2017) Interference of non-native species with fisheries and aquaculture. In: Impact of biological invasions on ecosystem services, pp 119–137

Gupta DS, Kumar MS (2022) The implications of quorum sensing inhibition in bacterial antibiotic resistance-with a special focus on aquaculture. J Microbiol Methods 203:106602

Hansen S, Abd El Wahed A (2020) Point-of-care or point-of-need diagnostic tests: time to change outbreak investigation and pathogen detection. Trop Med Infect Dis 5(4):151

Haq IU, Chaudhry WN, Akhtar MN, Andleeb S, Qadri I (2012) Bacteriophages and their implications on future biotechnology: a review. Virol J 9(1):1–8

Hargreaves JA (1998) Nitrogen biogeochemistry of aquaculture ponds. Aquaculture 166(3–4):181–212

Hasan KN, Banerjee G (2020) Recent studies on probiotics as beneficial mediator in aquaculture: a review. J Basic Appl Zool 81(1):1–16

Hasan M, Faruk M, Anka I, Azad M (2013) Investigation on fish health and diseases in rural pond aquaculture in three districts of Bangladesh. J Bangladesh Agric Univ 11(2):377–384

Hersoug B (2021) Why and how to regulate Norwegian salmon production?—the history of maximum allowable biomass (MAB). Aquaculture 545:737144

Hill B, Berthe F, Lightner D, Sais R (2013) Methods for disinfection of aquaculture establishments. In: Manual of diagnostic tests for aquatic animals, pp 28–39

Hishamunda, N., Ridler, N., & Martone, E. (2014). Policy and governance in aquaculture: Lessons learned and way forward (FAO Fisheries and Aquaculture Technical Paper No. 590). Food and Agriculture Organization of the United Nations. https://doi.org/10.4060/i3530e

Huan J, Li H, Wu F, Cao W (2020) Design of water quality monitoring system for aquaculture ponds based on NB-IoT. Aquac Eng 90(3):102088. https://doi.org/10.1016/j.aquaeng.2020.102088

Huang F, Shan J, Liang K, Yang M, Zhou X, Duan X, Jia X, Zhao H, Qin Q, Wang Q (2022) A new method to detect red spotted grouper neuro necrosis virus (RGNNV) based on CRISPR/Cas13a. Aquaculture 555:738217. https://doi.org/10.1016/j.aquaculture.2022.738217

Ijah A, Oladele O, Ishola B, Ayodele J, Yahaya U, Omodona S, Olukotun O (2020) Awareness and level of adoption of aquaculture management techniques in Igabi Local Government Area of Kaduna State, Nigeria. Asian J Fish Aquat Res 8(1):14–29

Inglis GJ, Morrisey D, Woods C, Sinner J, Newton M (2013) Managing the domestic spread of harmful marine organisms. Part A: operational tools for management. Ministry for Primary Industries, Wellington

Ivanitskaya O, Kalinina N, Kravets P, Shoshina E, Malaeva A (2022) Development of safety system in salmon fish farms of the Murmansk region. Paper presented at the BIO Web of Conferences

Iwama GK (2009) Interactions between aquaculture and the environment. Crit Rev Environ Control 21(3):177–216. https://doi.org/10.1080/10643389109388413

Jahangiri L, Shum S, Pfeiffer DU, Leung CF, Marques ARP, St-Hilaire S (2023) An economic model to assess biosecurity management strategies for marine fish farms in Hong Kong SAR. Aquaculture 567:739294

Jansen PA, Kristoffersen AB, Viljugrein H, Jimenez D, Aldrin M, Stien A (2012) Sea lice as a density-dependent constraint to salmonid farming. Proc R Soc B Biol Sci 279(1737):2330–2338

Jayaraman A, Wood TK (2008) Bacterial quorum sensing: signals, circuits, and implications for biofilms and disease. Annu Rev Biomed Eng 10:145–167

Jeyavani J, Sibiya A, Sivakamavalli J, Divya M, Preetham E, Vaseeharan B, Faggio C (2022) Phytotherapy and combined nanoformulations as a promising disease management in aquaculture: a review. Aquac Int 30(2):1071–1086

Jha B, Kavita K, Westphal J, Hartmann A, Schmitt-Kopplin P (2013) Quorum sensing inhibition by Asparagopsis taxiformis, a marine macro alga: separation of the compound that interrupts bacterial communication. Mar Drugs 11(1):253–265

Jiang G, Yu R, Zhou M (2004) Modulatory effects of ammonia-N on the immune system of Penaeus japonicus to virulence of white spot syndrome virus. Aquaculture 241(1–4):61–75

Johansen L-H, Jensen I, Mikkelsen H, Bjørn P-A, Jansen P, Bergh Ø (2011) Disease interaction and pathogens exchange between wild and farmed fish populations with special reference to Norway. Aquaculture 315(3–4):167–186

Jørgensen LG (2020) Zebrafish as a model for fish diseases in aquaculture. Pathogens 9(8):609

Kai Y-H, Chi S-C (2008) Efficacies of inactivated vaccines against betanodavirus in grouper larvae (Epinephelus coioides) by bath immunization. Vaccine 26(11):1450–1457

Kang B, Vitule JR, Li S, Shuai F, Huang L, Huang X et al (2023) Introduction of non-native fish for aquaculture in China: a systematic review. Rev Aquac 15(2):676–703

Kawato Y, Mekata T, Inada M, Ito T (2021) Application of environmental DNA for monitoring Red Sea bream Iridovirus at a fish farm. Microbiol Spectr 9(2):e00796-00721

Kennedy DA, Kurath G, Brito IL, Purcell MK, Read AF, Winton JR, Wargo AR (2016) Potential drivers of virulence evolution in aquaculture. Evol Appl 9(2):344–354

Kim MS, Kim KH (2019) Genetically engineered viral hemorrhagic septicemia virus (VHSV) vaccines. Fish Shellfish Immunol 95:11–15

Kim HJ, Park JS, Choi MC, Kwon SR (2016) Comparison of the efficacy of Poly (I: C) immunization with live vaccine and formalin-killed vaccine against viral hemorrhagic septicemia virus (VHSV) in olive flounder (Paralichthys olivaceus). Fish Shellfish Immunol 48:206–211

Kim H, Baek K-W, Ko E-J, Luan NT, Lim Y, Roh HJ et al (2020) Genome based quantification of VHSV in multiple organs of infected olive flounder (Paralichthys olivaceus) using real-time PCR. Genes Genomics 42:773–777

Kong J, Liu P, Shi T, Xiang J (2003) A baculovirus DNA fragment sequencing from Penaeus chinensis. Acta Oceanol Sin 25:186–189

Krkošek M (2017) Population biology of infectious diseases shared by wild and farmed fish. Can J Fish Aquat Sci 74(4):620–628

Kyule-Muendo D, Otachi E, Awour F et al (2022) Status of fish health management and biosecurity measures in fish farms, cages and hatcheries in Western Kenya. CABI Agric Biosci 3:18. https://doi.org/10.1186/s43170-022-00086-7

Lafferty KD, Harvell CD, Conrad JM, Friedman CS, Kent ML, Kuris AM et al (2015) Infectious diseases affect marine fisheries and aquaculture economics. Annu Rev Mar Sci 7:471–496

Lazado CC, Good C (2021) Survey findings of disinfection strategies at selected Norwegian and North American land-based RAS facilities: A comparative insight. Aquaculture 532:736038. https://doi.org/10.1016/j.aquaculture.2020.736038

Li Z, Li X, Zhang J, Wang X, Wang L, Cao Z, Xu Y (2016) Use of phages to control Vibrio splendidus infection in the juvenile sea cucumber Apostichopus japonicus. Fish Shellfish Immunol 54:302–311

Li D, Li X, Wang Q, Hao Y (2022) Advanced techniques for the intelligent diagnosis of fish diseases: a review. Animals 12(21):2938

Lin Y, Gao Z, Zhan A (2015) Introduction and use of non-native species for aquaculture in China: status, risks and management solutions. Rev Aquac 7(1):28–58

Liu C-H, Wu K, Chu T-W, Wu T-M (2018) Dietary supplementation of probiotic, Bacillus subtilis E20, enhances the growth performance and disease resistance against Vibrio alginolyticus in parrot fish (Oplegnathus fasciatus). Aquac Int 26:63–74

Liu L, Lu DQ, Xu J, Luo HL, Li AX (2019) Development of attenuated erythromycin-resistant Streptococcus agalactiae vaccine for tilapia (Oreochromis niloticus) culture. J Fish Dis 42(5):693–701

Liu S, Wang D, Cao Y, Lu T, Liu H, Li S (2020) Effects of propolis on the immune enhancement of the formalin-inactivated Aeromonas salmonicida vaccine. Aquac Res 51(11):4759–4770

Liu R, Han G, Li Z, Cun S, Hao B, Zhang J, Liu X (2022) Bacteriophage therapy in aquaculture: current status and future challenges. Folia Microbiol 67(4):573–590

Luo H, Li G, Weifeng P, Jie S, Qiuwei B (2015) Real-time remote monitoring system for aquaculture water quality. Int J Agric Biol Eng 8(6):136–143

Lymbery AJ, Doupé RG, Bennett T, Starcevich MR (2006) Efficacy of a subsurface-flow wetland using the estuarine sedge Juncus kraussii to treat effluent from inland saline aquaculture. Aquac Eng 34(1):1–7

Ma J, Bruce TJ, Jones EM, Cain KD (2019) A review of fish vaccine development strategies: conventional methods and modern biotechnological approaches. Microorganisms 7(11):569

Ma K, Bao Q, Wu Y, Chen S, Zhao S, Wu H, Fan J (2020) Evaluation of microalgae as immunostimulants and recombinant vaccines for diseases prevention and control in aquaculture. Front Bioeng Biotechnol 8:590431

Madhuri S, Mandloi A, Govind P, Sahni Y (2012) Antimicrobial activity of some medicinal plants against fish pathogens. Int Res J Pharm 3(4):28–30

Mai TT, Kayansamruaj P, Taengphu S, Senapin S, Costa JZ, del-Pozo J et al (2021) Efficacy of heat-killed and formalin-killed vaccines against Tilapia tilapinevirus in juvenile Nile tilapia (Oreochromis niloticus). J Fish Dis 44(12):2097–2109

Malone RF, Beecher LE (2000) Use of floating bead filters to recondition recirculating waters in warmwater aquaculture production systems. Aquac Eng 22(1–2):57–73

Mondal H, Thomas J (2022) A review on the recent advances and application of vaccines against fish pathogens in aquaculture. Aquac Int 30:1971–2000. https://doi.org/10.1007/s10499-022-00884-w

Manyi-Loh C, Mamphweli S, Meyer E, Okoh A (2018) Characterisation and Antibiotic Resistance of Selected Bacterial Pathogens Recovered from Dairy Cattle Manure during Anaerobic Mono-Digestion in a Balloon-Type Digester. Appl Sci 8:2088. https://doi.org/10.3390/app8112088

Marsian J, Hurdiss DL, Ranson NA, Ritala A, Paley R, Cano I, Lomonossoff GP (2019) Plant-made nervous necrosis virus-like particles protect fish against disease. Front Plant Sci 10:880

Menanteau-Ledouble S, Gonçalves RA, El-Matbouli M (2020) Feed supplementation with a commercially available probiotic solution does not alter the composition of the microbiome in the biofilters of recirculating aquaculture systems. Pathogens 9(10):830

Meyers T (2010). Regulation changes, policies and guidelines for Alaska fish and shellfish health and disease control. Alaska Department of Fish and Game,. Regional Information Report 5J10-01. Juneau, Alaska. 57 pp

Mohamad N, Amal MNA, Yasin ISM, Saad MZ, Nasruddin NS, Al-saari N et al (2019) Vibriosis in cultured marine fishes: a review. Aquaculture 512:734289

Mohamad A, Zamri-Saad M, Amal MNA, Al-Saari N, Monir MS, Chin YK, Md Yasin I-S (2021) Vaccine efficacy of a newly developed feed-based whole-cell polyvalent vaccine against vibriosis, streptococcosis and motile aeromonad septicemia in Asian seabass, Lates calcarifer. Vaccines 9(4):368

Mohanty, R. K., Kumar, A., Mishra, A., & Patil, D. (2014). Water budgeting and management: Enhancing aquacultural water productivity (Research Bulletin No. 63). Directorate of Water Management (DWM), Indian Council of Agricultural Research (ICAR). Bhubaneswar, India.

Mohd Yazid SH, Mohd Daud H, Azmai MNA, Mohamad N, Mohd Nor N (2021) Estimating the economic loss due to vibriosis in net-cage cultured Asian seabass (Lates calcarifer): evidence from the east coast of peninsular Malaysia. Front Vet Sci 8:644009

Mohd-Aris A, Muhamad-Sofie MHN, Zamri-Saad M, Daud HM, Ina-Salwany MY (2019) Live vaccines against bacterial fish diseases: a review. Vet World 12(11):1806

Mordecai GJ, Miller KM, Bass AL, Bateman AW, Teffer AK, Caleta JM et al (2021) Aquaculture mediates global transmission of a viral pathogen to wild salmon. Sci Adv 7(22):eabe2592

Moss SM (2002) Marine shrimp farming in the western hemisphere: past problems, present solutions, and future visions. Rev Fish Sci 10(3–4):601–620

Moss SM, Moss DR, Arce SM, Lightner DV, Lotz JM (2012) The role of selective breeding and biosecurity in the prevention of disease in penaeid shrimp aquaculture. J Invertebr Pathol 110(2):247–250

Moussa M, Cauvin E, Le Piouffle A, Lucas O, Bidault A, Paillard C et al (2021) A MALDI-TOF MS database for fast identification of Vibrio spp. potentially pathogenic to marine mollusks. Appl Microbiol Biotechnol 105:2527–2539

Mnyoro MS, Munubi RN, Pedersen L-F, Chenyambuga SW (2022) Evaluation of biofilter performance with alternative local biomedia in pilot scale recirculating aquaculture systems. J Clean Prod 366:132929. https://doi.org/10.1016/j.jclepro.2022.132929

Muniesa A, Furones D, Rodgers C, Basurco B (2022) An assessment of health management and biosecurity procedures in marine fish farming in Spain. Aquac Rep 25:101199

Munro A, Waddell I (1984) Furunculosis: experience of its control in the sea water cage culture of Atlantic salmon in Scotland. Int Counc Explor Sea Coop Res Rep 32:1–9

Murray AG, Peeler EJ (2005) A framework for understanding the potential for emerging diseases in aquaculture. Prev Vet Med 67(2–3):223–235

Murwantoko M, Negoro SLC, Isnansetyo A, Zafran Z (2018) Identification of marine leech and assessment of its prevalence and intensity on cultured hybrid groupers (Epinephelus sp.). Biodivers J Biol Divers 19(5):1798–1804

Nam H, An S, Kim C-H, Park S-H, Kim Y-W, Lim S-H (2014) Remote monitoring system based on ocean sensor networks for offshore aquaculture. Paper presented at the 2014 Oceans-St. John's

Nasr-Eldahan S, Nabil-Adam A, Shreadah MA, Maher AM, El-Sayed Ali T (2021) A review article on nanotechnology in aquaculture sustainability as a novel tool in fish disease control. Aquac Int 29:1459–1480

Nayak SK (2020) Current prospects and challenges in fish vaccine development in India with special reference to Aeromonas hydrophila vaccine. Fish Shellfish Immunol 100:283–299

Naylor R, Fang S, Fanzo J (2023) A global view of aquaculture policy. Food Policy 116:102422

Nguyen HT, Nguyen TTT, Tsai M-A, Ya-Zhen E, Wang P-C, Chen S-C (2017) A formalin-inactivated vaccine provides good protection against Vibrio harveyi infection in orange-spotted grouper (Epinephelus coioides). Fish Shellfish Immunol 65:118–126

Ninawe A, Sivasankari S, Ramasamy P, Kiran GS, Selvin J (2020) Bacteriophages for aquaculture disease control. Aquac Int 28:1925–1938

Nuraini D, Lutfiyah L (2020) Growth of cantang hybrid grouper juvenile (Epinephelus fuscoguttatus x Epinephelus lanceolatus) with different feeding frequency. Paper presented at the IOP Conference Series: Earth and Environmental Science

Nwadukwe FO, Arimoro FO (2012) The use of plastlon net fence as predator control in catfish nursery ponds. World J Fish Mar Sci 4(3):232–236

Oidtmann B, Johnston C, Klotins K, Mylrea G, Van P, Cabot S et al (2013) Assessment of the safety of aquatic animal commodities for international trade: the OIE aquatic animal health code. Transbound Emerg Dis 60(1):27–38

Onada OA, Akinwole AO, Ajani EK (2015) Study of interrelationship among water quality parameters in earthen pond and concrete tank. PeerJ 3:e845v841

Opiyo MA, Marijani E, Muendo P, Odede R, Leschen W, Charo-Karisa H (2018) A review of aquaculture production and health management practices of farmed fish in Kenya. Int J Vet Sci Med 6(2):141–148

Otieno NE, Shidavi E (2022) Effectiveness of physical barriers and enhanced fertilization in controlling predation on tilapia and catfish aquaculture systems by four piscivorous water bird families. Front Sustain Food Syst 6:1018064

Otta SK, Hoque F (2020) Health management in coastal aquaculture system. J Indian Soc Coastal Agric Res 38(2):38–42

Pal S (2015) Phage therapy an alternate disease control in aquaculture: a review on recent advancements. J Agric Vet Sci 8:68–81

Paladini G, Longshaw M, Gustinelli A, Shinn AP (2017) Parasitic diseases in aquaculture: their biology, diagnosis and control. In: Diagnosis and control of diseases of fish and shellfish, pp 37–107

Parker R (1974) Probiotics, the other half of the antibiotics story. Anim Nutr Health 29:4–8

Perelberg A, Ronen A, Hutoran M, Kotler M (2005) Protection of cultured Cyprinus carpio against lethal disease by attenuated virus. Vaccine 23:3396–3403

Perera, R.P., Jones, B., Beers, P., Kleeman, S., McGladdery, S. (2005) Maintaining biosecurity in aquaculture systems: a constraint or a challenge. In 6th Symposium on Diseases in Asian Aquaculture. VI p 3–20

Pérez MJ, Falqué E, Domínguez H (2016) Antimicrobial action of compounds from marine seaweed. Mar Drugs 14(3). https://doi.org/10.3390/md14030052

Phu TM, Phuong NT, Dung TT, Hai DM, Son VN, Rico A et al (2016) An evaluation of fish health-management practices and occupational health hazards associated with Pangasius catfish (Pangasianodon hypophthalmus) aquaculture in the Mekong Delta, Vietnam. Aquac Res 47(9):2778–2794

Porubcan R (1991) Reduction in chemical oxygen demand and improvement in Penaeus monodon yield in ponds inoculated with aerobic Bacillus bacteria. Paper presented at the program and abstract of the 22nd annual conference and exposition of the World Aquaculture Society, 1991

Pradeep PJ, Suebsing R, Sirithammajak S, Kampeera J, Turner W, Jeffs A et al (2017) Vertical transmission and concurrent infection of multiple bacterial pathogens in naturally infected red tilapia (Oreochromis spp.). Aquac Res 48(6):2706–2717

Preena PG, Swaminathan TR, Kumar VJR, Singh ISB (2020) Antimicrobial resistance in aquaculture: a crisis for concern. Biologia 75:1497–1517

Pruder G (1995) High health shrimp systems: seed supply-theory and practice. Paper presented at the swimming through troubled water, proceedings of the special session on shrimp farming, Aquaculture'95

Pruder GD (2004) Biosecurity: application in aquaculture. Aquac Eng 32(1):3–10

Raman RP, Prakash C, Marappan M, Pawar NA (2013) Environmental stress mediated diseases of fish: An overview. In: Advances in Fish Research, vol V. Narendra Publishing House, Delhi, India

Reverter M, Sarter S, Caruso D, Avarre J-C, Combe M, Pepey E et al (2020) Aquaculture at the crossroads of global warming and antimicrobial resistance. Nat Commun 11(1):1870

Rico A, Geng Y, Focks A, Van den Brink PJ (2013) Modeling environmental and human health risks of veterinary medicinal products applied in pond aquaculture. Environ Toxicol Chem 32(5):1196–1207

Ridzuan MSM, Abdullah A, Ramly R, Mansor NN, Ramli N, Firdaus-Nawi M (2022) Current status and advances of fish vaccines in Malaysia. Vet World 15(2):465

Robertsen B (2011) Can we get the upper hand on viral diseases in aquaculture of Atlantic salmon? Aquac Res 42:125–131

Rohani MF, Islam SM, Hossain MK, Ferdous Z, Siddik MA, Nuruzzaman M et al (2022) Probiotics, prebiotics and synbiotics improved the functionality of aquafeed: upgrading growth, reproduction, immunity and disease resistance in fish. Fish Shellfish Immunol 120:569–589

Ronen A, Perelberg A, Abramowitz J, Hutoran M, Tinman S, Bejerano I et al (2003) Efficient vaccine against the virus causing a lethal disease in cultured Cyprinus carpio. Vaccine 21(32):4677–4684

Rowan NJ (2023) The role of digital technologies in supporting and improving fishery and aquaculture across the supply chain–Quo Vadis? Aquac Fish 8(4):365–374

Rushton J, Bruce M, Bellet C, Torgerson P, Shaw A, Marsh T et al (2018) Initiation of global burden of animal diseases programme. Lancet 392(10147):538–540

Salin KR, Arome Ataguba G (2018) Aquaculture and the environment: towards sustainability. In: Sustainable aquaculture, pp 1–62

Sanawar H, Xiong Y, Alam A, Croué J-P, Hong P-Y (2017) Chlorination or monochloramination: balancing the regulated trihalomethane formation and microbial inactivation in marine aquaculture waters. Aquaculture 480:94–102

Santos L, Ramos F (2018) Antimicrobial resistance in aquaculture: current knowledge and alternatives to tackle the problem. Int J Antimicrob Agents 52(2):135–143

Schmidt W, Raymond D, Parish D, Ashton IG, Miller PI, Campos CJ, Shutler JD (2018) Design and operation of a low-cost and compact autonomous buoy system for use in coastal aquaculture and water quality monitoring. Aquac Eng 80:28–36

Schulz C, Gelbrecht J, Rennert B (2003) Treatment of rainbow trout farm effluents in constructed wetland with emergent plants and subsurface horizontal water flow. Aquaculture 217(1–4):207–221

Sekaran HR (2014) Biosecurity system in Malaysian fisheries: gearing up for safe and quality seafood. Fish People 12:33–37

Semenov A, Semenova K (2022) Ultraviolet disinfection of water in recirculating aquaculture system: a case study at sturgeon caviar fish farm. Acta Agric Slov 118(3):1–4

Senarathna DD, Abeysooriya K, Vithushana T (2021) Veterinary pharmaceuticals in aquaculture wastewater as emerging contaminant substances in aquatic environment and potential treatment methods. MOJ Eco Environ Sci 6:98–102

Sepúlveda D, Lorenzen E, Rasmussen JS, Einer-Jensen K, Collet B, Secombes C, Lorenzen N (2019) Time-course study of the protection induced by an interferon-inducible DNA vaccine against viral haemorrhagic septicaemia in rainbow trout. Fish Shellfish Immunol 85:99–105

Serrano PH (2005) Responsible use of antibiotics in aquaculture, vol 469. Food & Agriculture Org

Sidoruk M, Cymes I (2018) Effect of water management technology used in trout culture on water quality in fish ponds. Water 10(9):1264

Silverstein JT, Vallejo RL, Palti Y, Leeds TD, Rexroad CE III, Welch TJ, Wiens GD, Ducrocq V (2009) Rainbow trout resistance to bacterial cold-water disease is moderately heritable and is not adversely correlated with growth. J Anim Sci 87(3):860–867. https://doi.org/10.2527/jas.2008-1157

Steinberg KM, Okou DT, Zwick EM (2008) Applying rapid genome sequencing technologies to characterize pathogen genomes. ACS Publications

Su G, Logez M, Xu J, Tao S, Villéger S, Brosse S (2021) Human impacts on global freshwater fish biodiversity. Science 371(6531):835–838

Subasinghe RP (2005) Epidemiological approach to aquatic animal health management: opportunities and challenges for developing countries to increase aquatic production through aquaculture. Prev Vet Med 67(2–3):117–124

Subasinghe, R.P, Bondad-Reantaso, M.G., McGladdery, S.E. (2001). Aquaculture development, health and wealth. In: Subasinghe, R.P., Bueno, P., Phillips, M.J., Hough, C., McGladdery, S.E., Arthur, J.R. (Eds.). Aquaculture in the Third Millenium.Technical Proceedings of the Conference on Aquaculture in theThird Millenium, Bangkok, Thailand, 20–25 February 2000. http://www.fao.org/DOCREP/003/AB412E/ab412e09.htm)

Subasinghe R, Alday-Sanz V, Bondad-Reantaso MG, Jie H, Shinn AP, Sorgeloos P (2023) Biosecurity: Reducing the burden of disease. J World Aquacult Soc. https://doi.org/10.1111/jwas.12966

Swain PP, Sahoo L, Kumar R, Sundaray JK (2022) Applications of next-generation sequencing in aquaculture and fisheries. In: Advances in fisheries biotechnology. Springer, Singapore, pp 41–64

Tang K, Bondad-Reantaso M (2019) Impacts of acute hepatopancreatic necrosis disease on commercial shrimp aquaculture. Rev Sci Tech 38:477–490

Tang X, Gong J, Zeng C, Sheng X, Xing J, Zhan W (2019) Dynamic distribution of formalin-inactivated Edwardsiella tarda in flounder (Paralichthys olivaceus) post intraperitoneal vaccination. Fish Shellfish Immunol 89:393–402

Thanigaivel S, Thomas J, Chandrasekaran N, Mukherjee A (2023) Treating bacterial infections in fishes and shrimps using seaweed extracts. In: Aquaculture microbiology. Springer, pp 141–144

The World Bank (2013) The world bank annual report 2013. The World Bank, Washington, DC

Thiermann AB (2005) Globalization, international trade and animal health: the new roles of OIE. Prev Vet Med 67(2–3):101–108

Varela J, Ruiz-Jarabo I, Vargas-Chacoff L, Arijo S, León-Rubio J, García-Millán I et al (2010) Dietary administration of probiotic Pdp11 promotes growth and improves stress tolerance to high stocking density in gilthead seabream Sparus auratus. Aquaculture 309(1–4):265–271

Verschuere L, Rombaut G, Sorgeloos P, Verstraete W (2000) Probiotic bacteria as biological control agents in aquaculture. Microbiol Mol Biol Rev 64(4):655–671

Vora NM, Hannah L, Walzer C, Vale MM, Lieberman S, Emerson A et al (2023) Interventions to reduce risk for pathogen spillover and early disease spread to prevent outbreaks, epidemics, and pandemics. Emerg Infect Dis 29(3):1

Wang Y, Qi C, Pan H (2012) Design of remote monitoring system for aquaculture cages based on 3G networks and ARM-android embedded system. Procedia Eng 29:79–83

Warren JR, Marshall B (1983) Unidentified curved bacilli on gastric epithelium in active chronic gastritis. Lancet 321(8336):1273–1275

Watts JE, Schreier HJ, Lanska L, Hale MS (2017) The rising tide of antimicrobial resistance in aquaculture: sources, sinks and solutions. Mar Drugs 15(6):158

Wetten M, Aasmundstad T, Kjøglum S, Storset A (2007) Genetic analysis of resistance to infectious pancreatic necrosis in Atlantic salmon (Salmo salar L.). Aquaculture 272(1–4):111–117. https://doi.org/10.1016/j.aquaculture.2007.08.029

Wei G, Cai S, Wu Y, Ma S, Huang Y (2020) Immune effect of Vibrio harveyi formalin-killed cells vaccine combined with chitosan oligosaccharide and astragalus polysaccharides in♀ Epinephelus fuscoguttatus×♂ Epinephelus lanceolatus. Fish Shellfish Immunol 98:186–192

Wei LS, Goh KW, Abdul Hamid NK, Abdul Kari Z, Wee W, Van Doan H (2022) A mini-review on co-supplementation of probiotics and medicinal herbs: application in aquaculture. Front Vet Sci 9:869564

Wei J, Guo S, Wang Q, Xu H (2023) Present situation and countermeasure of animal quarantine in animal husbandry and veterinary. International Journal of Food Science and Agriculture 7(3):123–134. https://doi.org/10.26855/ijfsa.2023.06.008

Westers H (1983) Considerations in hatchery design for the prevention of diseases. In: A guide to integrated fish health management in the Great Lakes Basin. Great Lakes Fishery Commission, Ann Arbor, MI, pp 29–35

Wyban JA, Swingle J, Sweeney J, Pruder G (1993) Specific pathogen free Penaeus vannamei. World Aquac 24:39

Xu Z, Jin P, Zhou X, Zhang Y, Wang Q, Liu X et al (2021) Isolation of a virulent Aeromonas salmonicida subsp. masoucida bacteriophage and its application in phage therapy in turbot (Scophthalmus maximus). Appl Environ Microbiol 87(21):e01468–e01421

Zachritz W II, Hanson A, Sauceda J, Fitzsimmons K (2008) Evaluation of submerged surface flow (SSF) constructed wetlands for recirculating tilapia production systems. Aquac Eng 39(1):16–23

Zeng W, Wang Q, Wang Y, Zhao C, Li Y, Shi C et al (2016) Immunogenicity of a cell culture-derived inactivated vaccine against a common virulent isolate of grass carp reovirus. Fish Shellfish Immunol 54:473–480

Zhang S-Y, Li G, Wu H-B, Liu X-G, Yao Y-H, Tao L, Liu H (2011) An integrated recirculating aquaculture system (RAS) for land-based fish farming: the effects on water quality and fish production. Aquac Eng 45(3):93–102

Zhang H-P, Chen M-Y, Xu Y-X, Xu G-Y, Chen J-R, Wang Y-M et al (2020a) An effective live attenuated vaccine against Aeromonas veronii infection in the loach (Misgurnus anguillicaudatus). Fish Shellfish Immunol 104:269–278

Zhang Z, Liu G, Ma R, Qi X, Wang G, Zhu B, Ling F (2020b) The immunoprotective effect of whole-cell lysed inactivated vaccine with SWCNT as a carrier against Aeromonas hydrophila infection in grass carp. Fish Shellfish Immunol 97:336–343

Zhao J, Chen M, Quan C, Fan S (2015) Mechanisms of quorum sensing and strategies for quorum sensing disruption in aquaculture pathogens. J Fish Dis 38(9):771–786

Zulbainarni N, Megawati L (2019) Cost of biosecurity application: comparing aquaculture system and fish health in traditional fish farm. Paper presented at the International Society for Economics and Social Sciences of Animal Health-South East Asia 2019 (ISESSAH-SEA 2019)

Exploring the Depths: Social-Economic Mapping for Sustainable Aquaculture Communities and Resilience

9

Fittrie Meyllianawaty Pratiwy,
Asep Agus Handaka Suryana, and Rafidah Othman

Abstract

This insightful work delves into the realm of sustainable aquaculture communities, offering a comprehensive exploration of social-economic mapping and resilience strategies. The relevance of aquaculture is underscored by its vital role in supporting global food security, especially in developing nations. Social mapping techniques are a central theme, enabling the identification of needs, resources, and challenges within aquaculture communities. By harnessing the power of social-economic mapping, communities can gain a deeper understanding of their unique circumstances, facilitating effective decision-making and development planning. Several case studies illustrate the benefits of social-economic mapping, showcasing its utility in identifying and addressing various social and economic issues within coastal regions. It takes a close look at the significance of local culture, tradition, and gender roles within aquaculture communities. The impact of culture on gender roles is explored, emphasizing the variations in practices among different cultures and groups. Recommendations are provided to encompass multifaceted strategies and the application of social mapping to empower communities. The text concludes by presenting a framework for social-economic mapping, underscoring its potential to foster sustainable and resilient aquaculture communities.

F. M. Pratiwy (✉) · A. A. H. Suryana
Department of Fisheries, Faculty of Fisheries and Marine Science, Universitas Padjadjaran, Bandung, Indonesia
e-mail: fittrie.pratiwy@unpad.ac.id

R. Othman
Higher Institution Centre of Excellence (HICoE), Borneo Marine Research Institute, Universiti Malaysia Sabah, Kota Kinabalu, Sabah, Malaysia

N. M. Faudzi et al. (eds.), *Essentials of Aquaculture Practices*,
https://doi.org/10.1007/978-981-97-6699-4_9

Keywords

Social-economic mapping · Gender roles · Empower communities · Sustainable aquaculture

9.1 Introduction

Aquaculture, also known as fish cultivation, is the activity of maintaining, growing, breeding, and harvesting fish in a controlled environment. This includes various activities such as loading, transporting, storing, cooling, handling, processing, and preserving fish using ships. Aquaculture has experienced rapid development and is commonly practiced in open waters such as rivers, reservoirs, lakes, and coastlines. The global demand for fish has increased significantly in recent decades, leading to the expansion of aquaculture to meet this demand, especially in developing countries. This is driven by the growing awareness of the health benefits of fish consumption and the need for healthier diets. In terms of capture fisheries, of the top ten producers for global capture production, China is the leading fishing country, followed by Indonesia, Peru, India, and the Russian Federation. In 2020, 13 countries had over 75% of total inland captures, compared with 20 countries for marine captures (FAO 2020). However, it is uncertain whether the existing measurement and reporting frameworks effectively encompass present-day circumstances and obstacles. A prominent challenge, internationally recognized, is the need to balance the conservation of marine biodiversity (aligned with SDG 14) with the utilization of marine zones, such as aquaculture, to guarantee food security, generate income opportunities, and sustain viable coastal communities. This is notably linked to SDGs 1, 2, 3, 8, 9, 11, and 12. The potential of aquaculture not only addresses the demand for fish but also aims to improve the livelihoods of its communities. The potentiality of aquaculture not only addresses the demand for fish but also improves the livelihoods of its communities.

Global aquaculture production continued to grow in 2020 despite the challenges posed by the COVID-19 pandemic. The increase in total output, including fish, seaweed, shells, and pearls, is significant (Fig. 9.1). This growth not only contributes to food production but also has economic implications with the estimated increase in on-farm value.

According to FAO (2020), fisheries and aquaculture provide sustenance and economic support to millions of inhabitants residing in rural and coastal regions worldwide, frequently playing a significant role in shaping a society's cultural and societal fabric. These communities recognize the intrinsic value of fish as a nourishing and healthful dietary component with the capacity to address the needs of an expanding global population. Nevertheless, as the world's population continues to burgeon and the demand for fish increases, our natural resources come under increasing pressure. Sustainable management and development are essential to conserving these resources for the world's future system.

One strategy employed in community development is the utilization of social mapping. Typically, social mapping produces a visual representation structured to

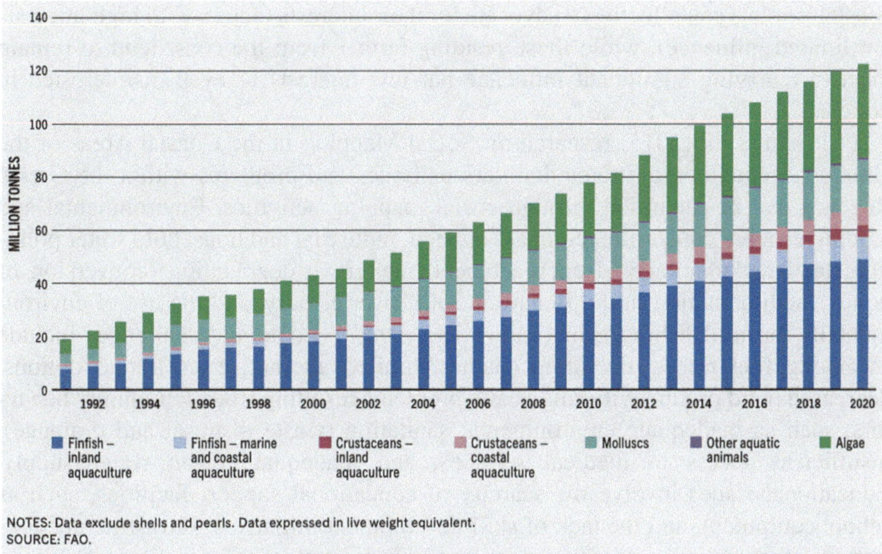

Fig. 9.1 World aquaculture production, 1991–2020 (FAO 2020)

provide a comprehensive overview of the characteristics or social challenges within a community. The underlying principle of social mapping is to gather a wealth of information to aid decision-making and facilitate the optimal development of specific regions (Moerad et al. 2014).

Earlier studies about social mapping analysis, such as Adibrata's (2012) research on Social Mapping of Coastal Communities to Support Sustainable Fisheries Management in Taman Wisata Perairan Padaido, Biak Numfor Regency, demonstrate the ability of social mapping to identify the needs and underlying issues perceived by coastal communities in their pursuit of improved well-being. Exploring the positive aspects embedded within the local socio-cultural context, such as traditional fishing permits or "*sasi*," can foster positive attributes like self-esteem, self-confidence, motivation for cooperation, mutual trust, and a shared commitment to addressing challenges and problems in TWP Padaido. The influential role of leadership triads serves as a social asset in mobilizing the communities toward progress.

Sukmayeti's (2019) study on Social Mapping of Resources and Fisher Accessibilities in Coastal Tourism Development Policies reveals that the distribution of social resources among fishermen in Meninting Village is uneven across five hamlets. Indicators include ownership of capital, possession of knowledge and skills, ownership of production tools, utilization of scientific knowledge and technology, diversification capabilities, socio-political relationships of fishermen, and their ability to recognize economic prospects associated with coastal tourism. Fishermen residing closer to the sea tend to possess more favorable social resources and exhibit a defined interest in coastal development. However, these development policies limit the accessibility of certain groups of fishermen. Only fishermen from coastal hamlets can effectively advocate for their interests (actors with high interests but limited influence), while those residing farther from the coast tend to remain inactive (carrying significant influence but low interest) or even disinterested in policies.

Fahrudin et al. (2015) research on Social Mapping in the Coastal Areas of the Gresik Regency demonstrates that various issues and problems within the Gresik Regency can be identified through social mapping activities. Environmental and natural resource concerns encompass erosion, industrial and household water pollution, land reclamation and encroachment, mangrove degradation (conversion of ponds), sedimentation in the Begawan Solo River estuary, and the use of environmentally harmful fishing gear (known as "cager"). Economic challenges include escalating fuel prices, declining catches, limited alternative livelihood options, decreased pond productivity, and health problems resulting from community behaviors, such as inadequate environmental sanitation (waste disposal and drainage), insufficient access to medical services, and inadequate clean water supply. Educational issues involve the scarcity of educational support facilities, such as school equipment, and the lack of skill development initiatives. Infrastructure concerns relate to limited public transportation and a scarcity of clean water. Institutional issues encompass power struggles among fishing groups, conflicts among members of the community, disputes with fishermen from other areas, and disagreements regarding claims to encroachments on water areas.

Infrastructure issues involve limited public transportation and scarce clean water. Institutional issues encompass power struggles among fishing groups, horizontal conflicts among community members, conflicts with fishermen from other areas, and disputes over claims to water area encroachments.

9.2 Aquaculture Fisheries Community

Indonesia boasts a wealth of marine resources that, if managed effectively and per the appropriate processes, can significantly boost the economic prospects of its citizens. According to the Food and Agriculture (2020), Indonesia was recognized as a primary global producer of seaweed and tuna. Shrimp and crabs held the second position. This data serves as a reminder of Indonesia's potential as a fisheries community with the capacity to make a global impact.

In a simplified description, a fisheries community can be depicted as a group of individuals whose livelihoods depend on fisheries resources, encompassing the processing, utilization, and marketing of these resources. The social and cultural aspects of fisheries communities are pivotal elements that can guide sustainable fisheries development (Sastrawidjaya 2002).

Fisheries communities can be categorized into several groups based on their direct contributions to the fisheries sector as a source of livelihood. These groups include fishermen and those involved in the collection and processing of catches. Additionally, some fisheries communities are not directly engaged in fisheries activities, such as craftsmen, employees, and others. These individuals fall under the category of fisheries communities due to their social ties with the groups directly engaged in fisheries activities, often residing in areas dominated by fisheries communities (Hartono and Purnomo 2019).

Fisheries communities are groups of individuals directly involved in fisheries activities, including fish capture, aquaculture, fish processing, fish trade, and related endeavors. They live and work in coastal areas, rivers, and other water bodies that serve as fisheries resources. Fisheries communities can be categorized into two main types. The first type is the capture fisheries community, which relies on the exploitation of fisheries resources through fish capture activities. The second type is the aquaculture fisheries community, whose livelihoods revolve around the cultivation of fish through environmentally engineered processes, utilizing the natural surroundings to generate products and services derived from fisheries resources (Hartono and Purnomo 2019).

Communities within the fisheries sector comprise individuals actively engaged in fisheries-related activities, including fish capture, aquaculture, fish processing, fish commerce, and other related pursuits. They inhabit and operate in coastal regions, rivers, and other aquatic environments that function as fisheries resources. Communities within the fisheries sector can be divided into two primary types. The first type is the *capture fisheries community*, which depends on the utilization of fisheries resources through fish capture activities. The second type is the *aquaculture fisheries community* (Fig. 9.2), whose means of living are centered around the farming of fish through environmentally designed processes, leveraging the natural environment to produce goods and services derived from fisheries resources (Hartono and Purnomo 2019).

Fig. 9.2 Aquaculture fisheries community

9.3 Social Mapping

Social mapping is a systematic process of depicting a community by collecting data and information regarding its profile and social issues. Social mapping is crucial to comprehend the social conditions of local communities since each community exhibits distinct social conditions, resulting in diverse problems and needs (Handoyo and Sudrajat 2016). As outlined by Meilantina (2013), social mapping can be conducted at a micro level, such as within a village, to understand the village, its potential, issues, efforts to address these problems, and action plans.

According to Prasetyo and Irwansyah (2020), a community is a group of individuals in a specific geographical area who interact within a social framework. On the other hand, fish farmers are individuals whose livelihood involves activities related to the care, cultivation, and harvesting of fish, as well as the collection of their yield within a controlled environment (Government Regulation of Republic Indonesia 2017).

Social mapping combines social and spatial aspects simultaneously and is closely related. The socio-spatial method aids research endeavors related to the governance of areas to comprehend facets of equality, efficiency, and transparency (Alemie et al. 2015). According to Safiullin et al. (2015), social mapping encompasses several dimensions, including statistical, anthropological, and network-oriented aspects.

The research conducted by Safriyawan et al. (2017) focused on the social mapping of Coastal Communities in Teluk Semanting Village, Pulau Derawan Subdistrict, Berau Regency, West Kalimantan Province. The study revealed that the coastal resources in Teluk Semanting Village have historically been utilized to sustain the livelihoods of the local population. The livelihoods of the fishing community are closely associated with traditional fishing practices based on local knowledge, the quality of coastal and marine resources, social institutions about fisheries, and mangrove ecosystems. Furthermore, the research highlighted the diversity of livelihoods based on fishing activities and agricultural land development, leading to a diversified community. The fishermen in Teluk Semanting Village engage in a range of activities, including fishing, farming, serving as village officials, trading, acting as leaders within their communities, and processing fishery products. The key issues identified by the community during the focus group discussions (FGD) are primarily related to the availability of clean water, which is impacted by land clearance activities. Land clearance has resulted in deforestation and forest fires, leading to environmental challenges in the area.

According to Sukari (2019), social mapping is one of the systematic processes for depicting communities, involving the collection of data and information about the community, such as their profiles and social issues. Social mapping is also carried out to identify the significance of the presence and activities of community empowerment actors. The results of social mapping can serve as the foundation for planning sustainable community development programs.

Bahruddin et al. (2013) stated that social mapping can provide a comprehensive picture of the mapped location, which includes actors who play a role in the process of social relations, social networks, and the strengths and interests of each actor in community life, especially in improving community living conditions. The mapping also identifies existing social problems, including the existence of vulnerable groups, as well as the available potential, both natural, human, financial, infrastructure, and social capital. This is related to agricultural development strategies which must accommodate existing heterogeneity in biophysical, economic, socio-cultural, institutional, and environmental aspects (Notenbaert et al. 2013).

9.4 Local Culture, Tradition, and Gender Roles

Culture is a phenomenon that examines the patterns and forms of behavior and activities of a community, with its elements being taken and transferred by individuals within the community. Culture also reflects the quality of society (Masinambow 1980 in Fahrudin et al. 2015). Culture is considered something inseparable from humans, formed within the life of a society. Human society is the creator, developer, and preserver of a culture. As expressed by Kistanto (2015), there is no society without culture, and there is no culture without society. The relationship between culture and society is explained by Anthony Giddens (1991) in Kistanto (2015) as follows:

> When we use the term in ordinary daily conversation, we often think of culture as equivalent to the 'higher things' of the mind—art, literature, music and painting... the concept includes such activities, but also far more. Culture refers to the whole way of life of the members of a society. It includes how they dress, their marriage custom and family life, their patterns of work, religious ceremonies, and leisure pursuits. It also covers the goods they create, and which become meaningful for them—bows and arrows, ploughs, factories and machines, computers, books, dwellings.

Interpreted as follows, when we use this term in everyday conversation, we often associate it with "high" things such as art, literature, music, and painting. However, this concept encompasses more than that. Culture refers to the entire lifestyle of community members. It includes clothing, marriage traditions, family life, work patterns, religious ceremonies, and recreational activities. It also encompasses items created by them and holds meaning for them—such as bows and arrows, plows, factories and machines, computers, books, and dwellings.

Traditions are practices that develop within a society, seen as values respected and adhered to by the supporting community. Traditions are considered eternal procedures passed down from one generation to the next as heritage. Customs are also defined as rules applied within the community environment and become distinctive features of a region that are attached to individuals in society. As stated by Koentjaraningrat (2003), customs are defined as a complex concept and firm rules strongly integrated into a cultural system, including the culture that governs human behavior in social life.

Furthermore, the gender system within society is influenced by the presence of culture deeply rooted in a specific region. According to a statement by Eckert

(1998), there is variation in gender practices among cultures from one place to another, as well as among different groups.

Gender roles arise from an individual's gender identity, and there are several factors contributing to the formation of one's overall identity (Ramli and Basri 2021). Gender identity is an integral part of a person's identity, involving the connection between an individual's perception of their gender in the past. Gender, in this context, becomes a cultural embodiment in categorizing men and women with characteristics of masculinity and femininity (Nurohim 2018).

There are ten roles in productive gender-based stratification among fish breeders, with six of them predominantly performed by men. These roles include seed distribution, fishpond preparation, pest and disease control, harvesting, fish sorting for sale, and fish packaging for sale. Meanwhile, the remaining four roles are mainly carried out by men, with limited female participation. These roles encompass fish seed provision, seed selection, fish preparation and sales, and fish weighing for sale.

9.5 Social Organization

An organization is defined as a unit formed by individuals with specific tasks. Organizations permeate and encompass all aspects of society, from the economic sphere to personal life. In a social context, the term "social organization" refers to the interpersonal relationships within a community. According to Dirdjosisworo (1985), social organization is defined as a social environment that comprises groups with clearly defined roles and tasks, striving to achieve specific goals. The UN social dimension classifications are described (Table 9.1), along with their fundamental elements and variables tailored to the context of aquaculture. Potential indicators, both quantitative and qualitative, might be the key point to portray the diverse social aspects of these key elements.

9.6 Socio-Economic Analysis for Sustainable Aquaculture

Social mapping is a systematic process of depicting society and involves collecting data and information about society, including the profile and social problems that exist in the society concerned. Netting et al. (1993) explained that social mapping is also called social profiling or creating a profile of a society, which is useful for helping and understanding changes in society.

For example, the livelihoods of fisheries communities in certain regions have been adversely impacted by dwindling fish stocks and competition for catch allocations from sportfishing groups. The decline in fishing activities across numerous communities has resulted in economic difficulties that undermine their overall resilience. To address these challenges, the establishment of aquaculture enterprises, such as shellfish and seaweed farms, becomes crucial. These enterprises not only support existing fisheries supply chains but also contribute to the sustainability of

Table 9.1 UN social dimension classifications in the context of aquaculture (United Nations 1989)

UN social dimension category	Key components
Population	Demographic change
Health	Community structure
	Nutrition/food
	Mental well-being/satisfaction
Education	Informed decision-making
	Formal education (training/skills)
	Informal education (experience)
Work	Income levels
	Justice
	Equality
	Livelihood protection
	Quality of employment
Housing	Type
	Proximity
	Social mobility
	Ownership

working waterfronts by driving the demand for boats, buoys, ropes, docks, and cleaning and processing facilities, as well as cold storage.

In cases where employing participatory approaches is challenging or when the results are unpredictable, the utilization of national social-economic statistics, commonly known as population censuses, along with other extensive, non-fishery-related data sources, can play a significant role in evaluating the spatial patterns of fisheries, as mentioned by Thiault et al. in 2017. These statistics help identify various conditions, including social, economic, and cultural aspects of aquaculture communities, as well as the potential problems existing in the community (Table 9.2).

This activity can be carried out using social mapping techniques with selected variables originating from theories in sociology. Social mapping of aquaculture communities aims to map the potential problems that exist in the community so that they can be used to prepare community development plans. Apart from that, social mapping of aquaculture communities can also be used to increase community capacity in managing waste independently, developing the potential of natural and human resources to support the achievement of Village SDGs and their resilience (Fig. 9.3).

9.7 Recommendations

1. Eco-friendly farming practices: Promote and incentivize the adoption of eco-friendly aquaculture practices, such as recirculating aquaculture systems (RAS) and integrated multi-trophic aquaculture (IMTA), which reduce environmental impact and enhance sustainability.

Table 9.2 Fish farmer characteristics

Characteristics of fish farmer	Category
Gender	Man Woman
Age	Early adulthood (26–35) Late adulthood (36–45) Early elderly (46–55) Late elderly (56–65) >65
Formal education level	Didn't pursue formal education Elementary school Junior high school Senior high school Higher education (level 1) Higher education (level 2)
Work experience	Low (1–4 years) Medium (5–10 years) High (> 10 years)
Number of respondent members	1–3 people 4–6 people >7 people
Number of respondents who work	Doesn't work 1–3 people > 4 people
Livelihood as an aquaculture farmer	Main jobs Side jobs
Aquaculture farmer business experience (years)	Low (1–4 years) Medium (5–10 years) High (> 10 years)
Number of family members	Small (1–3 people) Medium (4–6 people) Big (>6 people)

2. Community-driven initiatives: Encourage the establishment of community-driven aquaculture cooperatives, where local fishermen and fish farmers collectively manage and benefit from sustainable aquaculture projects, fostering economic and social resilience.
3. Preservation of cultural traditions: Integrate local cultural traditions and indigenous knowledge into aquaculture practices, ensuring that these traditions are respected and preserved while empowering communities to develop their aquaculture potential.

9.8 Conclusion

In conclusion, this work sheds light on the intricate world of sustainable aquaculture communities, with a focus on social-economic mapping. Furthermore, it underscores the significance of local culture and gender roles within these communities,

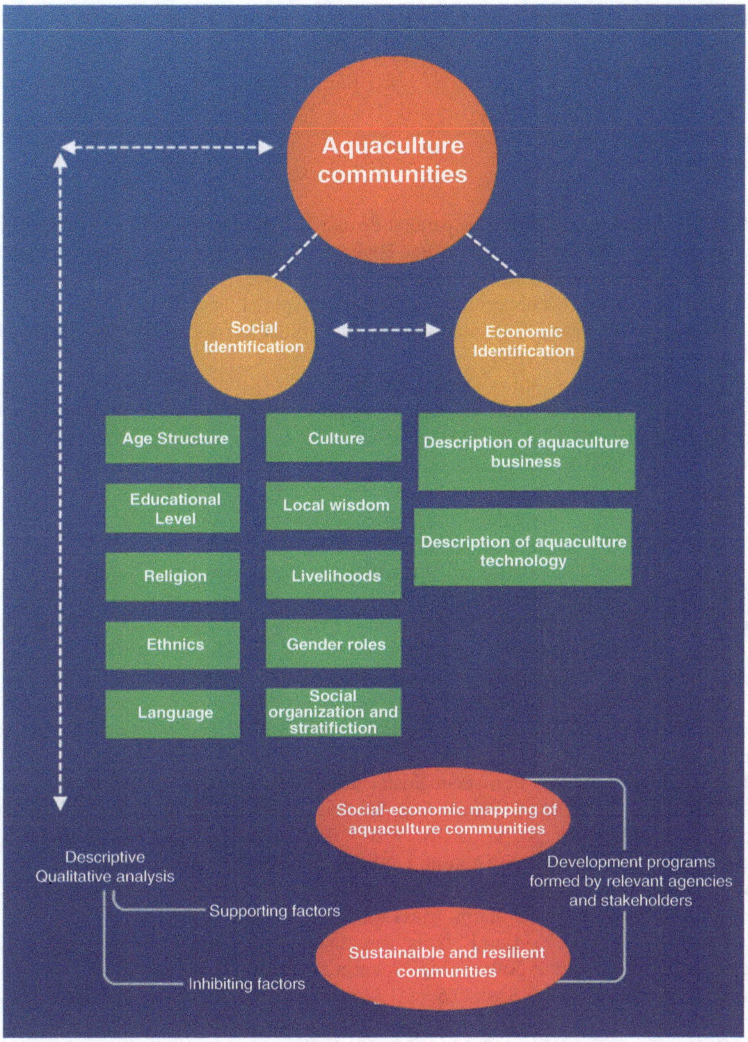

Fig. 9.3 Framework of social-economic mapping for aquaculture communities to achieve sustainable and resilient communities

recognizing the diversity of practices and values among different groups. This comprehensive exploration of aquaculture and resilience strategies is a valuable resource for policymakers, researchers, and community leaders striving to enhance the sustainability and cultural preservation of coastal communities worldwide. It provides a roadmap for addressing key challenges and fostering sustainable practices within the aquaculture sector.

Acknowledgement The authors want to express their sincere appreciation to Thalita Nazwa Faddhilah, M. Fachry Maula, and Mayapada Fajarwati for their insightful contributions to the writing process and data support.

References

Adibrata S (2012) Pemetaan Sosial Masyarakat Pesisir Untuk Mendukung Pengelolaan Perikanan Berkelanjutan Di TWP Padaido Kabupaten Biak Numfor. Akuatik: Jurnal Sumberdaya Perairan 6(1):1–6. https://journal.ubb.ac.id/akuatik/article/view/895

Alemie BK, Bennett RM, Zevenbergen J (2015) A socio-spatial methodology for evaluating urban land governance: the case of informal settlements. J Spat Sci 60(2):289–309. https://doi.org/10.1080/14498596.2015.1004654

Anthony G (1991) Sociology. Polity Press, Cambridge

Bahruddin, Krisdyatmiko, Darmawan DA, Soetomo (2013) Indikator proper Hijau Aspek Pengembangan Masyarakat (*Community Development*) Peraturan Menteri Lingkungan Hidup Republik Indonesia Nomor 06 Tahun 2013 Tentang Proper. Deputi Pengendalian dan Pencemaran Kementrian Lingkungan Hidup RI. Jurusan Pembangunan Sosial dan Kesejahteraan Fakultas Sosial dan Ilmu Politik Universitas Gadjah Mada

Dirdjosisworo S (1985) Asas-asas Sosiologi. Penerbit Armico, Bandung

Eckert P (1998) Gender and sociolinguistic variation. In: Coates J (ed) Language and gender. Blackwell Publisher, Ltd, Massachusetts

Fahrudin AMA, Al Amin T, Kodiran A, Hamdani A, Afandy, Trihandoyo (2015) Pemetaan Sosial (*Social Mapping*) di Wilayah Pesisir Kabupaten Gresik. Working Paper PKSPL-IPB

FAO (2020) The state of world fisheries and aquaculture 2020. Sustainability in action. FAO, Rome

Government Regulation of Republic Indonesia (2017) Government regulation of Republic Indonesia Number 28 of 2017 concerning fish cultivation. https://peraturan.bpk.go.id/Details/51590

Handoyo P, Sudrajat A (2016) Pemetaan Sosial untuk Perencanaan Pemberdayaan Masyarakat Desa Kemantren, Lamongan. Proceeding Seminar Nasional Mengawal Pelaksanaan SDGs (sustainable development goals). ISBN: 978- 979-028-859-1. Prodi Sosiologi Jurusan Ilmu Sosial Fakultas Ilmu Sosial dan Hukum. UNESA University Press, Surabaya

Hartono TT, Purnomo AH (2019) Modul 1: Konsep Dasar Kehidupan Sosial Masyarakat Perikanan. Universitas Terbuka, Sosial Ekonomi Masyarakat Perikanan, p 73

Kistanto NH (2015) Tentang Konsep Kebudayaan. Sabda: Jurnal Kajian Kebudayaan 10(2):1–11

Koentjaraningrat (2003) Kamus Antropologi. Jakarta Progress

Meilantina M (2013) Pemetaan Sosial (*Social Mapping*): Studi Di Wilayah Kabupaten Kapuas-Provinsi Kalimantan Tengah. J-SEA (J Soc Econ Agric) 8(1):32–43

Moerad SK, Windiani W, Mukhtar S, Nurif M (2014) Pemetaan Sosial Budaya Masyarakat Desa Sumber Agung Kecamatan Pesanggaran Kabupaten Banyuwangi. Jurnal Sosio Humaniora 7(1):61–74. https://doi.org/10.12962/j24433527.v7i1.596

Netting FE, Kettner PM, McMurtry S (1993) Social work macro practice. Longman, New York. ISBN-10 0133948528

Notenbaert A, Herrero M, Groote HD, You L, Gonzalez-Estrada E, Blummel M (2013) Identifying recommendation domains for targetting dual-purpose maize-based interventions in crop livestock systems in East Africa. Land Use Policy 30:834–846

Nurohim S (2018) Identitas dan Peran Gender Pada Masyarakat Suku Bugis. Sosietas 8(1):457–461. https://ejournal.upi.edu/index.php/sosietas/article/view/12499/7408

Prasetyo D, Irwansyah (2020) Memahami Masyarakat dan Perspektifnya. Jurnal Manajemen Pendidikan dan Ilmu Sosial 1(1):163–175. https://doi.org/10.38035/jmpis.v1i1.253

Ramli U, Basri L (2021) Peran Gender Pada Masyarakat Bugis. Jurnal Noken: Ilmu-Ilmu Sosial 7(1):78–89

Safiullin MR, Ermolaeva PO, Yermolaev OP, Selivanov RN (2015) Current perspectives on social mapping of urban territories. Asian Soc Sci 11(6):207. https://doi.org/10.5539/ass.v11n6p207

Safriyawan D, Gunawan BI, Juliani (2017) Study of sosial mapping coastal communities Kampung Teluk Semanting in Derawan Island sub-disrict Berau district. J Fish Agribusiness Dev 4(1):1–7. https://doi.org/10.30872/jppa.v4i1.148

Sastrawidjaya (2002) Nelayan dan Kemiskinan. Penerbit Pradnya Paramita Jakarta

Sukari S (2019) Sosial-Mapping Sebagai Landasan Perencanaan Pemberdayaan Masyarakat Berkelanjutan. J Riset Entrepr 2(1):52–61. https://doi.org/10.30587/jre.v2i1.796

Sukmayeti E (2019) Pemetaan Sosial terhadap Sumberdaya dan Aksesibilitas Nelayan dalam Kebijakan Pembangunan Wisata Pesisir. https://doi.org/10.33019/society.v7i2.101

Thiault L (2017) Mapping social-ecological vulnerability to inform local decision making. Conserv Biol 32:447. https://doi.org/10.1111/cobi.12989

United Nations (1989) Handbook on social indicators. United Nations Publication, Department of International Economic and Social Affairs, p 154. ST/ESA/STAT/SER.F/49

Caulerpa lentillifera Sea Grape: A Promising Source of Bioactive Metabolites for Aquaculture and Health

10

Muhammad Dawood Shah, Ching Fui Fui, Wahidatul Husna Zuldin, Fikri Akmal Khodzori, Wei Sheng Chong, Nurzafirah Mazlan, and Rossita Shapawi

Abstract

Caulerpa lentillifera, also known as sea grape, is a species of macroalgae belonging to the phylum Chlorophyta and the family Caulerpaceae. This species is found in Sabah, Malaysia, and other parts of the world. It thrives in shallow seagrass beds and sandy bottoms, especially in tropical and subtropical regions with warm water temperatures. This chapter aims to highlight the bioactivity and metabolite profile of *C. lentillifera*. Studies have shown that this species possesses various beneficial properties, such as antioxidant, anti-inflammatory, antimicrobial, antiparasitic, anti-cancer, anti-obesity, and hepatoprotective properties. These protective properties of *C. lentillifera* can be attributed to the presence of various bioactive metabolites of different natures, including carotenoids, fatty acids, alkanes, terpenoids, and alkaloids. The beneficial effects of *C. lentillifera* extracts and metabolites can be utilized for both human and animal health, especially in aquaculture, which relies heavily on chemical intervention. The extraction, isolation, and purification of these secondary metabolites from *C. lentillifera* could play a crucial role in the prevention and treatment of various diseases associated with oxidative stress, cytotoxicity, inflammation, microbial infections, and parasitic infestations in human and animal populations, especially in the aquaculture sector. Research into these organic metabolites is a long-term and practical means of improving health and sustainable aquaculture practices.

M. D. Shah (✉) · C. F. Fui · W. H. Zuldin · F. A. Khodzori · W. S. Chong · N. Mazlan · R. Shapawi
Higher Institution Centre of Excellence (HICoE), Borneo Marine Research Institute, Universiti Malaysia Sabah, Kota Kinabalu, Sabah, Malaysia
e-mail: dawoodshah@ums.edu.my

N. M. Faudzi et al. (eds.), *Essentials of Aquaculture Practices*,
https://doi.org/10.1007/978-981-97-6699-4_10

Keywords

Macroalgae · Antioxidant · Anti-obesity · Anti-cancer · Cytotoxicity

10.1 Introduction

Macroalgae, commonly known as seaweed, are found in the saltwater environment. Due to their chlorophyll coloring, they are categorized into brown algae (Phaeophyceae), green algae (Chlorophyceae), and red algae (Rhodophyceae). A variety of metabolic products can be identified within the macroalgae, including carotenoids, fatty acids, polysaccharides, phycobilins, vitamins, phycocyanins, sterols, and tocopherol (Dmytryk et al. 2017; Kadam and Prabhasankar 2010; Lordan et al. 2011; Murugan and Iyer 2013; Saravanan et al. 2008; Shah et al. 2022).

The extract and bioactive metabolites of macroalgae have been shown to exhibit various protective effects, including antioxidant, cytotoxic, antimicrobial, anti-inflammatory, and antiparasitic, making them potentially useful for the healthcare and aquaculture industries. They are also highly nutritious as they contain a variety of vitamins (A, C, and K) and minerals (iodine, calcium, iron, and magnesium) (Saravanan et al. 2008; Kadam and Prabhasankar 2010; Lordan et al. 2011; Murugan and Iyer 2013; Dmytryk et al. 2017; Shah et al. 2022).

Macroalgae are widespread in various regions of the world and colonize different horizontal zones in seas and oceans, namely, the supratidal (supra-littoral), intertidal (littoral), and subtidal (sub-littoral) zones. The intertidal zone is the most typical habitat for green macroalgae. A notable example is *C. lentillifera*, which not only occurs naturally but is also cultivated commercially. This particular species is characterized as a good source of bioactive compounds with significant protective potential (Hussin and Khoso 2017; Shah et al. 2022).

The study of the bioactivity and metabolite diversity of *C. lentillifera* is of great ecological and biomedical value. Knowledge of the bioactive metabolites found in these microalgae not only expands knowledge of marine biodiversity but also highlights their potential application in various fields. The extract and metabolites of *C. lentillifera* offer numerous benefits to the healthcare and aquaculture sectors. These include antimicrobial, anti-inflammatory, antioxidant, anti-obesity, anti-cancer, and hepatoprotective properties. Research into these aspects can lead to further developments in the pharmacology of natural products and the development of new therapeutic agents (Matanjun et al. 2009; Zhang et al. 2020; Shah et al. 2022).

The focus of this chapter is on a comprehensive analysis of *C. lentillifera*. This primarily involves an in-depth study of its bioactivity, metabolite profiling, and potential applications in the treatment and prevention of various diseases.

10.2 *Caulerpa lentillifera*: The Nutritious Sea Grape of the Sea

C. lentillifera, commonly known as sea grape or green caviar, is a well-known species of green seaweed recognized in Southeast Asia for its culinary versatility (Fig. 10.1). Its characteristic features include a bright green color, rounded branches,

Fig. 10.1 Image of *C. lentillifera*, also known as sea grape or green caviar, a nutritious and versatile seaweed used as a fresh vegetable in Southeast Asia

and a grape-like structure instead of traditional leaves. The seaweed belongs to the phylum Chlorophyta and the family Caulerpaceae, whose detailed classification includes the kingdom Plantae, division Chlorophyta, class Ulvophyceae, order Bryopsidales, family Caulerpaceae, genus *Caulerpa*, and species *Caulerpa lentillifera* (Matanjun et al. 2008; Guiry and Guiry 2020). This introductory classification forms the basis for research into the various aspects of *C. lentillifera.*

10.2.1 Nutritional Benefits and Tolerance to Ocean Acidification

Research has shown that *C. lentillifera* is a nutritional powerhouse, rich in essential nutrients and phytochemicals that provide numerous health benefits. Its composition contains high levels of protein, fiber, vitamins, and minerals such as calcium, magnesium, and iron. Seaweed also contains antioxidants, including carotenoids and flavonoids, which are known for their anti-inflammatory and anti-cancer properties (Syakilla et al. 2022). Apart from its nutritional profile, *C. lentillifera* has shown remarkable tolerance to ocean acidification, which is a major problem for seaweed aquaculture. Studies suggest that seaweed can increase its growth by absorbing carbon even when exposed to pH levels below 8. This unique adaptability makes *C. lentillifera* a promising candidate for sustainable aquaculture in the face of the challenges posed by climate change (Hengjie et al. 2023).

In addition, the *Caulerpa* spp., including *C. lentillifera* has remarkable pharmacological potential. *Caulerpa* spp. have been identified for their nociceptive and anti-inflammatory effects. Sulfated polysaccharides isolated from *Caulerpa* species offer potential biomedical applications as antinociceptive and anti-inflammatory agents (Rodrigues et al. 2012). The diverse properties of *C. lentillifera*, ranging from its nutritional richness to its environmental tolerance and pharmacological potential, emphasize its importance as an intriguing topic for further scientific exploration and research.

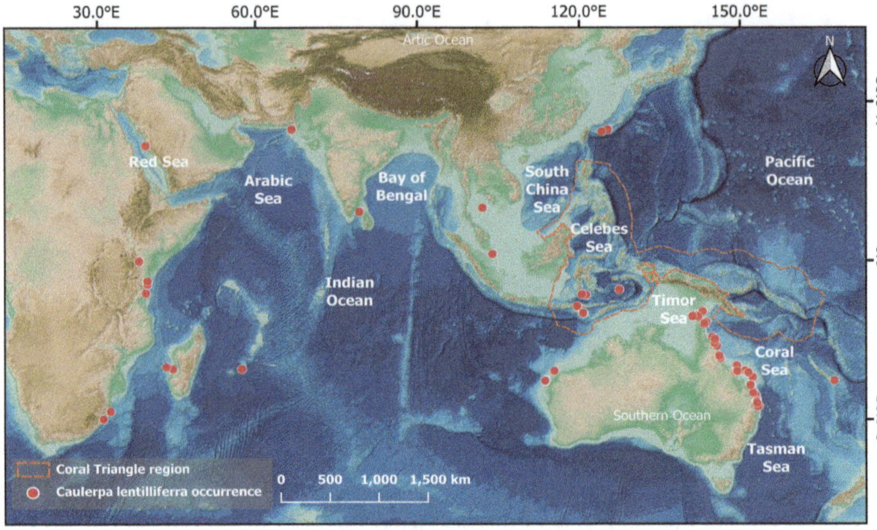

Fig. 10.2 The occurrence of *C. lentillifera* is indicated on the map. Data for *C. lentillifera* (red dots) are taken from the Ocean Biodiversity Information System (OBIS) (https://obis.org/; accessed on 01 January 2024) and recreated as a distribution map from the GEBCO World Map 2014 (www.gebco.net; accessed on 01 January 2024)

10.2.2 Distribution and Habitat of *C. lentillifera*

C. lentillifera is a green seaweed naturally distributed in tropical and subtropical regions such as the South China Sea, Southeast Asia, Japan, Okinawa, Taiwan, and Oceania (Chen et al. 2019) (Fig. 10.2). The distribution of *C. lentillifera* is influenced by factors such as water temperature, salinity, and nutrient concentration. This species thrives in shallow seagrass beds and sandy bottoms, especially in tropical and subtropical regions with warm water temperatures of 25–30 °C and salinity levels of 20–50 ppt (Chen et al. 2019; Stuthmann et al. 2023). In Malaysia, specifically in Sabah, *C. lentillifera* is commonly found in shallow waters along the coasts of Kota Kinabalu, Tawau, and Sandakan and is also cultivated in ponds in the districts of Kota Belud and Kuala Penyu (Syakilla et al. 2022). This species is also found in other coastal regions of the Asia-Pacific region, including the Philippines, Japan, Vietnam, Taiwan, and China (Syakilla et al. 2022).

10.2.3 Protective Effects of *C. lentillifera*

Antioxidant

Antioxidant metabolites play a crucial role in protecting cells from damage by neutralizing reactive oxygen species (ROS) and preventing them from causing damage. The antioxidant activity of these metabolites can be of great benefit in a number of degenerative diseases caused by oxidative stress, such as cardiovascular diseases, cancer, neurodegenerative diseases, respiratory diseases, and kidney and liver diseases (Shah and Iqbal 2010, 2018; Pizzino et al. 2017; Shah et al. 2020).

C. lentillifera has been reported to have antioxidant activity due to the presence of polyphenolic compounds. Studies have shown that *C. lentillifera* exhibits significant free radical scavenging and reducing power activity, with values of 2.16 TEAC and 362.11 FRAP mM mg/dry extract, respectively. In addition, the total phenolic content of *C. lentillifera* was reported to be 42.85 mg PGE/g dry extract (Matanjun et al. 2008). In another study, the fresh and dry extracts of *C. lentillifera* were analyzed for their antioxidant properties using the linoleic acid emulsion system method. The results showed that both extracts exhibited significant antioxidant activity with values of 54% and 79%, respectively. However, it is worth noting that the antioxidant activity of both extracts was significantly lower ($p < 0.05$) than that of the positive control, i.e., vitamin C with a value of 81% (Nofiani et al. 2018).

Anti-inflammatory

Inflammation is the body's natural reaction to tissue damage, injury, or infection. It is categorized into two types: acute and chronic inflammation. Acute inflammation is a short-term reaction that plays a decisive role in the body's defense mechanism. Chronic inflammation, on the other hand, lasts for a longer period and can lead to various diseases such as cardiovascular diseases and cancer. Chronic inflammation is harmful to the body because the excessive production of cytokines (interleukins (IL-1, 2 and 6), tumor necrosis factor-alpha (TNF-α), interferons (IFNs), etc.) and chemokines can cause tissue damage and trigger a cascade of harmful effects that contribute to disease progression. It is therefore important to control chronic inflammation to prevent the development of severe disease (Medzhitov 2008; Fleit 2014; Kotas and Medzhitov 2015).

Polysaccharides from *C. lentillifera* were found to exhibit anti-inflammatory activity by inhibiting the overproduction of TNF-α and IL-1β in lipopolysaccharide (LPS)-induced HT29 colorectal carcinoma cells (Sun et al. 2020). In another study, caulerpin extracted from *C. lentillifera* was reported to exhibit anti-inflammatory properties by inhibiting the overproduction of TNF-α and IL-1β in LPS-activated macrophages (Nagappan and Vairappan 2014).

Antimicrobial

Different microorganisms such as bacteria, fungi, viruses, and parasites cause different diseases (NIH 2007). *C. lentillifera* has the potential to act as an antimicrobial agent against microbes such as bacteria, viruses, and fungi.

In one study, both fresh and dry extracts of *C. lentillifera* were reported to have broad-spectrum antimicrobial activities against various microorganisms including *Escherichia coli*, *Staphylococcus aureus*, *Bacillus subtilis*, *Vibrio cholerae*, *Citrobacter freundii*, *Vibrio harveyi*, *Pseudomonas aeruginosa*, *Aeromonas hydrophila*, *Klebsiella pneumoniae*, and *Candida albicans*. The extract values for the minimum inhibitory concentration (MIC) ranged from 200 to 500 μg/well. However, the antimicrobial activities were classified as weak (Nofiani et al. 2018). The antibacterial potential of *C. lentillifera* chloroform extracts was tested for its efficacy in inhibiting MRSA (methicillin-resistant *S. aureus*) and *E. coli*

K1 (neuropathogenic *Escherichia coli* K1). The extract showed moderate efficacy against MRSA (62.17%) and weak activity against *E. coli* K1 (12.42%) at a concentration of 250 μg/mL (Yap et al. 2019). The antimicrobial properties of *C. lentillifera* extracts were investigated against four foodborne pathogens (*E. coli*, *S. aureus*, *Streptococcus* sp., and *Salmonella* sp.). The results showed that the crude extract of *C. lentillifera* effectively inhibited the growth of all organisms tested, with MIC values ranging from 136.50 to 175 mg/mL (Nagappan and Vairappan 2014).

Research studies have shown the antiviral potential of *C. lentillifera* extract to combat WSSV (white spot syndrome virus) infection. This is a highly infectious virus that causes a high mortality rate (80–100%) in farmed shrimp. The results of the administration of *C. lentillifera* extract were very promising. Shrimps injected with WSSV and pre-incubated with *C. lentillifera* (1–10 mg/mL) showed a significantly lower mortality rate of 0.0–6.7% compared to the positive control (100% mortality in shrimps injected with WSSV only) (Klongklaew et al. 2020).

In addition, *C. lentillifera* has also been reported to have fungicidal properties against infection with *L. thermophilum* and *H. sabahensis*. The ethanol extract of *C. lentillifera* can inhibit the growth of hyphae of *L. thermophilum* IPMB 1401, *L. thermophilum* IPMB 1601, and *H. sabahensis* IPMB 1603 (Saito and Lal 2019).

Anti-cancer

Cancer is a series of diseases characterized by the uncontrolled growth and spread of abnormal cells in the body. With an estimated 9.6 million deaths in 2018, cancer is one of the leading causes of death worldwide. However, mortality rates vary considerably depending on the type of cancer and the geographical region The causes of cancer are complex and multifactorial. Many cases of cancer are thought to be caused by environmental factors such as exposure to carcinogens, poor diet, and infections. Some cases of cancer can be caused by genetic factors or inherited mutations (Siegel et al. 2021; American Cancer Society 2023; World Health Organization 2023).

In one study, the hexane extract of *C. lentillifera* was reported to have anti-cancer properties against A172 human glioblastoma cells. In particular, the extract was found to inhibit cell proliferation and induce apoptosis, thus inhibiting growth. Additionally, the extract was observed to induce ROS accumulation and alter mitochondrial membrane potential in A172 cells (Tanawoot et al. 2021). In another study, oligosaccharides, particularly β-1,3-xylooligosaccharides, extracted from *C. lentillifera* were shown to have promising anti-tumor properties against MCF-7 (human breast adenocarcinoma cell line). Exposure to these oligosaccharides was observed to reduce the development of MCF-7 cells and promote apoptosis, as evidenced by induced chromatin condensation and degradation of poly-ADP-ribose polymerase (Maeda et al. 2012).

Anti-obesity

Obesity is a major public health problem that can lead to non-communicable diseases such as type 2 diabetes and cardiometabolic syndrome. As obesity develops, oxidative stress can occur in the liver, leading to an increase in ROS production and a decrease in the antioxidant potential of the system. This in turn can lead to the development of degenerative diseases associated with oxidative stress (Masschelin et al. 2020; Manoppo et al. 2022).

The study investigated the effects of *C. lentillifera* administered at a dose of 450 mg/kg body weight for 4 weeks to rats fed a cholesterol and fat-enriched diet (CFED). The results showed that the sea grape extract–treated rats had an 11% lower change in body weight compared to the CFED group (Manoppo et al. 2022). Another study investigated the effects of a 5% *C. lentillifera* extract on rats fed a CFED diet for 16 weeks and found a 39.5% reduction in body weight. In addition, a significant increase of 48.7% in HDL-C levels and reductions of 18.4%, 34.6%, and 33.7% in plasma TC, LDL-C, and TG levels were observed. In addition, treatment resulted in a decrease in plasma MDA levels and antioxidant enzymes and reduced glutathione peroxidase and catalase by 9%, 31.8%, and 3.14%, respectively, compared to rats fed a high-cholesterol diet (Matanjun et al. 2010).

Hepatoprotection

Hepatic disorders are diseases that affect the liver, which is responsible for many vital processes, including the production of bile to aid digestion, the storage of energy in the form of glycogen, the removal of toxins from the blood, and the metabolization of drugs and hormones. Hepatic diseases cause millions of deaths worldwide every year. There are a variety of forms, from acute, short-term to chronic, long-term diseases. Typical examples of liver diseases are cirrhosis, liver cancer, liver inflammation, etc. (Asrani et al. 2013; Mokdad et al. 2014). Alcohol consumption, obesity, medication use, viral infections, and non-alcoholic fatty liver disease all contribute to the prevalence of liver disease (Morgan 1994; Rehm et al. 2009; Björnsson et al. 2013; World Health Organization 2018; Asrani et al. 2019). Hepatotoxicity involves several processes, one of which is oxidative stress. This process triggers the overproduction of reactive oxygen species (ROS) and leads to changes in the levels of serum liver marker enzymes such as SGOT (serum glutamic-oxaloacetic transaminase), SGPT (serum glutamate pyruvate transaminase), and ALP (alkaline phosphatase) as well as antioxidant enzymes and lipid peroxidation in liver tissue (Shah et al. 2015, 2017; Wardani et al. 2017).

Studies have shown that the extract of *C. lentillifera* can protect the liver from damage caused by oxidative stress by increasing the activity of antioxidant enzymes and reducing lipid peroxidation. In a study, the potential hepatoprotective properties of the methanol extract of *C. lentillifera* against acetaminophen (APAP)-induced liver damage in juvenile zebrafish were investigated. In the control group treated with 10 mM APAP, the fish died. However, when the methanol extract of *C. lentillifera* was introduced at concentrations of 10–30 g/L to the tanks with the 10 mM APAP-exposed groups, there was a significant decrease in fish mortality. In addition, a dose-dependent reduction in hepatocyte swelling, leukocyte infiltration, hepatocyte vacuolization, and liver necrosis was observed under the light microscope in the liver tissue of zebrafish exposed to 10 mM APAP and simultaneously administered *C. lentillifera* extract (Codorniz et al. 2020).

To summarize, *C. lentillifera* is a remarkable natural resource with a spectrum of protective effects that demonstrates its potential as a therapeutic agent (Table 10.1). The seaweed exhibits remarkable antioxidant properties due to its polyphenolic compounds, which have significant free radical scavenging and reducing power. These antioxidant abilities are promising in mitigating degenerative diseases caused by oxidative stress, including cardiovascular diseases, cancer, neurodegenerative diseases, respiratory diseases, and kidney and liver diseases. In addition, *C. lentillifera* exhibits anti-inflammatory properties, as evidenced by its ability to inhibit the overproduction of cytokines, particularly TNF-α and IL-1β, in various cell models. The antimicrobial potential of *C. lentillifera* is evident from its broad-spectrum activity against bacteria, viruses, and fungi, indicating its role as a potential agent to combat microbial infections.

The anti-cancer properties of *C. lentillifera* are emphasized by its effect on various cancer cell lines, such as A172 human glioblastoma cells and MCF-7 human breast cancer cells. The anti-obesity effect of the algae has been demonstrated in

Table 10.1 Summary of the protective effects reported in *C. lentillifera*

Protective effects	*C. lentillifera* extract	Model of study	References
Antioxidant	Solvent extract	2,2′-azino-bis (3-ethylbenzothiazoline-6-sulfonate, linoleic acid emulsion system	Nofiani et al. (2018)
Anti-inflammatory	Sulfated polysaccharide, caulerpin metabolite	HT29 colorectal carcinoma cells, RAW 264.7 cells	Nagappan and Vairappan (2014), Sun et al. (2020)
Anti-cancer	Solvent extract, oligosaccharides	A172 human glioblastoma. Human breast adenocarcinoma cell line MCF-7	Maeda et al. (2012), Tanawoot et al. (2021)
Anti-obesity	Solvent extracts	Wistar albino rats, Sprague-Dawley rats	Matanjun et al. (2010), Manoppo et al. (2022)
Hepatoprotective	Solvent extract	Juvenile zebrafish	Codorniz et al. (2020)
Antimicrobial	Solvent extract	*Escherichia coli, Staphylococcus aureus, Bacillus subtilis, Vibrio cholerae, Citrobacter freundii, Vibrio harveyi, Pseudomonas aeruginosa, Aeromonas hydrophila, Klebsiella pneumoniae, Candida albicans.* MRSA and *E. coli* K1, *Streptococcus* sp., and *Salmonella* sp.	Nagappan and Vairappan (2014), Nofiani et al. (2018), Yap et al. (2019)

studies showing a reduction in body weight and an improvement in lipid profile in rats fed a high-cholesterol and high-fat diet. In addition, *C. lentillifera* exhibits hepatoprotective properties that protect the liver from damage caused by oxidative stress by increasing the activity of antioxidant enzymes and reducing lipid peroxidation. These multiple protective properties make *C. lentillifera* a promising candidate for further research and development in the fields of medicine and functional foods, offering potential solutions to a range of health problems.

10.2.4 Metabolite Profiling of *C. lentillifera* and Its Bioactivity

Analysis of metabolites in *C. lentillifera* has revealed the presence of a variety of bioactive compounds, including carotenoids, fatty acids, alkanes, terpenoids, and alkaloids. These metabolites exhibit various bioactive properties such as antioxidant, anti-cancer, anti-inflammatory, anti-obesity, anti-diabetic, neuroprotective, cardiovascular, and antiparasitic properties. These compounds were identified in the extracts using advanced analytical techniques such as liquid chromatography-mass spectrometry (LCMS), gas chromatography-mass spectrometry (GCMS), and nuclear magnetic resonance (NMR). Some of the metabolites and their activities are listed below.

Carotenoid

The species is a good source of carotenoids. Some of the important carotenoids are described below.

Astaxanthin

Astaxanthin is a red-pigmented carotenoid that is said to have antioxidant and anti-inflammatory, anti-apoptotic, anti-proliferative, neuroprotective, anti-diabetic, and kidney and hepatoprotective properties (Fig. 10.3). The positive effects of astaxanthin are achieved by influencing various molecular signaling pathways in the body. Certain pathways are triggered and inhibited, such as PI3K/AKT, JAK/STAT-3 signaling pathways, Nrf2 (NF-E2-related factor 2), the NF-κB family (nuclear factor kappa-light-chain-enhancer of activated B cells), MARKs (mitogen-activated protein kinases), and PPARγ (peroxisome proliferator–activated receptor-gamma) (Kohandel et al. 2022).

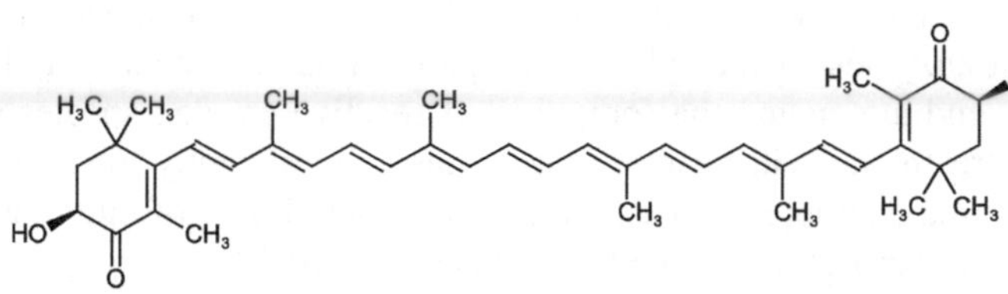

Fig. 10.3 The structure of astaxanthin

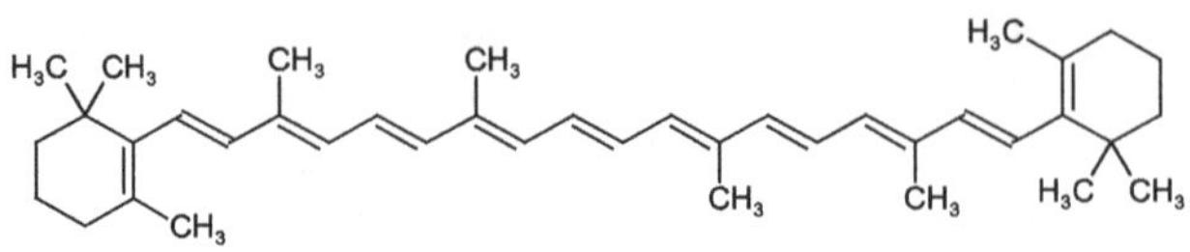

Fig. 10.4 The structure of β-carotene

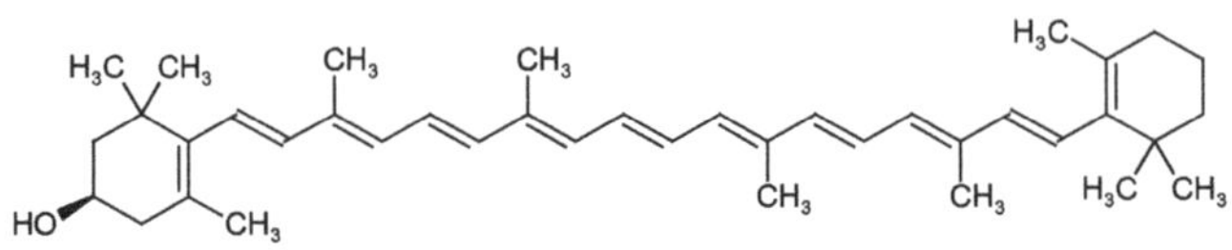

Fig. 10.5 The structure of β-cryptoxanthin

β-Carotene

β-Carotene is a red-orange-pigmented carotenoid characterized by beta rings at both ends of the molecule (Fig. 10.4). It is said to have an antioxidant and anti-inflammatory effect. β-Carotene at a concentration of 100 μM up-regulated heme oxygenase 1 (Hmox1) mRNA in RAW264.7 cells (Kohandel et al. 2022).

β-Cryptoxanthin

It is a pigmented carotenoid that is closely related to β-carotene and differs by the addition of a hydroxyl group (Fig. 10.5). It has been reported to have antioxidant and anti-cancer activities. In vitro experiments have shown that exposure to β-cryptoxanthin significantly inhibits the proliferation of human cervical cancer cells (HeLa) (Lorenzo et al. 2009; Balasubramaniam et al. 2020; Gansukh et al. 2020).

Canthaxanthin

It is a red-orange-pigmented carotenoid. It has been associated with antioxidant, anti-inflammatory, and neuroprotective properties (Fig. 10.6). Antioxidant and anti-inflammatory effects of canthaxanthin were observed at concentrations of 20 μM in nerve growth factor–differentiated PC12 cells injured by hydrogen peroxide (H_2O_2) or 1-methyl-4-phenylpyridinium ions (MPP^+). Pretreatment with canthaxanthin showed a concentration-dependent attenuation of $H_2O_2^-$ or MPP^+-induced cell death, lactate dehydrogenase release, increased DNA fragmentation, and decreased mitochondrial membrane potential (Chan et al. 2009).

Further pretreatment with canthaxanthin significantly maintained the activities of antioxidant enzymes such as glutathione peroxidase and catalase, reduced lipid peroxidation, and attenuated the formation of reactive oxygen species (ROS). In addition, pretreatment with canthaxanthin also significantly restored Na^+-K^+-ATPase activity and suppressed caspase-3 activity and levels of interleukin (IL)-1, IL-6, and tumor necrosis factor (TNF)-α (Chan et al. 2009).

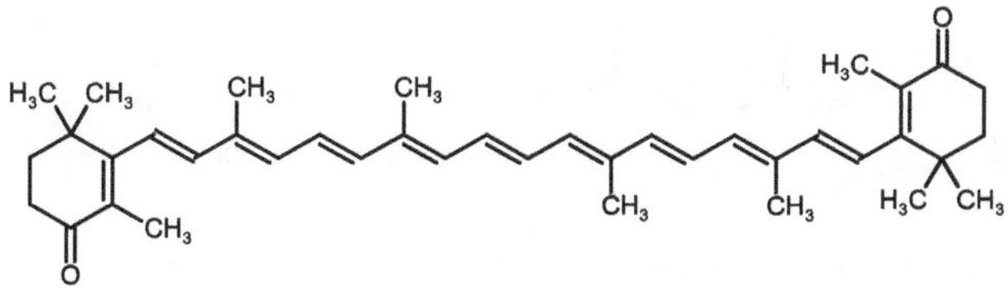

Fig. 10.6 The structure of canthaxanthin

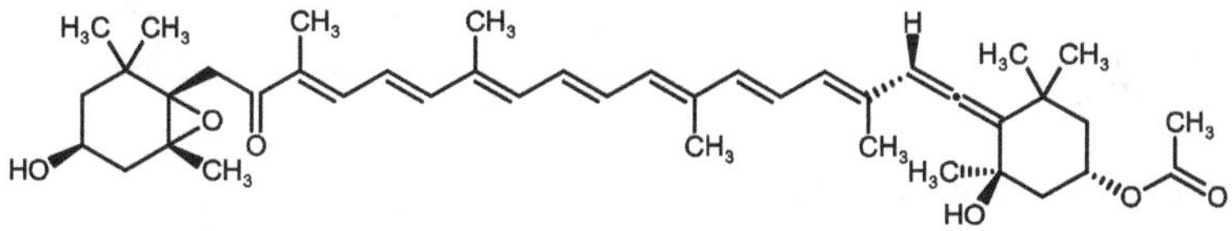

Fig. 10.7 The structure of fucoxanthin

Fucoxanthin

The pharmacologically active pigmented carcinoid has a distinctive molecular structure characterized by a unique mixture of chemical properties (Fig. 10.7). These include, above all, an allenic bond, which contributes to its particular reactivity and adaptability in various functional contexts. In addition to the allenic bond, the compound has a variety of oxygen-containing functional groups, each of which imparts a unique set of properties to the molecule.

Antioxidant, anti-cancer, and anti-inflammatory properties have been reported. Fucoxanthin inhibited the proliferation of melanoma cell lines (B16F10 cells) in a dose-dependent manner, accompanied by the induction of cell cycle arrest and apoptosis. The anti-inflammatory effect of fucoxanthin was demonstrated in the inflammatory bowel disease ulcerative colitis using mice with dextran sulfate sodium (DSS)-induced colitis. At a dosage of 50 and 100 mg/kg/day, it significantly protected against DSS-induced body weight loss, suppressed the increase in the Disease Activity Index, and counteracted the shortening of the colon. In addition, it improved histological damage in the colon, reduced prostaglandin levels in the colon, and attenuated the overexpression of inflammation-related molecules, including COX-2 and NF-κB. Beyond its anti-cancer and anti-inflammatory properties, fucoxanthin also exhibits anti-angiogenic, neuroprotective, cytoprotective, and anti-obesity properties (Kumar et al. 2013; Lourenço-Lopes et al. 2021).

Violaxanthin

The pigment carotenoid belongs to the xanthophylls (Fig. 10.8). The molecular structure of the metabolites consists of several conjugated double bonds and oxygen-containing functional groups. Its antioxidant properties have been reported. In particular, it has shown a strong inhibition of lipid peroxidation (Araki et al. 2016; Syakilla et al. 2022).

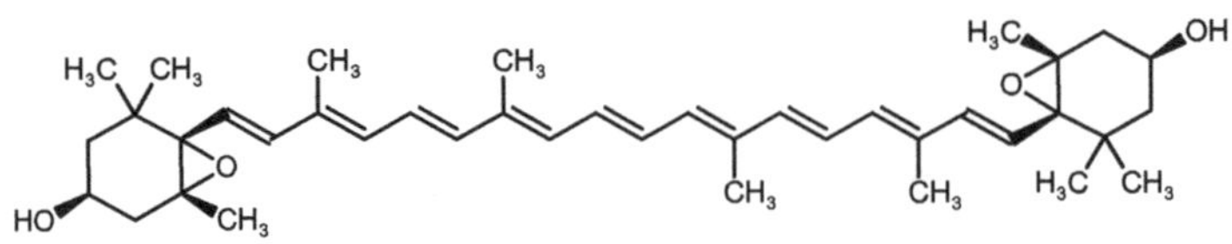

Fig. 10.8 The structure of violaxanthin

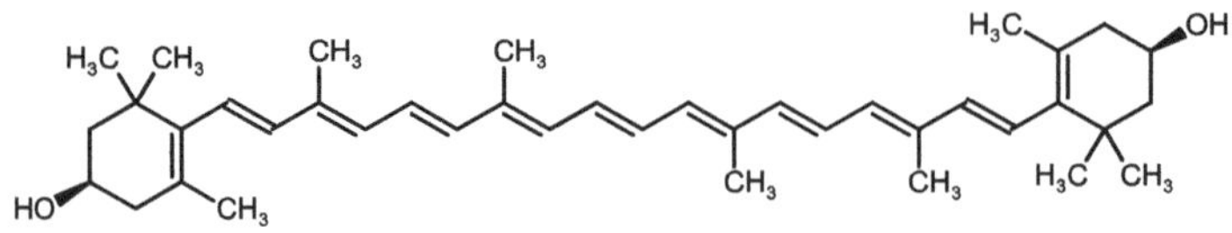

Fig. 10.9 The structure of zeaxanthin

Zeaxanthin

It is a xanthophyll carotenoid with promising nutraceutical effects (Fig. 10.9). Its antioxidant properties have been reported (Ngginak et al. 2017; Balasubramaniam et al. 2020).

10.2.5 Fatty Acid

Docosahexaenoic Acid and Eicosapentaenoic Acid

Both are long-chain omega-3 fatty acids. The number and position of their double bonds and the length of their carbon chains determine the basic structural differences between docosahexaenoic acid and eicosapentaenoic acid. Docosahexaenoic acid contains 22 carbon atoms (Fig. 10.10), while eicosapentaenoic acid contains 20 (Fig. 10.11) and is said to have antioxidant, anti-inflammatory, and anti-cancer properties.

Docosahexaenoic acid and eicosapentaenoic acid have been shown to efficiently inhibit the formation of reactive oxygen species from arachidonic acid, which is mediated by the COX-2 and XOD pathways. In addition, glutathione (GSH) levels were increased, and the activities of antioxidant enzymes were improved. They also reduced nitrite levels and inflammatory mediators such as thromboxane B2, prostaglandin E2, and 6-keto-prostaglandin F1α. The administration of docosahexaenoic acid and eicosapentaenoic acid led to a significant reduction in paw edema and the inflammatory markers tumor necrosis factor-alpha (TNF-α) (You and Hae 2007; Nobre et al. 2013). Docosahexaenoic acid has shown anti-cancer effects by inducing oxidative stress and apoptosis, inhibiting Wnt/β-catenin signaling (controls important processes in cells such as apoptosis, differentiation, migration, proliferation, and stem cell renewal), and suppressing the degradation of extracellular cellular matrix proteins and the expression of growth factors in pancreatic cancer cells (Park and Kim 2017).

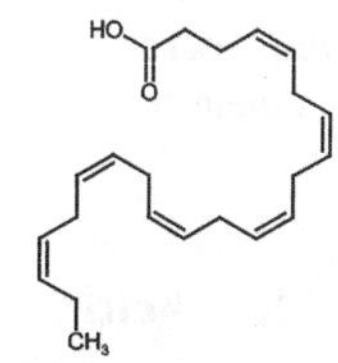

Fig. 10.10 The structure of docosahexaenoic acid

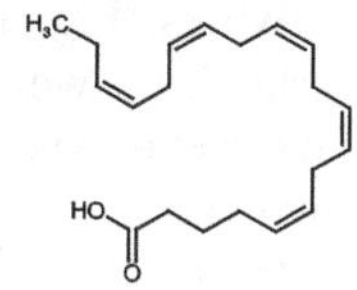

Fig. 10.11 The structure of eicosapentaenoic acid

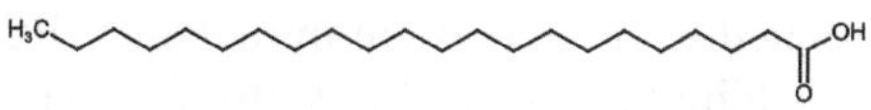

Fig. 10.12 The structure of behenic acid

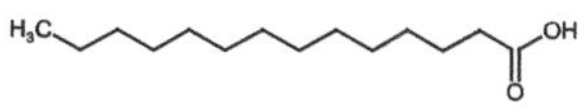

Fig. 10.13 The structure of myristic acid

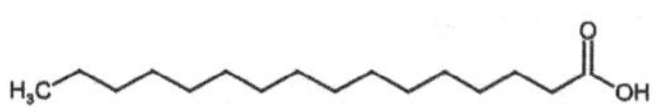

Fig. 10.14 The structure of palmitic acid

Behenic Acid

It is a saturated fatty acid (Fig. 10.12). It has been reported to have antimicrobial activity against *Agrobacterium tumefaciens* and *Ralstonia solanacearum* strains with a minimum inhibitory concentration (MIC) of 250 μg/mL (Xie et al. 2021).

Myristic Acid

The saturated fatty acid is said to have anti-diabetic and anti-inflammatory properties (Fig. 10.13). Administration of myristic acid at a concentration of 300 mg/kg led to an improvement in glucose tolerance and a reduction in insulin-dependent blood glucose levels, as well as an increase in diacylglycerol kinase (DGK) δ levels (a key enzyme involved in the exacerbation of type 2 diabetes) in the skeletal muscle of Nagoya-Shibata-Yasuda (NSY) mice. These mice are used to study type 2 diabetes in the context of obesity [64]. In addition, myristic acid downregulated the expression of interleukin (IL)-1β, IL-6, and TNF-α in the BV-2 mouse microglia cell line (Huang et al. 2023).

Palmitic Acid

Saturated fatty acid is known for its anti-tumor effect (Fig. 10.14). Palmitic acid shows its anti-tumor effect by regulating the PI3K/AKT signaling pathway, which is essential for tumor formation and metastasis (Wang et al. 2023).

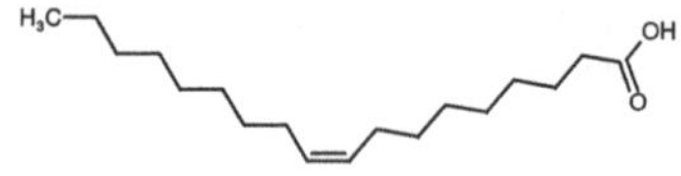

Fig. 10.15 The structure of oleic acid

Oleic Acid

It is a monounsaturated fatty acid. Oleic acid has been reported to have anti-tumor activities (Fig. 10.15). It has the potential to inhibit the overexpression of HER2 (erbB-2), a well-studied oncogene that is important for the development, spread, and metastasis of various human malignancies. In addition, oleic acid may be involved in calcium signaling pathways in cells associated with the proliferation process. Oleic acid has been shown to induce apoptosis, or cell death, in cancer cells. The mechanisms behind the apoptotic event triggered by oleic acid could be related to an increase in intracellular ROS formation or the activation of caspase-3. In addition, oleic acid was also found to have an antibacterial effect. The metabolite inhibited the growth of three gram-positive bacteria at a MIC of 1.0 mg/mL (Dilika et al. 2000; Carrillo et al. 2012).

10.2.6 Alkane

Eicosane

It is categorized as an alkane. Eicosane has a remarkable potential to promote wound healing and exhibits antioxidant activities (Fig. 10.16). The metabolite showed IC_{50} values of 56 (μg/mL). In particular, in a diabetic rat wound model, the administration of eicosane proved to be effective in accelerating the wound healing process. This observed improvement in wound healing is attributed to the strong free radical scavenging properties of eicosane (Balachandran et al. 2023). Eicosane has been reported to exhibit very potent anti-inflammatory, analgesic, and anti-pyretic effects when acting on formalin, carrageenan, acetic acid, and Brewer's yeast–induced inflammation in rats. The activities may be due to inhibition of the release of histamine, bradykinin, prostaglandins, thromboxanes, and leukotrienes as well as decreased stimulation of TNF, IL-1b, IL-6, and interferon-α and interaction with the Cox enzyme (Okechukwu 2020).

10.2.7 Terpenoid

Azulene

It is classified as a terpenoid and has reportedly shown anti-inflammatory properties in the rat model (Fig. 10.17). After exposure, the compound showed a significant reduction in erythema on the skin of irradiated rats compared to the control group (Guarrera et al. 2001).

Limonene

It is a terpenoid compound that has shown antiparasitic activities (Fig. 10.18). Its efficacy was observed in the inhibition of troublesome *Leishmania amazonensis*

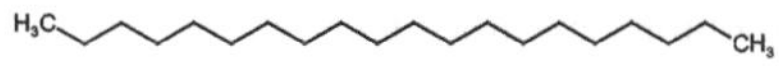

Fig. 10.16 The structure of eicosane

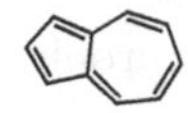

Fig. 10.17 The structure of azulene

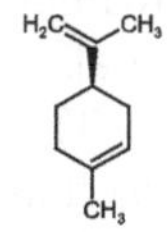

Fig. 10.18 The structure of limonene

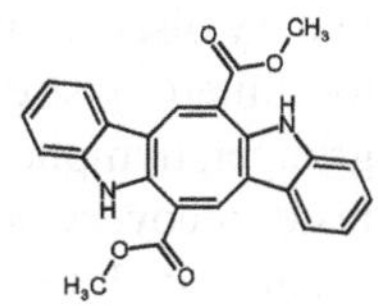

Fig. 10.19 The structure of caulerpin

promastigotes and *amastigotes* with 50% inhibitory doses of 252.0 ± 49.0 and 147.0 ± 46.0 μM, respectively. This inhibitory effect also extends to other Leishmania species. Treatment of macrophages infected with *L. amazonensis* with 300 μM Limonene resulted in a significant reduction in infection rates by 78%. In addition, topical or intrarectal application of Limonene to *L. amazonensis*-infected mice resulted in a considerable reduction in lesion diameters, which was accompanied by a significant decrease in parasite burden within the treated lesions, as confirmed by histological examinations (Arruda et al. 2009).

10.2.8 Alkaloid

Caulerpin

It is an alkaloid that has shown antinociceptive and anti-inflammatory properties in various tests, including convulsions, hot plate, formalin-induced pain, capsaicin-induced ear edema, and carrageenan-induced peritonitis (Fig. 10.19). It reduced nociception, inhibited the effect of formalin, and showed anti-inflammatory properties when administered orally at a concentration of 100 mol/kg (De Souza et al. 2009). Caulerpin also showed antiviral activity against Chikungunya virus infections. The isolated compound showed high virucidal activity and inhibited about 90% of Chikungunya virus infectivity at a concentration of 5 μM (Esteves et al. 2019).

10.3 Recommendation and Future Vision of *C. lentillifera* Metabolites for Human Health and Sustainable Aquaculture

As mentioned earlier, *C. lentillifera* is a good source of metabolites. The beneficial effects of *C. lentillifera* metabolites can be extended to both human and animal health, especially in the aquaculture industry, which is heavily dependent on chemicals. The antioxidant metabolites such as astaxanthin, β-carotene, β-cryptoxanthin, violaxanthin, zeaxanthin, docosahexaenoic acid, eicosapentaenoic acid, and eicosane can help defend against oxidative stress and reduce the risk of chronic disease in humans, while in the aquaculture industry, the inclusion of these metabolites in fish feed can increase their resistance to environmental stress, leading to improved growth and health (Chan et al. 2009; Nobre et al. 2013; Balasubramaniam et al. 2020; Lourenço-Lopes et al. 2021; Kohandel et al. 2022).

The anti-inflammatory and anti-cancer properties of *C. lentillifera* metabolites such as astaxanthin, β-carotene, canthaxanthin, fucoxanthin, myristic acid, palmitoleic acid, oleic acid, and azulene may have potential for the treatment of inflammatory diseases in humans. In aquaculture, where fish can suffer inflammation due to numerous stresses, integrating these metabolites in the diet of species such as grouper, humphead wrasse, and sea bass may help to reduce inflammation, promote faster recovery, and improve the overall well-being of the fish (Arruda et al. 2009; Chan et al. 2009; Nobre et al. 2013; Balasubramaniam et al. 2020; Lourenço-Lopes et al. 2021; Kohandel et al. 2022; Huang et al. 2023).

The antipathogenic effects of *C. lentillifera* metabolites, such as behenic acid and limonene, could have potential applications in human health and contribute to the development of innovative therapeutic agents against infectious diseases. The antibacterial and antiparasitic effects are particularly important in aquaculture. Aquaculture operators can explore natural ways to treat bacterial infections and parasites in grouper, humphead wrasse, sea bass, etc. by incorporating these metabolites into their feed, thus reducing reliance on drugs and supporting more sustainable aquaculture techniques (Arruda et al. 2009; Xie et al. 2021).

Further research in this area is needed to better understand the molecular mechanisms underlying these benefits. This could lead to creative applications that benefit both human consumers and the aquaculture industry in terms of health, nutrition, and sustainable practices.

10.4 Conclusion

In conclusion, *C. lentillifera* is a promising resource with important implications in various fields. The species contains a variety of bioactive metabolites that are antioxidant, anti-inflammatory, antibacterial, anti-cancer, anti-obesity, and hepatoprotective, including carotenoids, omega-3 fatty acids, and alkaloids. According to these data, *C. lentillifera* is a promising source of valuable natural medicine. Both the healthcare and aquaculture industries can benefit from its applications, which include the prevention and treatment of a wide range of diseases, including microbial infections, cytotoxicity, parasitic infestations, oxidative stress, and inflammation. Further research is needed to fully understand the potential of *C. lentillifera* and to extract and purify its bioactive metabolites for drug development and other applications.

Acknowledgement The study was funded by the Ministry of Higher Education Malaysia through the Higher Institutions Centre of Excellence (HiCoE) Research Grant (Grant code: HIC2404).

References

American Cancer Society (2023). What is cancer? https://www.cancer.org/research/cancer-facts-statistics/all-cancer-facts-figures/2023-cancer-facts-figures.html. Accessed on March 27, 2023.

Araki M, Kaku N, Harada M, Ando Y, Yamaguchi R, Shindo K (2016) Production of auroxanthins from violaxanthin and 9-cis-violaxanthin by acidic treatment and the antioxidant activities of violaxanthin, 9-cis-violaxanthin, and auroxanthins. J Agric Food Chem 64(49):9352. https://doi.org/10.1021/acs.jafc.6b04506

Arruda DC, Miguel DC, Yokoyama-Yasunaka JKU, Katzin AM, Uliana SRB (2009) Inhibitory activity of limonene against Leishmania parasites in vitro and in vivo. Biomed Pharmacother 63(9):643–649. https://doi.org/10.1016/j.biopha.2009.02.004

Asrani SK, Larson JJ, Yawn B, Therneau TM, Kim WR (2013) Underestimation of liver-related mortality in the United States. Gastroenterology 145(2):375–82.e2. https://doi.org/10.1053/j.gastro.2013.04.005

Asrani SK, Devarbhavi H, Eaton J, Kamath PS (2019) Burden of liver diseases in the world. J Hepatol 70(1):151–171. https://doi.org/10.1016/j.jhep.2018.09.014

Balachandran A, Choi SB, Beata MM, Małgorzata J, Froemming GRA, Lavilla CA, Billacura MP, Siyumbwa SN, Okechukwu PN (2023) Antioxidant, wound healing potential and in silico assessment of naringin, eicosane and octacosane. Molecules 28(3):1043. https://doi.org/10.3390/MOLECULES28031043/S1

Balasubramaniam V, June Chelyn L, Vimala S, Mohd Fairulnizal MN, Brownlee IA, Amin I (2020) Carotenoid composition and antioxidant potential of Eucheuma denticulatum, Sargassum polycystum and Caulerpa lentillifera. Heliyon 6(8):e04654. https://doi.org/10.1016/j.heliyon.2020.e04654

Björnsson ES, Bergmann OM, Björnsson HK, Kvaran RB, Olafsson S (2013) Incidence, presentation, and outcomes in patients with drug-induced liver injury in the general population of Iceland. Gastroenterology 144(7):1419–25.e3. https://doi.org/10.1053/j.gastro.2013.02.006

Carrillo C, Cavia MM, Alonso-Torre SR (2012) Antitumor effect of oleic acid; mechanisms of action: a review. Nutr Hosp 27(6):1860. https://doi.org/10.3305/NH.2012.27.6.6010

Chan KC, Mong MC, Yin MC (2009) Antioxidative and anti-inflammatory neuroprotective effects of astaxanthin and canthaxanthin in nerve growth factor differentiated PC12 cells. J Food Sci 74(7):H225–H231. https://doi.org/10.1111/j.1750-3841.2009.01274.x

Chen X, Sun Y, Liu H, Liu S, Qin Y, Li P (2019) Advances in cultivation, wastewater treatment application, bioactive components of Caulerpa lentillifera and their biotechnological applications. PeerJ 7:e6118. https://doi.org/10.7717/peerj.6118

Codorniz KD, Marquina REM, Nolasco ADG, Palencia PDD, Mata SB (2020) Evaluation of the hepatoprotective effect of methanolic extract of Caulerpa lentillifera against acetaminophen-induced liver toxicity in juvenile zebrafish (Danio rerio). Jurnal Ilmiah Farmasi 16(1):31–38. https://doi.org/10.20885/JIF.VOL16.ISS1.ART4

De Souza ÉT, De Lira DP, De Queiroz AC, Da Silva DJC, De Aquino AB, Campessato Mella EA, Lorenzo VP, De Miranda GEC, De Araújo-Júnior JX, De Oliveira Chaves MC, Barbosa-Filho JM, De Athayde-Filho PF, De Oliveira Santos BV, Alexandre-Moreira MS (2009) The antinociceptive and anti-inflammatory activities of caulerpin, a bisindole alkaloid isolated from seaweeds of the genus Caulerpa. Mar Drugs 7(4):689. https://doi.org/10.3390/md7040689

Dilika F, Bremner PD, Meyer JJM (2000) Antibacterial activity of linoleic and oleic acids isolated from Helichrysum pedunculatum: a plant used during circumcision rites. Fitoterapia 71(4):450–452. https://doi.org/10.1016/S0367-326X(00)00150-7

Dmytryk A, Tuhy Ł, Chojnacka K (2017) Algae as source of pharmaceuticals. In: Prospects and challenges in algal biotechnology. Springer, Singapore, pp 295–310. https://doi.org/10.1007/978-981-10-1950-0_11

Esteves PO, de Oliveira MC, de Souza Barros C, Cirne-Santos CC, Laneuvlille VT, Palmer Paixão IC (2019) Antiviral effect of caulerpin against Chikungunya. Nat Prod Commun 14(10):1934578X1987829. https://doi.org/10.1177/1934578X19878295/ASSET/IMAGES/LARGE/10.1177_1934578X19878295-FIG5.JPEG

Fleit HB (2014) Chronic inflammation. In: Pathobiology of human disease. Elsevier, pp 300–314. https://doi.org/10.1016/B978-0-12-386456-7.01808-6
Gansukh E, Nile A, Sivanesan I, Rengasamy KRR, Kim DH, Keum YS, Saini RK (2020) Chemopreventive effect of β-cryptoxanthin on human cervical carcinoma (HeLa) cells is modulated through oxidative stress-induced apoptosis. Antioxidants 9(1):28. https://doi.org/10.3390/antiox9010028
Guarrera M, Turbino L, Rebora A (2001) The anti-inflammatory activity of azulene. J Eur Acad Dermatol Venereol 15(5):486–487. https://doi.org/10.1046/j.1468-3083.2001.00340.x
Guiry MD, Guiry GM (2020) Algaebase: listing the world's algae. https://www.algaebase.org/
Hengjie T, Das SK, Zainee NFA, Yana R, Rozaimi M (2023) Ocean acidification and aquacultured seaweeds: progress and knowledge gaps. J Mar Sci Eng 11(1):78. https://doi.org/10.3390/jmse11010078
Huang Q, Chen C, Zhang Z, Xue Q (2023) Anti-inflammatory effects of myristic acid mediated by the NF-κB pathway in lipopolysaccharide-induced BV-2 microglial cells. Mol Omics 19(9):726. https://doi.org/10.1039/d3mo00063j
Hussin H, Khoso A (2017) Seaweed cultivation and coastal communities in Malaysia: an overview. Asian Fish Sci 30(2):87–100. https://doi.org/10.33997/j.afs.2017.30.2.003
Kadam SU, Prabhasankar P (2010) Marine foods as functional ingredients in bakery and pasta products. Food Res Int 43(8):1975. https://doi.org/10.1016/j.foodres.2010.06.007
Klongklaew N, Praiboon J, Tamtin M, Srisapoome P (2020) Antibacterial and antiviral activities of local Thai green macroalgae crude extracts in pacific white shrimp (Litopenaeus vannamei). Mar Drugs 18(3):140. https://doi.org/10.3390/md18030140
Kohandel Z, Farkhondeh T, Aschner M, Pourbagher-Shahri AM, Samarghandian S (2022) Anti-inflammatory action of astaxanthin and its use in the treatment of various diseases. Biomed Pharmacother 145:112179. https://doi.org/10.1016/j.biopha.2021.112179
Kotas ME, Medzhitov R (2015) Homeostasis, inflammation, and disease susceptibility. Cell 160(5):816–827. https://doi.org/10.1016/j.cell.2015.02.010
Kumar CH, Ramesh A, Kumar JNS, Pradesh A, Pradesh A, Pradesh A (2013) A review on hepatoprotective activity of medicinal plants. J Med Plants Stud 6(1):1–12
Lordan S, Ross RP, Stanton C (2011) Marine bioactives as functional food ingredients: potential to reduce the incidence of chronic diseases. Mar Drugs 9(6):1056. https://doi.org/10.3390/md9061056
Lorenzo Y, Azqueta A, Luna L, Bonilla F, Domínguez G, Collins AR (2009) The carotenoid β-cryptoxanthin stimulates the repair of DNA oxidation damage in addition to acting as an antioxidant in human cells. Carcinogenesis 30(2):308–314. https://doi.org/10.1093/carcin/bgn270
Lourenço-Lopes C, Fraga-Corral M, Jimenez-Lopez C, Carpena M, Pereira AG, Garcia-Oliveira P, Prieto MA, Simal-Gandara J (2021) Biological action mechanisms of fucoxanthin extracted from algae for application in food and cosmetic industries. Trends Food Sci Technol 117:163–181. https://doi.org/10.1016/j.tifs.2021.03.012
Maeda R, Ida T, Ihara H, Sakamoto T (2012) Induction of apoptosis in MCF-7 cells by β-1,3-xylooligosaccharides prepared from Caulerpa lentillifera. Biosci Biotechnol Biochem 76(5):1032–1034. https://doi.org/10.1271/bbb.120016
Manoppo JIC, Nurkolis F, Pramono A, Ardiaria M, Murbawani EA, Yusuf M, Qhabibi FR, Yusuf VM, Amar N, Karim MRA, Subali AD, Natanael H, Rompies R, Halim RF, Bolang ASL, Joey G, Novianto CA, Permatasari HK (2022) Amelioration of obesity-related metabolic disorders via supplementation of Caulerpa lentillifera in rats fed with a high-fat and high-cholesterol diet. Front Nutr 9:2124. https://doi.org/10.3389/fnut.2022.1010867
Masschelin PM, Cox AR, Chernis N, Hartig SM (2020) The impact of oxidative stress on adipose tissue energy balance. Front Physiol 10:1638. https://doi.org/10.3389/fphys.2019.01638
Matanjun P, Mohamed S, Mustapha NM, Muhammad K, Ming CH (2008) Antioxidant activities and phenolics content of eight species of seaweeds from North Borneo. J Appl Phycol 20(4):367–373. https://doi.org/10.1007/s10811-007-9264-6
Matanjun P, Mohamed S, Mustapha NM, Muhammad K (2009) Nutrient content of tropical edible seaweeds, Eucheuma cottonii, Caulerpa lentillifera and Sargassum polycystum. J Appl Phycol 21(1):75–80. https://doi.org/10.1007/s10811-008-9326-4

Matanjun P, Mohamed S, Muhammad K, Mustapha NM (2010) Comparison of cardiovascular protective effects of tropical seaweeds, Kappaphycus alvarezii, Caulerpa lentillifera, and Sargassum polycystum, on high-cholesterol/high-fat diet in rats. J Med Food 13(4):792–800. https://doi.org/10.1089/jmf.2008.1212

Medzhitov R (2008) Origin and physiological roles of inflammation. Nature 454(7203):428–435. https://doi.org/10.1038/nature07201

Mokdad AA, Lopez AD, Shahraz S, Lozano R, Mokdad AH, Stanaway J, Murray CJ, Naghavi M (2014) Liver cirrhosis mortality in 187 countries between 1980 and 2010: a systematic analysis. BMC Med 12(1):145. https://doi.org/10.1186/s12916-014-0145-y

Morgan MY (1994) The prognosis and outcome of alcoholic liver disease. Alcohol Alcohol Suppl 2:335–343

Murugan K, Iyer VV (2013) Differential growth inhibition of cancer cell lines and antioxidant activity of extracts of red, brown, and green marine algae. In Vitro Cell Dev Biol Anim 49(5):324–334. https://doi.org/10.1007/s11626-013-9603-7

Nagappan T, Vairappan CS (2014) Nutritional and bioactive properties of three edible species of green algae, genus Caulerpa (Caulerpaceae). J Appl Phycol 26(2):1019. https://doi.org/10.1007/s10811-013-0147-8

Ngginak J, Mangibulude JC, Rondonuwu FS (2017) The identification of carotenoids and testing of carotenoid antioxidants from sand lobster (Panulirus homarus). Egg Extract 22(3):155–160. https://doi.org/10.14710/ik.ijms.22.3.155-160

NIH (2007) Understanding emerging and re-emerging infectious diseases—National Institute of Health (NIH) curriculum supplement series—NCBI Bookshelf. https://www.ncbi.nlm.nih.gov/books/NBK20370/

Nobre MEP, Correia AO, Borges MB, Sampaio TMA, Chakraborty SA, Gonçalves DO, Brito GAC, Leal LKAM, Felipe CFB, Lucetti DL, Arida RM, De Barros Viana GS (2013) Eicosapentaenoic acid and docosahexaenoic acid exert anti-inflammatory and antinociceptive effects in rodents at low doses. Nutr Res 33(5):422. https://doi.org/10.1016/j.nutres.2013.02.011

Nofiani R, Hertanto S, Zaharah TA, Gafur S (2018) Proximate compositions and biological activities of Caulerpa lentillifera. Mol Ther 13(2):141. https://doi.org/10.20884/1.jm.2018.13.2.441

Okechukwu PN (2020) Evaluation of anti-inflammatory, analgesic, antipyretic effect of eicosane, pentadecane, octacosane, and heneicosane. Asian J Pharm Clin Res 13:29–35. https://doi.org/10.22159/ajpcr.2020.v13i4.36196

Park M, Kim H (2017) Anti-cancer mechanism of docosahexaenoic acid in pancreatic carcinogenesis: a mini-review. J Cancer Prev 22(1):1. https://doi.org/10.15430/JCP.2017.22.1.1

Pizzino G, Irrera N, Cucinotta M, Pallio G, Mannino F, Arcoraci V, Squadrito F, Altavilla D, Bitto A (2017) Oxidative stress: harms and benefits for human health. Oxidative Med Cell Longev 2017:1–13. https://doi.org/10.1155/2017/8416763

Rehm J, Mathers C, Popova S, Thavorncharoensap M, Teerawattananon Y, Patra J (2009) Global burden of disease and injury and economic cost attributable to alcohol use and alcohol-use disorders. Lancet 373:2223. https://doi.org/10.1016/S0140-6736(09)60746-7

Rodrigues JAG, Vanderlei ESO, Silva LMCM, Araújo IWF, Queiroz INL, Paula GA, Abreu TM, Ribeiro NA, Bezerra MM, Chaves HV, Lima V, Jorge RJB, Monteiro HSA, Leite EL, Benevides NMB (2012) Antinociceptive and anti-inflammatory activities of a sulfated polysaccharide isolated from the green seaweed Caulerpa cupressoides. Pharmacol Rep 64(2):282–292. https://doi.org/10.1016/s1734-1140(12)70766-1

Saito H, Lal TM (2019) Antimycotic activity of seaweed extracts (Caulerpa lentillifera and Eucheuma cottonii) against two genera of marine oomycetes, Lagenidium spp. and Haliphthoros spp. Biocontrol Sci 24(2):73–80. https://doi.org/10.4265/BIO.24.73

Saravanan KR, Ilangovan K, Khan AB (2008) Floristic and macro faunal diversity of Pondicherry mangroves, South India. Trop Ecol 49(1):91–94

Shah MD, Iqbal M (2010) Diazinon-induced oxidative stress and renal dysfunction in rats. Food Chem Toxicol 48(12):3345–3353. https://doi.org/10.1016/j.fct.2010.09.003

Shah MD, Iqbal M (2018) Antioxidant activity, phytochemical analysis and total polyphenolics content of essential oil, methanol extract and methanol fractions from Commelina nudiflora. Int J Pharm Pharm Sci 10(8):36. https://doi.org/10.22159/ijpps.2018v10i8.25922

Shah MD, Gnanaraj C, Haque AE, Iqbal M (2015) Antioxidative and chemopreventive effects of *Nephrolepis biserrata* against carbon tetrachloride (CCl_4)-induced oxidative stress and hepatic dysfunction in rats. Pharm Biol 53(1):31–39. https://doi.org/10.3109/13880209.2014.909502

Shah MD, D'Souza UJA, Iqbal M (2017) The potential protective effect of Commelina nudiflora L. against carbon tetrachloride (CCl_4)-induced hepatotoxicity in rats, mediated by suppression of oxidative stress and inflammation. Environ Health Prev Med 22(1):1–19. https://doi.org/10.1186/s12199-017-0673-0

Shah MD, Seelan Sathiya Seelan J, Iqbal M (2020) Phytochemical investigation and antioxidant activities of methanol extract, methanol fractions and essential oil of Dillenia suffruticosa leaves. Arab J Chem 13(9):7170–7182. https://doi.org/10.1016/j.arabjc.2020.07.022

Shah MD, Maran BAV, Shaleh SRM, Zuldin WH, Gnanaraj C, Yong YS (2022) Therapeutic potential and nutraceutical profiling of North Bornean seaweeds: a review. Mar Drugs 20(2):101. https://doi.org/10.3390/md20020101

Siegel RL, Miller KD, Fuchs HE, Jemal A (2021) Cancer statistics, 2021. CA Cancer J Clin 71(1):7. https://doi.org/10.3322/caac.21654

Stuthmann LE, Brix da Costa B, Springer K, Kunzmann A (2023) Sea grapes (Caulerpa lentillifera J. Agardh, Chlorophyta) for human use: structured review on recent research in cultivation, nutritional value, and post-harvest management. J Appl Phycol 35(6):2957–2983. https://doi.org/10.1007/s10811-023-03031-x

Sun Y, Liu Z, Song S, Zhu B, Zhao L, Jiang J, Liu N, Wang J, Chen X (2020) Anti-inflammatory activity and structural identification of a sulfated polysaccharide CLGP4 from Caulerpa lentillifera. Int J Biol Macromol 146:931–938. https://doi.org/10.1016/j.ijbiomac.2019.09.216

Syakilla N, George R, Chye FY, Pindi W, Mantihal S, Wahab NA, Fadzwi FM, Gu PH, Matanjun P (2022) A review on nutrients, phytochemicals, and health benefits of green seaweed, Caulerpa lentillifera. Food Secur 11(18):2832. https://doi.org/10.3390/foods11182832

Tanawoot V, Vivithanaporn P, Siangcham T, Meemon K, Niamnont N, Sobhon P, Tamtin M, Sangpairoj K (2021) Hexane extract of seaweed Caulerpa lentillifera inhibits cell proliferation and induces apoptosis of human glioblastoma cells. Sci Technol Asia 26(2):128. https://doi.org/10.14456/scitechasia.2021.32

Wang X, Zhang C, Bao N (2023) Molecular mechanism of palmitic acid and its derivatives in tumor progression. Front Oncol 13:1224125. https://doi.org/10.3389/fonc.2023.1224125

Wardani G, Farida N, Andayani R, Kuntoro M, Sudjarwo S (2017) The potency of red seaweed (Eucheuma cottonii) extracts as hepatoprotector on lead acetate-induced hepatotoxicity in mice. Pharm Res 9(3):282–286. https://doi.org/10.4103/pr.pr_69_16

World Health Organization (2018) Global burden of disease estimates for 2000–2012. WHO. pp 1–41. http://www.who.int/gho/mortality_burden_disease/en/index.html

World Health Organization (2023) Cancer. https://www.who.int/health-topics/cancer#tab=tab_1

Xie Y, Peng Q, Ji Y, Xie A, Yang L, Mu S, Li Z, He T, Xiao Y, Zhao J, Zhang Q (2021) Isolation and identification of antibacterial bioactive compounds from Bacillus megaterium L2. Front Microbiol 12:645484. https://doi.org/10.3389/fmicb.2021.645484

Yap WF, Tay V, Tan SH, Yow YY, Chew J (2019) Decoding antioxidant and antibacterial potentials of Malaysian green seaweeds: Caulerpa racemosa and Caulerpa lentillifera. Antibiotics 8(3):152. https://doi.org/10.3390/antibiotics8030152

You JK, Hae YC (2007) Antioxidative and anti-inflammatory actions of docosahexaenoic acid and eicosapentaenoic acid in renal epithelial cells and macrophages. J Med Food 10(2):225. https://doi.org/10.1089/jmf.2006.092

Zhang M, Ma Y, Che X, Huang Z, Chen P, Xia G, Zhao M (2020) Comparative analysis of nutrient composition of Caulerpa lentillifera from different regions. J Ocean Univ China 19(2):439. https://doi.org/10.1007/s11802-020-4222-x

Advances in Population Genetics and Aquaculture of Penaeid Shrimp: Integrating Morphometric Characters and Molecular Markers

11

Noora Barzkar, Saeid Tamadoni Jahromi, Rafidah Othman, Sitti Raehanah Muhamad Shaleh, Rossita Shapawi, and Ching Fui Fui

Abstract

This chapter delves into the use of morphometric features and molecular markers in population genetics and aquaculture, specifically focusing on crustacean species, particularly penaeid shrimps. Combining morphometric analysis and molecular techniques has become a potent method for addressing issues related to species identification, population structure, and genetic diversity within aquaculture systems. Recent research has employed morphometric features to distinguish between fishery stocks and species, emphasizing the importance of considering various factors that influence morphological variability. Molecular markers like mitochondrial DNA (mtDNA) and microsatellites have been extensively utilized to uncover genetic relationships and population structure in penaeid shrimps, aiding in phylogenetic studies, stock identification, and assessment of genetic diversity in both wild and cultured populations. This section also discusses the usefulness of principal component analysis (PCA) and discriminant function analysis (DFA) in morphometric studies, providing insights into patterns of covariation and shape variation among populations. Additionally, it explores the benefits of microsatellite markers, such as their high polymorphism and co-dominant nature, in population genetics research. Case studies showcase how morphometric analysis and molecular markers are applied in species delin-

N. Barzkar · R. Othman · S. R. Muhamad Shaleh · R. Shapawi · C. F. Fui
Higher Institution Centre of Excellence (HICoE), Borneo Marine Research Institute, Universiti Malaysia Sabah, Kota Kinabalu, Sabah, Malaysia

S. Tamadoni Jahromi (✉)
Persian Gulf and Oman Sea Ecology Research Centre, Iranian Fisheries Sciences Research Institute, Agricultural Research Education and Extension Organization (AREEO), Bandar Abbas, Iran

N. M. Faudzi et al. (eds.), *Essentials of Aquaculture Practices*,
https://doi.org/10.1007/978-981-97-6699-4_11

eation, phylogenetic reconstruction, and population genetic studies of penaeid shrimps. The integration of these approaches contributes to our understanding of evolutionary relationships, population dynamics, and aquaculture management strategies.

Keywords

mtDNA · Microsatellite markers · Genetic diversity · Population dynamics

11.1 Introduction

Identifying marine resource stocks is essential for sustainable fisheries (Barzkar et al. 2022, 2023a, b, 2024a, b). In addition to genetic and environmental methods, morphometric analysis plays a significant role in stock detection (Robalino et al. 2011; Rawat et al. 2017). Recent studies continue to stress the importance of stock identification in fisheries management. Cadrin (2000) research underscores the value of morphometric analysis in accurately identifying and distinguishing between different fish stocks, thereby contributing to sustainable fisheries practices. Tzeng (2004) suggested that animals with minor morphometric differences can also form a stock, an approach widely used in fishery stock differentiation studies. Morphological variability has been extensively used in recent years to explore taxonomic and stock relationships in marine and freshwater fishes (Prado et al. 2006).

Understanding genetic definitions of stocks is crucial for sustainably managing natural resources. Different broodstocks or seed stocks often exhibit varying growth rates, disease resistance, or other important characteristics (Sugama et al. 2002). Data on these factors and their associations with specific genotypes are critical for developing a systematic aquaculture program. Significant advances have been made in recent years to facilitate the ease and utilization of molecular markers (Ramesh et al. 2020). Such markers will assist in the development of penaeid broodstock selection. One of the most efficient current methods for determining differentiation between and within species is through the use of mitochondrial DNA (mtDNA) and microsatellites. Halim et al. (2021) studied the population genetics of the commercially important green tiger prawn (*Penaeus semisulcatus*) in the Indo-Pacific Ocean. In another study, Vu et al. (2021) investigated the use of both neutral and outlier single nucleotide polymorphism (SNP) datasets to determine the population structure, connectivity, and local adaptation of *Penaeus monodon* populations from throughout the Indo-Pacific region. Ren (2020) examined genetic diversity and relatedness among domesticated culture lines, generating information that was used to produce a base population with broad genetic variation.

The result of subsequent quantitative genetic analyses has shown that growth and several key reproductive traits can theoretically be exploited via selection in future breed improvement programs (Ren 2020). Baldwin et al. (1998) demonstrated the usefulness of analyzing mtDNA sequences in studying the phylogenetics of *Penaeus chinensis* populations in different locations, including the north coast of the

Shandong Peninsula, the west coast of the Korea Peninsula, 1182 sea area in the Yellow Sea, and an aquaculture farm where the parent shrimp were captured from wild populations at the Chinese coast of the Yellow Sea. Lavery et al. (2004) used mitochondrial DNA sequences to reconstruct the phylogeny of marine shrimp in the *Penaeus* sensu lato genus, testing hypotheses on species groupings and evolutionary history.

11.2 Application of Morphometric Characters in Shrimp Population Studies

Morphometric characters refer to quantitative measurements of lengths between defined anatomical features and are commonly used to distinguish fishery stocks and species. Various factors such as stress, food, or reproductive cycle can affect morphometric variation. Recent studies have advanced our understanding of morphometric analysis in shrimp species. For example, Carvalho-Batista et al. (2019) characterized the morphometrics of two species of the genus *Xiphopenaeus*. Similarly, Carvalho et al. (2019) investigated the morphometric differentiation of two sympatric species, *Farfantepenaeus aztecus* and *F. californiensis*, from the Gulf of California. Additionally, Ngrid et al. (2013) explored morphometric variation and sexual dimorphism in the white-leg shrimp, *Litopenaeus vannamei*, from Rio Grande do Norte State, Northeastern Brazil. Pérez-Castañeda and Defeo (2002) conducted a study on morphometric variation in pink shrimp populations of two sympatric species, *Farfantepenaeus brasiliensis* and *F. paulensis*, from the Rio de Janeiro coast of Brazil. Furthermore, Chow and Sandifer (1991) analyzed morphometric variation and sexual dimorphism of the Pacific white shrimp, *Litopenaeus vannamei*, from the coast of Jalisco, Mexico. Recent studies have enhanced our understanding of morphometric analysis and its application in differentiating shrimp species.

11.3 Variations in the Length-Weight Connections

Variations in the length-weight connections among sexes, species, and across different seasons and locations have been documented for both wild and cultured penaeid shrimps. Seasonal changes in the length-weight relationship of wild shrimp populations have been linked to seasonal fluctuations in reproductive biology, with mature shrimps exhibiting higher weight/length ratios than immature ones. Recent research has delved deeper into these associations. For instance, Sousa et al. (2018) examined the length-weight relationships of six commercially important penaeid shrimp species from the Madeira archipelago. Additionally, Amani et al. (2015) investigated the length-weight relationship and condition factor of commercially important penaeid shrimps from the west coast of Peninsular Malaysia. Kaka et al. (2019) provided length-weight relationships for three penaeid shrimp species from Malindi–Ungwana Bay in Kenya, while Mohale et al. (2024) explored length-weight relationships and the conditional factors in seven *Penaeus* shrimp populations on the southeast coast of Tamil Nadu, India.

Many studies have been conducted to elucidate evolutionary relationships in *Penaeus* species. Recent research indicated that *Penaeus* relationships based on morphological parameters can be misleading (Tsoi et al. 2005a, b). Tsoi et al. (2007) found significant genetic differences between two closely related species of kuruma shrimp, *Litopenaeus japonicas*. Although it can be easily bred, most hatcheries still depend on wild populations as the source of broodstock for generating larvae of higher value. Thus, determination of the source of wild broodstock is very important (Smith et al. 2017). Several reports on morphometric studies have been documented in shrimps.

11.4 Multivariate Analysis

An alternative approach for data summarization involves multivariate analyses (Yiğit and Mendeş 2018; Eyduran et al. 2019). Principal component (PC) analysis is an important tool in the multivariate methodology used for correlated traits. PC functions mainly to reduce data dimensionality and to explore the relationship between traits (Parés-Casanova et al. 2012; Verma et al. 2015). This technique has been applied in the inter-populational morphometric variability of the freshwater shrimp species. For example, Konan et al. (2014) verified the inter-populational morphometric variability using multivariate analysis (PCA) in four male populations (Bia, Comoé, Bandaman, and Sassandra) of the freshwater shrimp species *Macrobrachium vollenhovenii* from Côte d'Ivoire river in the western part of Africa. The study was aimed at determining whether *M. vollenhovenii* from the Bia and Comoe rivers, located in the eastern part of Côte d'Ivoire, constitute a homogeneous or heterogeneous stock. The results of the study indicated that *M. vollenhovenii* males from rivers in Côte d'Ivoire exhibit wide morphometric variability and therefore were heterogeneous. In another study on morphometric variability among the shrimp populations of the genus *Palaemonetes* (*P. carteri* and *P. ivonicus*), discriminant function analysis (DFA) was performed in order to assess their taxonomic status in the Amazon Basin and to verify the intra- and inter-populational morphometric variability (García-Dávila et al. 2005).

11.5 Morphological Characters Investigated for Two Morphotypes of *Penaeus semisulcatus* (Green Tiger Shrimp)

A morphometric analysis by Tamadoni Jahromi et al. (2019a, b) found significant differences in various variables between the banded and non-banded antenna types based on gender. The study followed the methods of Tzeng and Yeh (1999) and Tsoi et al. (2005a, b) for all collected specimens. Sixteen morphometric characters were examined in the study, and measurements were taken using a vernier caliper or ruler, depending on the specific character being measured (Tamadoni Jahromi et al. 2019a, b). Each original character was standardized by transforming it into a ratio of body length (BL), carapace length (CL), or telson length (TL), resulting in a total of 25 variables for further analysis. These generated a total of 25 variables that were subsequently analyzed (see Table 11.1 and Figs. 11.1 and 11.2).

A morphometric analysis by Tamadoni Jahromi et al. (2019a, b) found significant differences in various variables between the banded and non-banded antenna types based on gender. When grouped by sex, 10 out of the 25 measured variables showed significant disparities between the two types ($P < 0.05$) (refer to Table 11.1). Among males, four variables were significant for both banded and non-banded types, whereas among females, five variables were significant for both types.

Table 11.1 Description of traditional morphometric variables of a penaeid shrimp

No.	Variable	Code of variable	Description of variable
1	Antennal spine width	(ASW)	Width of flagellate appendage projecting from the front of the cephalothorax
2	Body length	(BL)	Total length is measured from the tip of the rostrum to the tip of the telson
3	Carapace length	(CL)	Carapace length is a measurement from the post-orbital notch to the posterior margin of the carapace
4	Diagonal of carapace length	(DCL)	The base from the eye stalk to the bottom of the posterior end of the carapace
5	First abdominal somite high	(ASH1)	The maximum height of the first abdominal somite
6	First abdominal somite length	(ASL1)	The maximum length of the first abdominal somite
7	First abdominal somite width	(ASW1)	The length of the maximum depth of the first abdominal somite
8	Second abdominal somite high	(ASH2)	The maximum height of the second abdominal somite
9	Second abdominal somite length	(ASL2)	The maximum length of the second abdominal somite
10	Second abdominal somite width	(ASW2)	The length of the maximum depth of the second abdominal somite
11	Sixth abdominal somite high	(ASH6)	The maximum height of the sixth abdominal somite
12	Sixth abdominal somite length	(ASL6)	The maximum length of the sixth abdominal somite
13	Sixth abdominal somite width	(ASW6)	The length of the maximum depth of the sixth abdominal somite
14	Hepatic spine width	(HSW)	The maximum depth of the lateral spine is situated near the anterior margin of the hepatic region of the carapace
15	Rostrum length	(RL)	The length from the tip to the end of the rostrum
16	Telson length	(TeL)	The maximum length of the telson

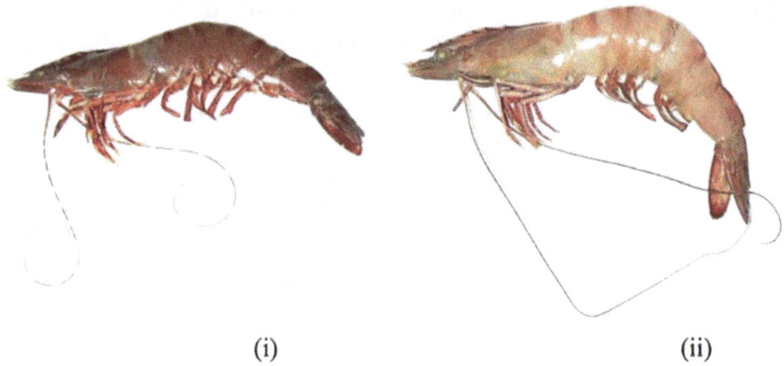

Fig. 11.1 The body color patterns for two types of *P. semisulcatus*: the banded and non-banded morphotypes

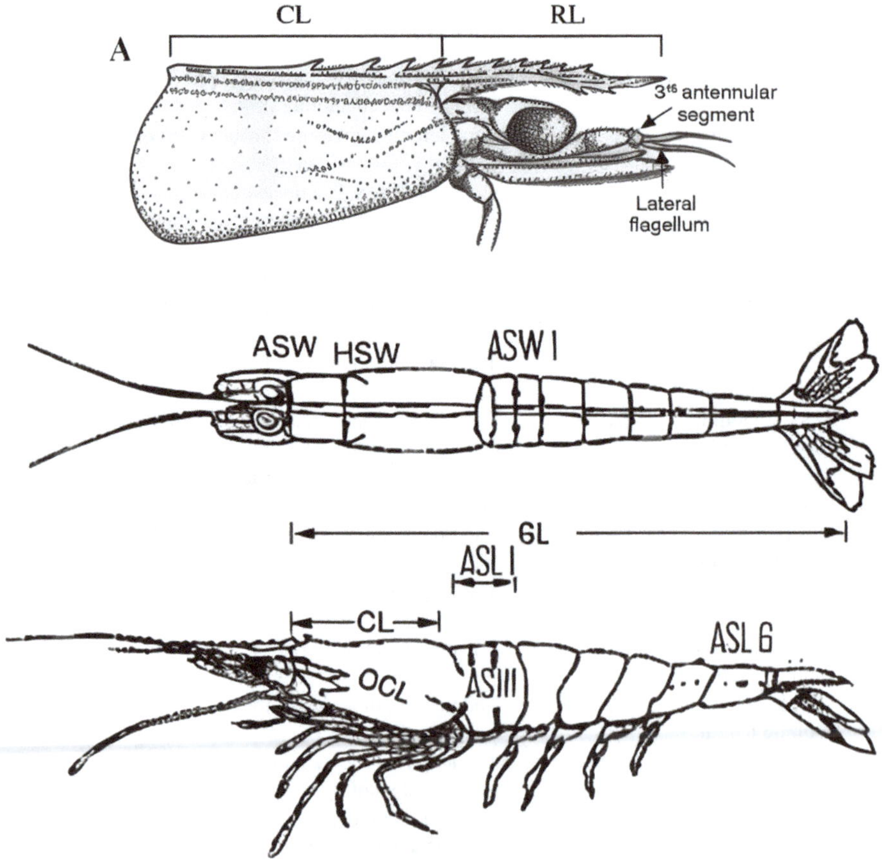

Fig. 11.2 Examples of morphometric characters investigated. The full name for the abbreviation is given in Table 11.1

11.6 Statistical Analyses in Multivariate Methods

Two of the most utilized statistics in multivariate analyses are principal component analysis (PCA) and discriminant function analysis (DFA). They are briefly described below.

11.6.1 Principal Component Analysis (PCA)

Principal component analysis (PCA) is a mathematical process that converts a set of observations of possibly correlated variables into a set of values of uncorrelated variables called principal components (PCs). PCA combines and reviews the connected measured variables into several principal components (PCs). It is widely used in exploratory data analysis to generate predictive models. In this analysis, a comparatively small number of components that group sets of many interrelated variables are identified.

From the dataset, a correlation or variance-covariance matrix is created. Munjal et al. (2018) stated that these principal components are referred to as patterns of covariation in size and shape particularly in morphometric analysis. The first principal component, often considered a size factor, shows a high correlation of characters and accounts for the major contribution of total variation. The second component (and the remaining principal components) explains the rest of the variation and is typically interpreted as a shape component (Dytham 2011). The first component is positively interrelated to all dimensions, but the second one and other components may be positively connected to some variables and negatively correlated with others (Rawat et al. 2017).

11.6.2 Discriminant Function Analysis (DFA)

Discriminant function analysis (DFA) is a statistical analysis to calculate a categorical dependent variable by one or more continuous or double independent variables. It is useful in determining whether a set of variables can predict membership.

Fisher (1936) stated that this analysis is regarded as a multivariate extension of the analysis of variance (ANOVA) to study variation and covariation among taxonomic groups. DFA is a technique that optimizes between samples in discrimination compared to within samples according to the variables. Discriminant functions (DFs) are designed by obtaining a weighted combination of variables which have maximal variances between groups compared to within groups. In the next stage, these discriminant functions can resort the individuals into selected groups (Stella 2019). The first function explains the highest amount of between-group covariance, and the second and also the rest of the functions clarify in decreasing order of importance group classification. In a total number of populations (n), the maximum number of functions in the analysis is $n - 1$. A scatter plot diagram can be designed to visualize the degree of division between groups. Furthermore, a matrix of generalized squared distances between populations can be created from the discriminant function analysis to assess the between-group relationship.

11.7 Molecular Markers in Population Genetics and Aquaculture

Mutations can occur in organisms due to normal cellular processes or interactions with the environment, leading to genetic variation or polymorphism, which can result from selection and genetic drift. Genetic variation can occur at various levels, such as among individuals, species, and higher taxonomic levels. For genetic studies, these variations must be heritable and recognizable by their phenotypic variations or through molecular techniques (Halaschek-Wiener et al. 2009). Before DNA markers were developed, allozyme markers were widely used in fish and other aquaculture species to identify broodstocks (Kucuktas and Liu 2007). Allozymes are unique allelic forms of similar enzymes located at the same locus (Okumuş and Çiftci 2003). However, with the advances of new techniques, notably the polymerase chain reaction (PCR), DNA markers have almost completely replaced allozymes as the marker of choice. Molecular markers can provide invaluable information for various aspects of aquaculture practices. These include genetic identification and discrimination of aquaculture stocks, controlling the end results of founding and spreading aquaculture stocks, assisting selective breeding programs, and measuring chromosomal and gene manipulations such as induction of polyploidy and gynogenesis (Mojekwu and Anumudu 2013).

Various types of markers are commonly used in aquaculture genetics. Allozyme and mitochondrial DNA (mtDNA) markers are extensively employed in genetic studies. Additional marker types include randomly amplified polymorphic DNA (RAPD), restriction fragment length polymorphism (RFLP), amplified fragment length polymorphism (AFLP), and microsatellite markers (Liu and Cordes 2004). Genetic differences among similar-looking species in the Penaeida family suggest a consistent pattern. Molecular genetics can help reevaluate the relationships of penaeid shrimp when such differences are observed (Baldwin et al. 1998). There are two main types of molecular markers: those associated with known gene functions (type I) and those associated with unknown functions (type II) (Liu and Cordes 2004).

According to this classification, RFLP and allozyme markers are categorized as type I markers, while RAPD and AFLP markers are classified as type II markers due to their amplification from unidentified genomic regions. AFLP is considered more effective in distinguishing between different genetic variations compared to RAPD. It is a PCR-based method that requires only a small amount of initial DNA, unlike RFLP. AFLP does not need prior genetic information or probes and resolves the issue of inconsistent reproducibility present in RAPD. Kai et al. (2002) utilized 14 AFLP markers to differentiate three color variations in the black rockfish, *Sebastes inermis*, identifying investigative AFLP loci as well as loci with major frequency divergences (Kai et al. 2002; Tan et al. 2019).

11.8 Applications of Mitochondrial DNA in Fisheries

Studies have shown that mitochondrial DNA (mtDNA) exhibits a high level of polymorphism and possesses maternal and clonal inheritance (Awadalla et al. 1999). These characteristics have made mtDNA one of the most valuable markers for genetic studies in aquaculture (Giuffra et al. 1994). The use of DNA markers has enabled rapid advancements in aquaculture through the examination of genetic variability and inbreeding (Mojekwu and Anumudu 2013), parentage assignments, species and strain identification, and the creation of high-resolution genetic linkage maps for aquaculture species which have been studied by researchers (Wang et al. 2024). Early research has shown that mtDNA displays a high level of sequence diversity at both species and infra-specific levels (Jørgensen et al. 2018; Tan et al. 2019). The completion of mtDNA genome sequencing in various organisms using PCR has provided valuable information in vertebrates, particularly in fishes (Kumar and Kocour 2017). MtDNA markers have been widely used to study stock structure in various fish species and shrimps, such as eels (Avise et al. 1986). Studies have extensively utilized sequence analyses of mtDNA to examine the phylogenetics of natural shrimp populations and to evaluate the genetic divergence among captive populations (Lin et al. 2024). This approach has been employed in multiple studies conducted by various researchers over the years.

High genetic variations have been found in the mitochondrial 16S rRNA gene within *P. notialis* and between *P. notialis* and *P. schmitti* (García-Dávila et al. 2005). Similarly, significant genetic divergence has been observed in *P. monodon* (Xu et al. 2001) and between two visually similar varieties of *P. japonicas* (kuruma shrimps) (Tsoi et al. 2005a, b). Magoulas et al. (1998) elucidated polymorphism in Greek shrimp broodstocks and carried on producing the first restriction map of the mtDNA molecule of this species. The authors, furthermore, investigated the nucleotide sequence of a 350 bp segment from the cytochrome b gene. Large genetic variations of the COI gene were detected among 15 species of the genus *Penaeus* (Palumbi and Benzie 1991). On the other hand, Baldwin et al. (1998) observed 0–3% COI variation within some *Penaeus* species.

11.9 Molecular Phylogeny Using mtDNA Sequencing

Phylogenetics is the science of estimating and analyzing evolutionary relationships among organisms. Over the years, molecular markers have been widely used for phylogenetic analysis. One of the important advantages of the molecular approach in detecting phylogenetic relationships over more classical approaches, such as morphological or life cycle traits, is that the dissimilarities are immediately quantifiable. Sequences can be compared from dissimilar organisms, which would eventually allow comparisons of nucleotide differences. These data are then translated into the form of trees. The positions and lengths of the branches are correlated to the relatedness among the organisms (Pace 1996).

In the process of identifying the relationships among species, the trees constructed can either be phenetic or phylogenetic. The phenetic models and algorithms are based on distance methods for the calculation of relationships in generating the trees. The phylogenetic tree is based on the cladistic method, which implies ancestor-descendant relationships (Hall 2004). A tree is reconstructed by considering the various possible pathways of evolution and selecting from among them the one which stands out to be the best tree. Trees reconstructed via these methods are known as "cladograms." Computer algorithms rely on parsimony or maximum probability methods for measuring relationships and building trees (Gill and Mooi 2002).

Molecular genetics techniques have been particularly useful for studying the population genetics and systematics of penaeids that are similar in appearance and ecology. These techniques have helped refine the taxonomic position of species (Maggioni et al. 2001), identify new species (Gusmão et al. 2000), and develop genetic markers for use in aquaculture (Moore et al. 1999; Pirian et al. 2016). Palumbi and Benzie (1991) demonstrated the phylogenetic differences between two morphological species of kuruma shrimps (*P. japonicus*) by analyzing 16S and COI genes of mtDNA. Mitochondrial DNA is typically a circular molecule about 16 kb in length (Fig. 11.3). According to Brown (1985), studies suggest that mutations

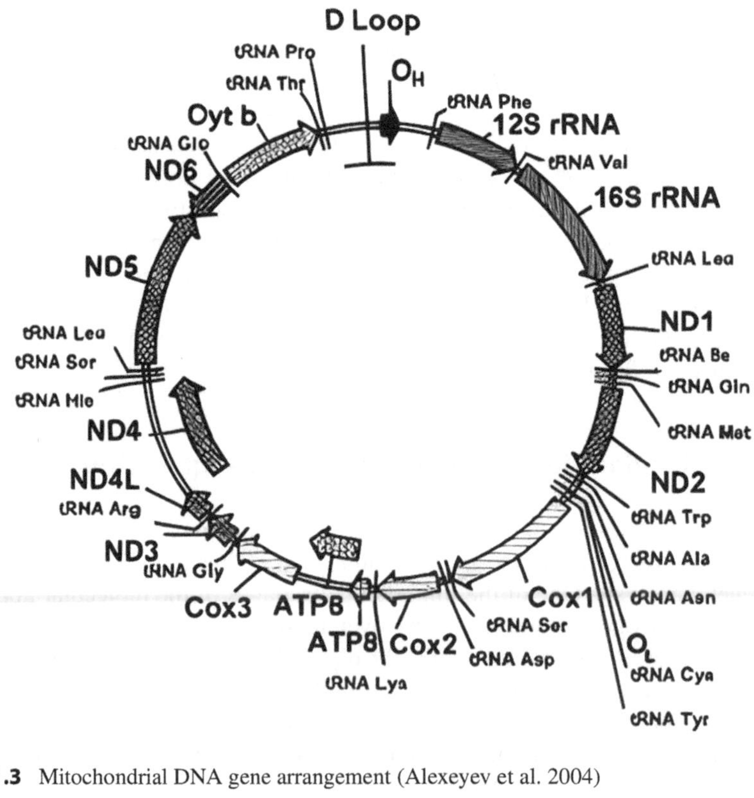

Fig. 11.3 Mitochondrial DNA gene arrangement (Alexeyev et al. 2004)

occur more rapidly in mitochondrial DNA than in nuclear DNA, possibly due to two factors: first, a higher mutation rate in mtDNA resulting from limited repair mechanisms during replication (Lax et al. 2011) and, second, a smaller effective population size due to the exclusive maternal inheritance of the haploid mitochondrial genome (Birky et al. 1989).

11.10 Applications of Mitochondrial DNA in Fisheries

Researchers have extensively employed mtDNA markers to investigate the stock structure of various fish species and shrimps (Avise et al. 1986). Analysis of mtDNA sequences has been widely used in studying the phylogeny of natural shrimp populations (Gusmão et al. 2000) and comparing genetic differences between captive populations (Lin et al. 2024). High genetic variations of the mitochondrial 16S rRNA gene have been observed within and between different shrimp species (García-Dávila et al. 2005). Significant divergence has been identified in one shrimp species, while genetic differences were reported between two similar varieties of another shrimp species (Khamnamtong et al. 2009).

Studies have explored polymorphism in Greek shrimp broodstocks and produced the first restriction map of the mtDNA molecule of this species, along with investigating the nucleotide sequence of a 350 bp segment from the cytochrome b gene (Magoulas et al. 1998). Navajas et al. (1996) found significant genetic variation in the mitochondrial cytochrome oxidase subunit I (COI) gene across various arthropod species. In their study, Palumbi and Benzie (1991) identified large genetic variations in the COI gene among 15 species of the *Penaeus* genus. Conversely, Baldwin et al. (1998) observed COI variations ranging from 0% to 3% within certain *Penaeus* species. Tamadoni Jahromi et al. (2019a, b) utilized two mtDNA genes, 16S rRNA and COI, to explore the taxonomic status of two *P. semisulcatus* morphotypes and their genetic relationships with related species (Table 11.2).

Tamadoni Jahromi et al. (2019a, b) reported the highest total genetic divergences for the 16S rRNA gene sequences between specific morphotypes from Jask and Penang, with a genetic distance of 6.6%. The lowest nucleotide divergence was recorded among individuals from Busher, Hormoz, and Jask, with a genetic distance of 3.9% (Table 11.2). Also, the results of that study revealed that banded and non-banded morphotypes clustered into distinct monophyletic groups. Seventeen (17) genetically divergent sites were recorded for the two morphotypes of *P. semisulcatus* (Fig. 11.4).

A comparison of all *Penaeus* species in this study revealed that the highest nucleotide divergence (0.127) was recorded between *P. monodon* (MO.CH49IR) and banded morphotype of *P. semisulcatus* (B.PE25 MA) and the lowest (0.011) between *P. merguiensis* (ME. HO44IR) and *P. indicus* (P.IND.AF335279), respectively. Amplification of the 16S rRNA gene produced a product of approximately 530 base pairs for most samples analyzed. After editing, a final consensus sequence of 457 characters was aligned and used for tree construction (Fig. 11.4).

Table 11.2 Sample size and sites of studied species

Species	N[a]	Site (area)
P. semisulcatus (banded type Iran)	2	Hormoz (Iran)
P. semisulcatus (banded type Iran)	2	Jask area Iran
P. semisulcatus (banded type Malaysia)	4	Penang (Malaysia)
P. semisulcatus (non-banded type Iran)	2	Bushehr Port (Iran)
P. semisulcatus (non-banded type Iran)	1	Jask area Iran
P. semisulcatus (non-banded type Iran)	1	Hormoz (Iran)
P. monodon (Iran)	1	Chabahar (Iran)
P. monodon (Malaysia)	1	Penang
P. merguiensis(Iran)	2	Hormoz (Iran)
P. indicus (Iran)	2	Jask area (Iran)
*P. indicus*16S & COI	2	NCBI
P. merguiensis 16S & COI	2	NCBI
M. affinis.16S & COI	2	NCBI
P. semisulcatus 16S &COI	2	NCBI
P. monodon 16S &COI	2	NCBI

[a] *N* = number of individuals

In a different study, researchers used mitochondrial DNA analysis of 16S ribosomal RNA to explore the evolutionary relationship of *Trichiurus lepturus* in the Persian Gulf compared to other areas. The study by Tamadoni et al. revealed that the examined strain belongs to *T. lepturus* and has 42 divergence sites compared to partial sequences of the 16S ribosomal RNA gene from other regions (West Atlantic and Indo-Pacific area). The phylogeny results indicated that all 18 haplotypes of the species were grouped into five clades with high bootstrap support. The observed structuring within the Persian Gulf and the Indo-Pacific individuals suggested a similarity between the two mentioned areas.

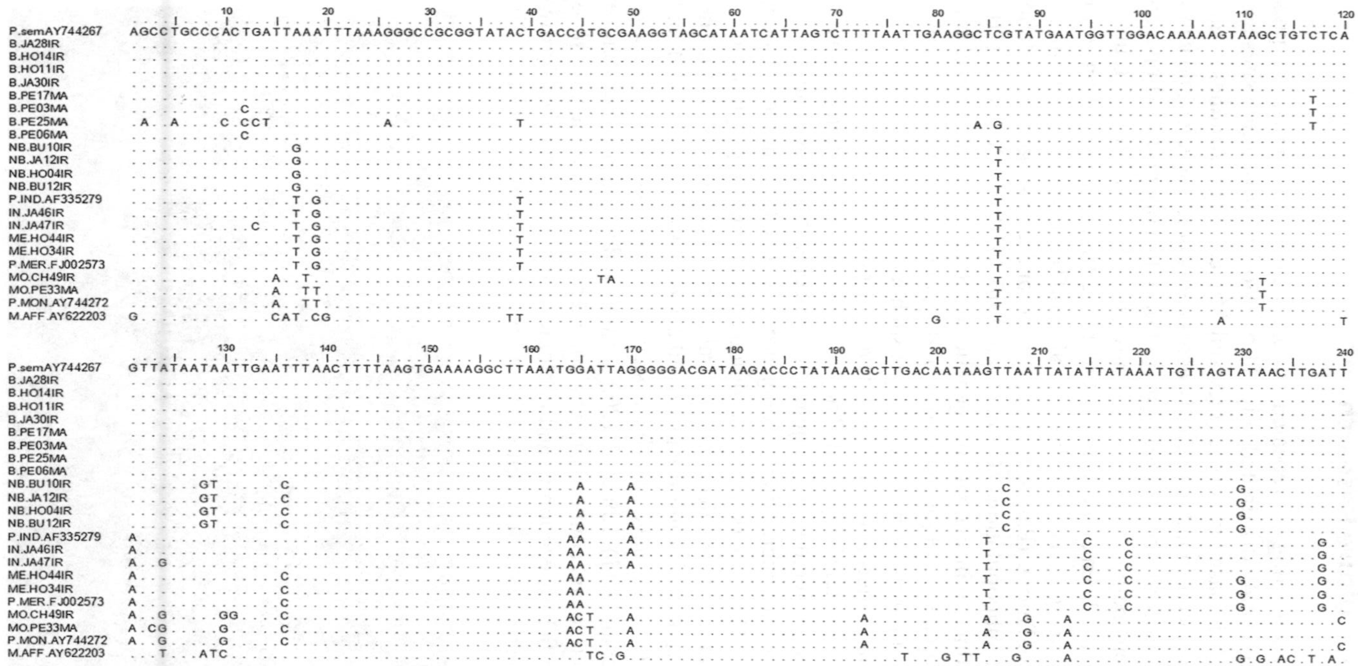

Fig. 11.4 Aligned sequences of 16S rRNA of 23 taxa including both morphotypes of *P. semisulcatus*, *P. merguiensis*, *P. indicus*, and *P. monodon* from Persian Gulf (IR) and Penang (PE). Dots denote identity to the top sequence. Details on the name of taxa are given in Table 11.2

11.11 Microsatellite Markers and Their Application in Aquaculture Research

Microsatellites, consisting of repeated simple sequence patterns, are present in genomes and are evenly distributed (Fig. 11.5). Most microsatellite sites are small and crucial for PCR-based genotyping (Abdul-Muneer 2014; Antunes et al. 2022). Researchers have been found abundantly across all studied species and can mutate at high rates (Crawford and Cuthbertson 1996), with polymerase slippage during DNA replication being a proposed cause for this variation (Tautz 1989). Microsatellites have been employed to estimate the effective population size in grey seals (*Halichoerus grypus*) (Palsbøll et al. 2007), determine migration levels between subpopulations (Palsbøll et al. 2007), and assess genetic relationships within different subpopulations (Cornuet and Luikart 1996). Xu et al. (2001) utilized microsatellites to investigate the genetic structure of wild *P. monodon* populations in the Philippines and its connection with mangrove status and shrimp culture systems. Due to their numerous advantages such as small locus size, co-dominant nature, Mendelian inheritance, and high levels of polymorphism (Xu et al. 2001), microsatellites are extensively used. They are present in abundance across various eukaryotic genomes, including all aquaculture species studied to date.

Microsatellites are increasingly being used as genetic markers in population genetics studies. These markers are co-dominant and highly polymorphic and are generally considered to be selectively neutral (Jarne and Lagoda 1996). They have been found in various marine organisms like mollusks (Wenne 2023) and squid (Shaw et al. 1997) and have also been studied in lobsters (Tan et al. 2019) and shrimps (Aziz et al. 2011). Liu et al. (2022) employed microsatellites to investigate population structure and connectivity in a commercially important *Konosirus punctatus*, a pelagic fish of economic and ecological importance in the Northwestern, providing valuable data for conservation and management efforts. Similarly,

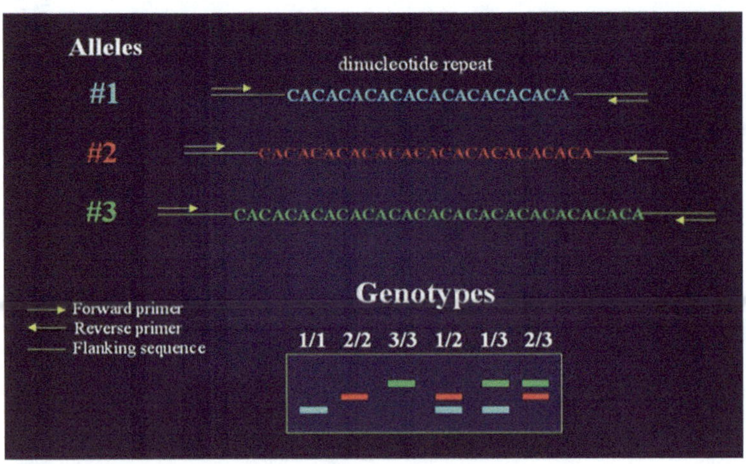

Fig. 11.5 Principle of microsatellite. (Modified from Meglécz et al. 2007)

Liu et al. (2022) used microsatellite markers to assess genetic diversity and population structure in squid populations, shedding light on their demographic history and connectivity patterns. In the field of aquaculture, microsatellites continue to play a crucial role. Jerry et al. (2004) utilized microsatellite markers for parentage analysis in a shrimp breeding program, demonstrating their usefulness in ensuring genetic traceability and optimizing breeding strategies.

Furthermore, the progress in genomic technologies has allowed for the creation of high-throughput microsatellite genotyping methods, as demonstrated by Pimentel et al. (2018), which have made it easier to conduct large-scale genetic studies in aquaculture species. In the field of aquaculture research, microsatellite markers are highly effective for identifying parentage and assessing genetic diversity in hatchery populations. Vanavichit et al. (1998) used microsatellites as genetic markers in *P. monodon* and found them suitable for studying the shrimp genome and managing broodstock. Factors like geographic location, marine currents, life cycle, and ecological characteristics have a significant impact on the genetic distance and differentiation between shrimp populations. These factors were investigated using microsatellites in the study of *Litopenaeus schmitti*, contributing to conservation and breeding efforts (Maggioni et al. 2001). Microsatellites also played a crucial role in evaluating genetic variability in both captive broodstock and wild populations of *Litopenaeus vannamei* (Cruz et al. 2004). Furthermore, microsatellites were used to examine genetic diversity within selective breeding families (Cao 2018).

In a study by Li et al. (2007), two microsatellite multiplex systems were developed for the black tiger shrimp *Penaeus monodon* to analyze the genetic diversity of two populations. In addition, there have been advancements in microsatellite isolation techniques such as the development of microsatellite enrichment to increase the quantity of colonies in a library containing the microsatellite motif of interest (Nunome et al. 2020). The traditional way to isolate microsatellite colonies involves creating a small-insert partial genomic library in a plasmid. Microsatellite enrichment has also been developed to increase the number of colonies in a given library containing the microsatellite motif of interest (Nunome et al. 2006). These recent studies emphasize the ongoing importance and usefulness of microsatellites in understanding the genetic structure, diversity, and dynamics of shrimp populations, as well as in supporting conservation and breeding programs.

11.12 Construction of Microsatellite Enriched Genomic Library

Polymorphic markers are widely used in studying the biological relationships of organisms, mapping important traits, and cloning genes (Gonzalez and Zardoya 2013). Microsatellite markers are considered one of the most efficient molecular marker systems, but their utility is often limited by the difficulty, cost, and time required for their acquisition (Gonzalez and Zardoya 2013). Developing primers for the DNA flanking specific microsatellite repeats is necessary to amplify microsatellite loci (Primmer et al. 1996; Zhu et al. 2000). Various microsatellite markers have been developed for fish species like seabass (*Lates calcarifer*) (Zhu et al. 2000), as well as for marine invertebrates like oysters (Naciri-Graven et al. 1998). Microsatellites have also been characterized for commercially important marine shrimps such as *Penaeus vannamei* (Mangabeira-Silva et al. 2024), *P. japonicus* (Zhang et al. 2023), and *P. monodon* (Xu et al. 2003).

A small number of microsatellites have also been developed for a variety of penaeid species, including cultured shrimp *P.* (*Litopenaeus*) *vannamei* (Maggioni et al. 2001), and have been used in population genetic investigations in white shrimp, *Litopenaeus setiferus* (Ball and Chapman 2003), genetic relationships (Khamnamtong et al. 2009), and gene mapping (Wuthisuthimethavee et al. 2005). Also, a small number of microsatellites have been developed for a variety of penaeid species for cultured shrimp *P.* (*Litopenaeus*) *vannamei* (Maggioni et al. 2001) and the population structure of *Litopenaeus schmitti* (Meehan et al. 2003) and have mainly been used in population genetic investigations in white shrimp, *Litopenaeus setiferus* (Ball and Chapman 2003), genetic relationships (Xu et al. 2003), and gene mapping as well (Wuthisuthimethavee et al. 2003).

11.13 Conclusion

In summary, the chapter emphasizes the importance of combining morphometric analysis and molecular markers to advance our understanding of population genetics and aquaculture, specifically focusing on penaeid shrimps. This combined approach provides researchers with valuable insights into species identification, population structure, and genetic diversity. By using statistical methods like PCA and DFA in conjunction with molecular techniques such as microsatellites and mitochondrial DNA analysis, we can distinguish fishery stocks, evaluate genetic variability, and guide aquaculture management practices. Examples from actual cases demonstrate the practical usefulness of these methods in species identification, phylogenetic reconstruction, and sustainable resource management. Ultimately, integrating morphometric and molecular approaches improves our ability to address important questions in evolutionary biology and contributes to the development of effective conservation and aquaculture strategies for penaeid shrimps and other related species.

References

Abdul-Muneer P (2014) Application of microsatellite markers in conservation genetics and fisheries management: recent advances in population structure analysis and conservation strategies. Genet Res Int 2014:691759

Alexeyev MF, LeDoux SP; Wilson GL (2004) Mitochondrial DNA and aging. Clin Sci (Lond) 107(4):355–364. https://doi.org/10.1042/CS20040148

Amani A, Arshad A, Yusoff F, Amin S (2015) Length-weight relationship and relative condition factor of Parapenaeopsis sculptilis (Heller, 1862) from the coastal waters of Perak, Peninsular Malaysia

Antunes AM, Nunes Stival JG, Targueta CP, de Campos Telles MP, Soares TN (2022) A pipeline for the development of microsatellite markers using next generation sequencing data. Curr Genomics 23(3):175–181. https://doi.org/10.2174/1389202923666220428101350. PMID: 36777003; PMCID: PMC9878831

Archak S, Meduri E, Sravana Kumar P, Nagaraju J (2007) A microsatellite database of fully sequenced insect genomes. Nucl Acids Res 35:D36–D39

Avise JC, Helfman GS, Saunders NC, Hales LS (1986) Mitochondrial DNA differentiation in North Atlantic eels: population genetic consequences of an unusual life history pattern. Proc Natl Acad Sci U S A 83(12):4350–4354

Awadalla P, Eyre-Walker A, Smith JM (1999) Linkage disequilibrium and recombination in hominid mitochondrial DNA. Science 286(5449):2524–2525

Aziz D, Siraj SS, Daud SK, Panandam JM, Othman MF (2011) Genetic diversity of wild and cultured populations of *Penaeus monodon* using microsatellite markers. J Fish Aquat Sci 6:614–623

Baldwin JD, Bass AL, Bowen BW, Clark WH Jr (1998) Molecular phylogeny and biogeography of the marine shrimp Penaeus. Mol Phylogenet Evol 10(3):399–407

Ball AO, Chapman RW (2003) Population genetic analysis of white shrimp, Litopenaeus setiferus, using microsatellite genetic markers. Mol Ecol 12:2319–2330. https://doi.org/10.1046/j.1365-294X.2003.01922.x

Barzkar NO, Babich R, Das S, Sukhikh ST, Jahromi MS (2022) Marine Bacterial dextranases: fundamentals and Applications. Molecules 27(17):5533

Barzkar, N. S. Sukhikh, O. Babich, B.A. Venmathi Maran, S. Tamadoni Jahromi (2023a). Marine collagen: purification, properties and application, Front Mar Sci 10 1245077.

Barzkar NV, Rungsardthong ST, Jahromi Q, Laraib R, Das O, Babich SS (2023b) A recent update on fucoidonase: source, isolation methods and its enzymatic activity. Front Mar Sci 10:1129982

Barzkar N, Sukhikh S, Babich O (2024a) Study of marine microorganism metabolites: new resources for bioactive natural products. Front Microbiol 14:1285902

Barzkar N, Babich O, Sukhikh S, Maran BAV, Jahromi ST, Luwor RB, Sorsa T, Das R (2024b) Exploring the sources and potential applications of marine collagenases, Biocatalysis and Agricultural. Biotechnology 103150

Birky CW Jr (1995) Uniparental inheritance of mitochondrial and chloroplast genes: mechanisms and evolution. Proc Natl Acad Sci 92(25):11331–11338. https://doi.org/10.1073/pnas.92.25.11331

Birky CW, Fuerst P, Maruyama T (1989) Organelle gene diversity under migration, mutation, and drift: equilibrium expectations, approach to equilibrium, effect of heteroplasmic cells, and comparison to nuclear genes. Genetics 121:613–627

Brown WM (1985) The mitochondrial genome of animals. In: Molecular evolutionary genetics. Plenum, New York, pp 95–130

Cadrin SX (2000) Advances in morphometric identification of fishery stocks. Rev Fish Biol Fish 10:91–112. https://doi.org/10.1023/A:1008939104413

Cao TG (2018) Selective breeding to develop a genetically improved strain of Pacific whiteleg shrimp (Litopenaeus vannamei). University of the Sunshine Coast

Carvalho CD, Keunecke KA, Lavrado HP (2019) Morphometric variation in pink shrimp populations at Rio de Janeiro coast (SE Brazil): are they really similar in closer areas? An Acad Bras Cienc 91:e20180252

Carvalho-Batista A, Terossi M, Zara FJ et al (2019) A multigene and morphological analysis expands the diversity of the seabod shrimp *Xiphopenaeus* Smith, 1869 (Decapoda: Penaeidae), with descriptions of two new species. Sci Rep 9:15281. https://doi.org/10.1038/s41598-019-51484-3

Chow S, Sandifer PA (1991) Differences in growth, morphometric traits, and male sexual maturity among Pacific white shrimp, Penaeus vannamei, from different commercial hatcheries. Aquaculture 92:165–178. https://doi.org/10.1016/0044-8486(91)90018-3

Cornuet JM, Luikart G (1996) Description and power analysis of two tests for detecting recent population bottlenecks from allele frequency data. Genetics 144(4):2001–2014. https://doi.org/10.1093/genetics/144.4.2001. PMID: 8978083; PMCID: PMC1207747

Crawford AM, Cuthbertson RP (1996) Mutations in sheep microsatellites. Genome Res 6(9):876–879

Cruz P, Ibarra AM, Mejia-Ruiz H et al (2004) Genetic variability assessed by microsatellites in a breeding program of Pacific white shrimp (*Litopenaeus vannamei*). Mar Biotechnol 6:157–164. https://doi.org/10.1007/s10126-003-0017-5

Dytham C (2011) Choosing and using statistics: a biologists guide. Wiley, Chichester

Eyduran E, Akin M, Eyduran SP (2019) Application of multivariate adaptive regression splines through R software. Nobel Academic Publishing, Ankara

Fisher RA (1936) The use of multiple measurements in taxonomic problems. Ann Eugenics 7(2):179–188

García-Dávila CR, Magalhães C, Guerrero JCH (2005) Morphometric variability in populations of Palaemonetes spp. (Crustacea, Decapoda, Palaemonidae) from the Peruvian and Brazilian Amazon Basin. Iheringia Sér Zool 95:327–334

Gill A, Mooi R (2002) Phylogeny and systematics of fishes. In: Handbook of fish biology and fisheries, vol 1, pp 15–42

Giuffra E, Bernatchez L, Guyomard R (1994) Mitochondrial control region and protein coding genes sequence variation among phenotypic forms of brown trout Salmo trutta from northern Italy. Mol Ecol 3(2):161–171

Gonzalez EG, Zardoya R (2013) Microsatellite DNA capture from enriched libraries. Methods Mol Biol 1006:67–87. https://doi.org/10.1007/978-1-62703-389-3_5. PMID: 23546784

Gusmão J, Lazoski C, Solé-Cava A (2000) A new species of Penaeus (Crustacea: Penaeidae) revealed by allozyme and cytochrome oxidase I analyses. Mar Biol 137:435–446

Halaschek-Wiener J, Amirabbasi-Beik M, Monfared N, Pieczyk M, Sailer C, Kollar A, Thomas R, Agalaridis G, Yamada S, Oliveira L, Collins JA, Meneilly G, Marra MA, Madden KM, Le ND, Connors JM, Brooks-Wilson AR (2009) Genetic variation in healthy oldest-old. PLoS One 4(8):e6641. https://doi.org/10.1371/journal.pone.0006641. PMID: 19680556; PMCID: PMC2722017

Halim SAAA, Othman AS, Akib NAM, Jamaludin NA, Esa Y, Nor SAM (2021) Mitochondrial markers identify a genetic boundary of the green tiger prawn (*Penaeus semisulcatus*) in the Indo-Pacific Ocean. Zool Stud 60:e8. https://doi.org/10.6620/ZS.2021.60-08. PMID: 34386093; PMCID: PMC8315928

Hall BG (2004) Phylogenetic trees made easy: a how-to manual. Sinauer Associates, Sunderland

Jarne P, Lagoda PJ (1996) Microsatellites, from molecules to populations and back. Trends Ecol Evol 11(10):424–429

Jerry DR, Preston NP, Crocos PJ, Keys S, Meadows JRS, Li Y (2004) Parentage determination of Kuruma shrimp Penaeus (Marsupenaeus) japonicus using microsatellite markers (Bate). Aquaculture 235(1):237–247. https://doi.org/10.1016/j.aquaculture.2004.01.019

Jørgensen TE, Karlsen BO, Emblem Å et al (2018) Mitochondrial genome variation of Atlantic cod. BMC Res Notes 11:397. https://doi.org/10.1186/s13104-018-3506-3

Kai Y, Nakayama K, Nakabo T (2002) Genetic differences among three colour morphotypes of the black rockfish, Sebastes inermis, inferred from mtDNA and AFLP analyses. Mol Ecol 11(12):2591–2598. https://doi.org/10.1046/j.1365-294x.2002.01628.x. PMID: 12453242

Kaka RM, Jung'a JO, Badamana M, Ruwa RK, Karisa HC (2019) Morphometric length-weight relationships of wild penaeid shrimps in Malindi-Ungwana Bay: implications to aquaculture development in Kenya. Egypt J Aquat Res 45(2):167–173. https://doi.org/10.1016/j.ejar.2019.06.003

Khamnamtong B, Klinbunga S, Menasveta P (2009) Genetic diversity and geographic differentiation of the giant tiger shrimp (*Penaeus monodon*) in Thailand analyzed by mitochondrial *COI* sequences. Biochem Genet 47:42–55. https://doi.org/10.1007/s10528-008-9205-3

Konan KM, Ouattara A, Da Costa KS, Adépo-Gourène AB, Gourène G (2014) Allometric growth and condition factor of West African shrimp, *Macrobrachium vollenhovenii* (Herklots, 1857), in the rivers of Côte d'Ivoire. Mar Freshw Res 65:849–856

Kucuktas H, Liu Z (2007) Allozyme and mitochondrial DNA markers. In: Aquaculture genome technologies, p 73

Kumar G, Kocour M (2017) Applications of next-generation sequencing in fisheries research: a review. Fish Res 186:11–22. https://doi.org/10.1016/j.fishres.2016.07.021

Lax NZ, Turnbull DM, Reeve AK. (2011). Mitochondrial mutations: newly discovered players in neuronal degeneration. Neuroscientist. Dec;17(6):645-658. https://doi.org/10.1177/1073858411385469

Lavery S, Snell H, Moritz C (2004) Phylogenetic relationships of the genus Penaeus s.l. based on mitochondrial DNA sequences. Mar Freshw Res 55(2):103–110

Li Y, Wongprasert K, Shekhar M, Ryan J, Dierens L, Meadows J et al (2007) Development of two microsatellite multiplex systems for black tiger shrimp Penaeus monodon and its application in genetic diversity study for two populations. Aquaculture 266(1):279–288. https://doi.org/10.1016/j.aquaculture.2007.01.038

Lin S, Han F, Li Z, Chang T, Xiao X, Wu Y et al (2024) Regulatory effects of myo-inositol on carbohydrate and lipid metabolism and low-salinity adaptation in Pacific white shrimp (Penaeus vannamei). Available at SSRN 4815816

Liu Z, Cordes JF (2004) DNA marker technologies and their applications in aquaculture genetics. Aquaculture 238(1–4):1–37

Liu B, Li J, Zhang K, Peng Y, Liu Y, Jin X, Zheng S, Wang Y, Liu L, Lü Z, Zhang S, Gong L (2022) Population structure and genetic diversity in wild dotted gizzard shad (*Konosirus punctatus*) revealed by microsatellite markers. Front Mar Sci 9:1048279. https://doi.org/10.3389/fmars.2022.1048279

Maggioni R, Rogers AD, Maclean N, D'Incao F (2001) Molecular phylogeny of western Atlantic Farfantepenaeus and Litopenaeus shrimp based on mitochondrial 16S partial sequences. Mol Phylogenet Evol 18(1):66–73. https://doi.org/10.1006/mpev.2000.0866

Magoulas A, Bartley D, Basurco B (1998) Application of molecular markers to aquaculture and broodstock management with special emphasis on microsatellite DNA. Cahiers Options Mediterr 34:153–168

Mangabeira-Silva IS, Soares PET, da Silva YTV, Rodrigues de Albuquerque BHD, Câmera de Oliveira MTF, Honorato Ferreira LA, Bezerra de Souza MF, Vieira de Lucena D, Paiva Pereira JM, Pinheiro e Silva RP, Ferreira Lanza DC (2024) Characterization of microsatellite markers in the coding regions of the Penaeus vannamei genome. PloS One 19(5):e0289351. https://doi.org/10.1371/journal.pone.0289351

Meehan D, Xu X, Zuniga G, Alcivar-Warren A (2003) High frequency and large number of polymorphic microsatellites in cultured shrimp Penaeus (Litopenaeus) vannamei. [Crustacea:Decapoda]. Mar Biotechnol 5:311–330. https://doi.org/10.1007/s10126-002-0092-z

Meglécz E, Anderson SJ, Bourguet D, Butcher R, Caldas A, Cassel-Lundhagen A, D'Acier AC, Dawson DA, Faure N, Fauvelot C, Franck P, Harper G, Keyghobadi N, Kluetsch C (2007) Microsatellite flanking region similarities among different loci within insect species. Insect Mol Biol 16(2):175–185. https://doi.org/10.1111/j.1365-2583.2006.00713.x

Mohale HP, Jawahar P, Jayakumar N, Oli GA, Ravikumar T, Bhosle R (2024) Length-weight relationship and condition factors of seven Penaeus shrimp species along the southeast coast of Tamil Nadu (Southern India). Indian J Anim Res 1:10

Mojekwu T, Anumudu C (2013) Microsatellite markers in Aquaculture: Application in Fish population genetics. IOSR J Environ Sci Toxicol Food Technol 5:43–48

Moore SS, Barendse W, Berger KT, Armitage SM, Hetzel DJ (1999) Microsatellite mapping of a gene affecting horn development in Bos taurus. Nat Genet 21(4):305–306

Munjal G, Hanmandlu M, Srivastava S (2018) Phylogenetics algorithms and applications. Ambient Commun Comput Syst 904:187–194. https://doi.org/10.1007/978-981-13-5934-7_17. PMCID: PMC7123334

Naciri-Graven Y, Martin AG, Baud JP, Renault T, Gérard A (1998) Selecting flat oyster Ostrea edulis for survival when infected by the parasite Bonamia ostreae. J Exp Mar Biol Ecol 224:91–107

Navajas M, Gutierrez J, Lagnel J, Boursot P (1996) Mitochondrial cytochrome oxidase I in tetranychid mites: a comparison between molecular phylogeny and changes of morphological and life history traits. Bull Entomol Res 86(4):407–417. https://doi.org/10.1017/S0007485300034994

Ngrid, V., A., Ccioly, Aulo, A., L., Ima, -, F., Ilho, Erezinha, L., S., Antos, Na, c.A., B., arbosa, Ourdès, B., S., Ampos, Osemairy, V, S., ouza, Ashington, C., A., raujo, Agner, F, M.C., & Olina (2013). Sexual dimorphism in Litopenaeus vannamei (Decapoda) identified by geometric morphometrics. Pan-American Journal of Aquatic Sciences, 8(4):276–281

Nunome M, Yoshioka R, Shinkai T, Kino K, Matsuda Y (2020) Assessment of genetic diversity and genetic relationships of farm and laboratory quail populations in Japan using microsatellite DNA markers. Vet Med Sci 6:1000–1008

Nunome T, Negoro S, Miyatake K, Yamaguchi H, Fukuoka H (2006) A Protocol for the Construction of Microsatellite Enriched Genomic Library. Plant Mol Biol Report 24:305–312. https://doi.org/10.1007/BF02913457

Okumuş İ, Çiftci Y (2003) Fish population genetics and molecular markers: II - Molecular markers and their applications in fisheries and aquaculture. Turk J Fish Aquat Sci 3(1):51–79

Pace NR (1996) New perspective on the natural microbial world: molecular microbial ecology. ASM News 62(10):463–470

Palsbøll PJ, Berube M, Allendorf FW (2007) Identification of management units using population genetic data. Annu Rev Ecol Evol Syst 49:405–430

Palumbi SR, Benzie JAH (1991) Large mitochondrial DNA differences between morphologically similar penaeid shrimp. Mol Mar Biol Biotechnol 1(1):27–34

Pérez-Castañeda R, Defeo O (2002) Morphometric relationships of penaeid shrimps in a coastal lagoon: Spatiotemporal variability and management implications. Estuaries 25:282–287. https://doi.org/10.1007/BF02691315

Parés-Casanova PM, Sinfreu I, Villalba D (2012) Principal component analysis of cephalic morphology to classify some Pyrenean cattle. Anim Genet Res 50:59–64

Pimentel JSM, Carmo AO, Rosse IC, Martins APV, Ludwig S, Facchin S, Pereira AH, Brandão-Dias PFP, Abreu NL, Kalapothakis E (2018) High-throughput sequencing strategy for microsatellite genotyping using neotropical fish as a model. Front Genet 9:73. https://doi.org/10.3389/fgene.2018.00073. PMID: 29593777; PMCID: PMC5855144

Pirian K, Piri K, Sohrabipour J, Jahromi ST, Blomster J (2016) Nutritional and phytochemical evaluation of the common green algae, Ulva spp. (Ulvophyceae), from the Persian Gulf. Fund Appl Limnol 188(4):315–327

Prado PRR, Fontoura Costa L, Moraes EM, Manfrin MH, Sene FM (2006) Curvature Measurement As A Tool For The Analysis Of Morphometric VariationUsing Drosophila Wings As A Model. Brazilian Journal of morphologycal Sciences 23(3-4):333–342

Primmer CR, Moller AP, Ellengren H (1996) A wide-range survey of cross species microsatellite amplifications in birds. Molecular Ecology 5:365–378

Ramesh P, Mallikarjuna G, Sameena S, Kumar A, Gurulakshmi K, Reddy BV et al (2020) Advancements in molecular marker technologies and their applications in diversity studies. J Biosci 45:1–15

Rawat S et al (2017) Identification of fish stocks based on truss morphometric: a review. J Fish Life Sci 2(1):9–14

Ren S (2020) Genetic improvement of the Pacific white shrimp (Penaeus vannamei) in China. Queensland University of Technology. https://doi.org/10.13140/RG.2.2.23346.63685

Robalino J, Chapman RW, de la Vega E, O'Leary NA, Gorbach DM, Du Z-Q, Rothschild MF, Browdy CL, Warr GW, Labreuche Y (2011) Advances in genomics and genetics of penaeid shrimp. In: Fletcher GL, Rise ML (eds) Aquaculture biotechnology. https://doi.org/10.1002/9780470963159.ch2

Shaw PW, Arkhipkin AI, Al-Khairulla H, Yoon JH (1997) Genetic structuring of Patagonian toothfish populations in the Southwest Atlantic Ocean: the effect of the Antarctic Polar Front and deep-water troughs as barriers to genetic exchange. Mol Ecol 6(6):583–596

Smith EG, Hume BCC, Delaney P, Wiedenmann J, Burt JA (2017) Genetic structure of coral-Symbiodinium symbioses on the world's warmest reefs. PLoS One 12(6):e0180169. https://doi.org/10.1371/journal.pone.0180169. PMID: 28666005; PMCID: PMC5493405

Sousa R, Gouveia L, Pinto A, Timóteo V, Delgado J, Henriques P (2018) Weight-length relationships of six shrimp species caught off the Madeira Archipelago, Northeastern Atlantic. Braz J Biol 79:133–138

Stella O (2019) Discriminant analysis: an analysis of its predictship function. J Educ Pract 10(5):50–57

Sugama K, Haryanti, Benzie JAH, Ballment E (2002) Genetic variation and population structure of the giant tiger prawn, Penaeus monodon, in Indonesia. Aquaculture 205(1):37–48. https://doi.org/10.1016/S0044-8486(01)00662-7

Tamadoni Jahromi S, Othman AS, Rosazlina R (2019a) Morphometrics and mitochondrial DNA genes analysis suggest a new species of Penaeus (Crustacea: Penaeidae) from the Persian Gulf. Biochem Genet 57:193–213

Tamadoni Jahromi S, Hosseini SA, Mohammadinasab A, Ahmadi M, Mokhtari M (2019b) Morphometric analysis of two morphotypes of Penaeus semisulcatus (De Haan, 1844) in the Persian Gulf. Iran J Fish Sci 18(1):115–129

Tan MP, Wong LL, Razali SA, Afiqah-Aleng N, Mohd Nor SA, Sung YY, Van de Peer Y, Sorgeloos P, Danish-Daniel M (2019) Applications of next-generation sequencing technologies and computational tools in molecular evolution and aquatic animals conservation studies: a short review. Evol Bioinformatics Online 15:117693431989228. https://doi.org/10.1177/1176934319892284

Tautz D (1989) Hypervariability of simple sequences as a general source for polymorphic DNA markers. Nucleic Acids Res 17(16):6463–6471

Tsoi KH, Chan TY, Chu KH (2005a) Genetic divergence between two morphologically similar varieties of Penaeus japonicus. J Crustac Biol 45(3):112–125

Tsoi KH, Wang ZY, Chu KH (2005b) Genetic divergence between two morphologically similar varieties of the kuruma shrimp *Penaeus japonicus*. Mar Biol 147:367–379. https://doi.org/10.1007/s00227-005-1585-x

Tsoi KH, Chan TY, Chu KH (2007) Molecular population structure of the kuruma shrimp *Penaeus japonicus* species complex in western Pacific. Mar Biol 150:1345–1364. https://doi.org/10.1007/s00227-006-0426-x

Tzeng T-D (2004) Morphological variation between populations of spotted mackerel (Scomber australasicus) off Taiwan. Fish Res 68(1–3):45–55. https://doi.org/10.1016/j.fishres.2004.02.011

Tzeng TD, Yeh SZ (1999) Stock discrimination of three penaeid shrimp species in Taiwan waters by traditional morphometric characters and linear discriminant analysis. Zool Stud 38(3):259–269

Vanavichit A, Klinbunga S, Kamonrat W, Menasveta P (1998) Characterization of microsatellites in the black tiger shrimp Penaeus monodon. Aquaculture 165(3–4):283–292

Verma D, Sankhyan V, Katoch S, Thakur YP (2015) Principal component analysis of biometric traits to reveal body confirmation in local hill cattle of Himalayan state of Himachal Pradesh, India. Vet World 8(12):1453–1457

Vu NTT, Zenger KR, Silva CNS, Guppy JL, Jerry DR (2021) Population structure, genetic connectivity, and signatures of local adaptation of the giant black tiger shrimp (Penaeus monodon) throughout the Indo-Pacific region. Genome Biol Evol 13(10):evab214. https://doi.org/10.1093/gbe/evab214. PMID: 34529049; PMCID: PMC8495139

Wang L, Sun F, Yang Z, Lee M, Yeo S, Wong J, Wen Y, Yue GH (2024) Mapping the genetic basis for sex determination and growth in hybrid tilapia (Oreochromis mossambicus × O. niloticus). Available at SSRN: https://ssrn.com/abstract=4758748 or https://doi.org/10.2139/ssrn.4758748

Wenne R (2023) Microsatellites as molecular markers with applications in exploitation and conservation of aquatic animal populations. Genes 14:808. https://doi.org/10.3390/genes14040808

Wilson AC, Cann RL, Carr SM, George M, Gyllensten UB, Helm-Bychowski KM, Higuchi RG, Palumbi SR, Prager EM (1985) Mitochondrial DNA and two perspectives on evolutionary genetics. Biol J Linnean Soc 26(4):375–400

Wuthisuthimethavee S, Lumubol P, Toojinda T, Tragoonrung S, Vanavichit A (2005) SSLP-based linkage map construction in black tiger shrimp (Penaeus monodon Fabricius). ScienceAsia 31(2):91–97

Wuthisuthimethavee, S., Lumubol, P., Vanavichit A. and Tragoonrung, S. (2003). Linkage map construction in the black tiger shrimp (Penaeus monodon Fabricius) by SSLP markers. Book of Abstracts. Plant and animal genome (PAG XI) conference, Town and Country Hotel, San Diego, Jan 11–15. Abstr. P645. pp. 235.

Xu Z, Primavera J, de la Peña LD, Pettit P, Belak J, Alcivar-Warren A (2001) Genetic diversity of wild and cultured black tiger shrimp (Penaeus monodon) in the Philippines using microsatellites. Aquaculture 199(1–2):13–40. https://doi.org/10.1016/S0044-8486(00)00535-4

Xu Z, Dhar AK, Wyrzykowski J, Alcivar-Warren A (2001) Identification of abundant and informative microsatellites from shrimp (Penaeus monodon) genome. Anim Genet 32(3):255–262. https://doi.org/10.1046/j.1365-2052.1999.00458.x

Xu Z, Wyrzykowski J, Alcivar-Warren A, Argue BJ, Moss JM, Arce SM, Traub M, Calderon FR, Lotz OJ, Breland V (2003) Genetic analyses for TSV-susceptible and TSV-resistant shrimp (Litopenaeus vannamei) using M1 microsatellite. J World Aquacult Soc 34:332–343

Yiğit S, Mendeş M (2018) Multivariate perspective for investigating relations among some behavioral traits and feed programs. J Anim Plant Sci 24(4):1186–1188

Zhang M, Zhao Y, Shan B, Liu Y, Yang C, Wang L, Liu M, Xie Q, Li Y, Zou J et al (2023) Microsatellite-Marker-Based Evaluation of Stock Enhancement for Kuruma Prawn Penaeus japonicus in Beibu Gulf. South China Sea Fishes 8(12):568. https://doi.org/10.3390/fishes8120568

Zhu Y, Queller D, Strassmann J (2000) A Phylogenetic Perspective on Sequence Evolution in Microsatellite Loci. J Mol Evol 50:324–338. https://doi.org/10.1007/s002399910036

Coral Aquaculture: A Review of In Situ and Ex Situ Culture Systems, Conditions, Applications, and Challenges

12

Fikri Akmal Khodzori, Nur Angelie Agnezara Roger, Aslam Zuhairi Nor'ashikin, Aiman Azseri, Lucyie Liana Misi, Muhammad Addin Mazni, Hariz Khairul Hisham, Muhammad Dawood Shah, and Wei Sheng Chong

Abstract

Coral reefs serve as vital ecosystems, playing a crucial role in supporting diverse marine life and sustaining human livelihoods. However, human-induced pressures and climate change disturbances have led to coral decline, necessitating concerted efforts and innovative solutions. Coral aquaculture, also known as coral farming or coral gardening, has emerged as a focal point in marine conservation initiatives, aiming to reduce the growing threats to coral reefs worldwide. Innovative techniques in coral aquaculture encompass both in situ and ex situ cultivation methods. In situ coral aquaculture involves cultivating corals within their natural habitats to enhance coral cover and health conditions. Conversely, ex situ coral aquaculture entails controlled coral cultivation in aquarium environments, aiming to optimize coral production. This industry has evolved into large-scale systems, meeting the demands of various sectors, including marine aquariums, pharmaceuticals, and reef restorations. The applications of coral aquaculture extend beyond environmental conservation, contributing significantly to economic growth by generating employment opportunities for the local communities. Additionally, the diverse array of secondary bioactive metabolites found within corals has led to the discovery of novel drugs, enhancing the pharmaceutical industry. Despite its promising potential, coral aquaculture faces challenges such as sediment impact from dredging activities as well as various

F. A. Khodzori (✉) · N. A. A. Roger · A. Z. Nor'ashikin · A. Azseri · L. L. Misi · M. D. Shah · W. S. Chong
Higher Institution Centre of Excellence (HICoE), Borneo Marine Research Institute, Universiti Malaysia Sabah, Kota Kinabalu, Sabah, Malaysia
e-mail: mfakmal@ums.edu.my

M. A. Mazni · H. K. Hisham
WWF-Malaysia Sabah Office 6th Floor, CPS Tower, Centre Point Complex No. 1, Jalan Centre Point, Kota Kinabalu, Sabah, Malaysia

N. M. Faudzi et al. (eds.), *Essentials of Aquaculture Practices*,
https://doi.org/10.1007/978-981-97-6699-4_12

types of emerging diseases and compromised health. By exploring these aspects, the review emphasizes the potential for coral aquaculture to make a sustainable contribution to both marine conservation and human being.

Keywords

Coral farming · Culture systems · Pharmaceutical · Marine conservation

12.1 Introduction

Coral reefs are among the most productive marine ecosystems in the world (Wilkinson 2008). They serve as habitats for approximately 25% of the total global marine biodiversity (Hoegh-Guldberg et al. 2017; Souter et al. 2020). The biological and ecological services provided by coral reefs offer invaluable benefits such as food subsistence and cash-crop fisheries, which are crucial for supporting and sustaining human livelihoods (Cabral and Geronimo 2018; Hoegh-Guldberg et al. 2019).

However, coral populations at local and regional scales face multiple human stressors such as sedimentation and water quality degradation (De'ath and Fabricius 2010; Duprey et al. 2016), as well as industrial, sewage, and microplastic pollution (Abaya et al. 2018; Tan et al. 2020). On a global scale, corals are vulnerable to climate change disturbances such as sea surface warming and ocean acidification (Rinkevich 2015, 2021). In addition, wild harvesting and direct collection for the ornamental market, reef restoration enterprises, and the pharmaceutical industry exert substantial pressure on coral communities worldwide (Delbeek 2001, 2008; Leal et al. 2012, 2013; Cooper et al. 2014).

Recognizing the increasing threats to coral reefs, coral aquaculture, also known as coral farming or coral gardening, serves as one of the initiatives for reef rehabilitation (Musco et al. 2017; Bradley et al. 2020). This industry has rapidly expanded over the past decades, with the aim to reduce depleted coral species caused by wild collection (Rinkevich 2008, 2014; Young et al. 2012). In general, coral aquaculture can be defined as a process of rearing or cultivating corals in both in situ and ex situ environments (Sheridan et al. 2013; Barton et al. 2017). In situ coral aquaculture involves culturing corals within their natural habitats by exposing them to conditions presumed to be optimal for coral growth (Gibbs and Newlands 2022). Meanwhile, ex situ coral aquaculture entails the controlled cultivation and propagation of corals in aquarium environments, commonly using a flow-through aquaculture system (FTAS) and recirculating aquaculture system (RAS) (Leal et al. 2016). These systems provide flexibility and precise control over culture system parameters such as water temperature, salinity, pH levels, and nutrient concentrations (Zaidi et al. 2023; Leal et al. 2016).

In recent years, researchers and aqua culturists have developed multiple advanced techniques, utilizing interdisciplinary approaches to maximize yield and minimize costs associated with coral production (Tagliafico et al. 2018a, b; Lin et al. 2021, 2023). Cutting-edge methodologies, including topological optimization, have been used to design efficient heat paths for heat conduction, ensuring that the corals are maintained at optimal temperatures (Li et al. 2017). The intricate coral structures influenced by wave hydrodynamics have also been explored to ensure the delivery of vital nutrients essential for coral growth (Lin et al. 2023).

Moreover, the scale of coral aquaculture has now expanded from a hobbyist pursuit to industrial-scale systems, driven by the increasing demand for cultured corals (Leal et al. 2016; Thatcher et al. 2022). From 2010 until 2014, this surge in demand reached around 1.5 million live corals per year, which has been attributed to the increased supply to the ornamental industry (Cooper et al. 2014; Leal et al. 2016; Tagliafico et al. 2018a, b). As of 2013, approximately 140 hard coral species were traded for the aquarium market (Rocha et al. 2013). Notably, coral aquaculture has focused on resilient, attractive, and fast-growing species like *Turbinaria*, *Plerogyra*, *Fimbriaphyllia*, and *Pocillopora* corals (Fig. 12.1), characterized by their robust growth rates. Meanwhile, slower-growing species like *Dipsastraea* and *Porites* corals (Fig. 12.2) are less preferred in aquaculture settings (Rocha et al. 2013).

Beyond its role in commercial interests for marine ornamental trade, coral aquaculture also stimulates economic growth by generating employment opportunities for the communities, thereby contributing significantly to local economies (Delbeek 2001, 2008). Additionally, the rich diversity of secondary bioactive metabolites extracted from corals, including terpenoids, alkaloids, peptides, and proteins, has captured the attention of pharmacists and natural product researchers worldwide,

Fig. 12.1 Certain types of hard corals preferred in marine aquariums due to their vibrant colors, hardiness, and adaptability to captive environments include (**a**) *Turbinaria*, (**b**) *Plerogyra*, (**c**) *Fimbriaphyllia*, and (**d**) *Pocillopora*

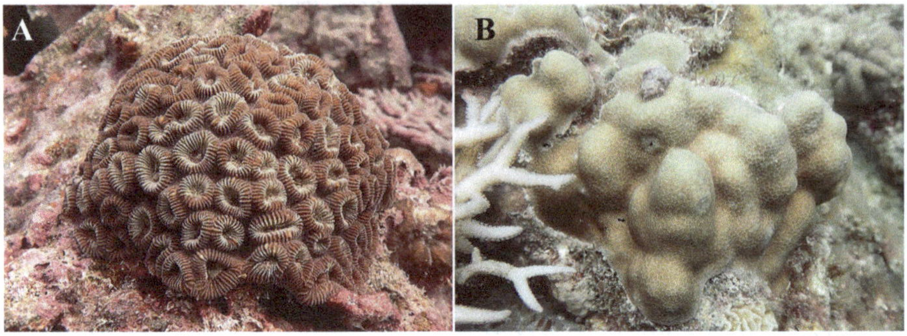

Fig. 12.2 Certain types of hard corals such as (**a**) *Dipsastraea* and (**b**) *Porites* exhibited relatively slow growth rates

leading to the discovery of novel drugs (Leal et al. 2012, 2013, 2016). Furthermore, coral aquaculture also aids in enhancing coral adaptation methods to stressors induced by climate change, thereby assisting ongoing efforts to restore and conserve coral reefs (Rinkevich 2015, 2021).

Nonetheless, coral aquaculture faces numerous challenges such as disease occurrence in the cultivation process. Coral disease is defined as any disorder in coral health, resulting in abnormalities in their morphology and physiology caused by biotic and abiotic stressors (Raymundo et al. 2008). These diseases can be characterized by observing and documenting changes in pigmentation of the coral tissue at the lesion area, the presence of microbial bands, and tissue loss revealing their bare skeleton (Moriarty et al. 2020).

The signs of diseases commonly observed in natural habitats include brown jelly syndrome, rapid tissue necrosis, skeleton eroding band, and white syndrome (Bruckner 2015; Akmal et al. 2023), all of which also manifest in cultured corals, demanding careful management strategies (Sheridan et al. 2013). Furthermore, the competition with algae and limited genetic diversity among surviving coral species during the nursery phase poses ongoing challenges for researchers and aqua culturists (Neil et al. 2021).

Accordingly, this review comprehensively elucidates various aspects of coral aquaculture, delving into culture systems, conditions, applications, and challenges. By gathering information from peer-reviewed articles from 2000 to 2023, the review offers a nuanced understanding of both in situ and ex situ coral aquaculture. The information was retrieved from Internet databases such as PubMed, Web of Science, Scopus, and registries housing dissertations and proceedings. The search criteria were tailored to include terms such as in situ and ex situ systems, culture systems and conditions, applications and challenges, and diseases and compromised coral health.

12.2 Coral Culture Systems

Coral aquaculture involves coral cultivation using in situ and ex situ systems. The in situ system refers to the practice of cultivating corals directly within their natural reefs. This involves transplanting coral fragments onto damaged reefs to aid in their recovery and restoration. Meanwhile, the ex situ system involves coral cultivation in controlled environments outside of their natural habitat. This approach is often used for research, conservation, and educational purposes.

The in situ system benefits from readily available resources in the natural environment such as food and water supplies. However, it lacks control over biotic and abiotic factors influenced by seasonal variations that potentially restrict optimal coral growth and reproduction. Moreover, factors like environmental changes, competition, and coral predators like the crown-of-thorns *Acanthaster* starfish (Fig. 12.3) impede coral survivability (Sheridan et al. 2013; Gibbs and Newlands 2022).

In the ex situ, the selection of a water flow system is crucial for maintaining the health of coral specimens. Two commonly used systems are the flow-through aquaculture system (FTAS) and the recirculating aquaculture system (RAS). FTAS involves the introduction of seawater into the culture system directly from the surrounding environment, which then undergoes various filtration processes (e.g., mechanical filters, protein skimmers, and biological filters) to remove impurities, excess nutrients, and potential pollutants and subsequently returns to the sea. Conversely, RASs use either artificial seawater or naturally filtered seawater with a relatively low rate of water replacement. Nevertheless, both systems often use advanced filtration techniques, including biological, mechanical, and chemical filtration to ensure the water remains clean and suitable for coral growth and reproduction (Sheridan et al. 2013; Leal et al. 2016; Barton et al. 2017).

Additionally, corals can be cultivated through asexual fragmentation and sexual propagation in the ex situ system. Asexual fragmentation is an easier method but results in limited viable propagules, leading to low genetic diversity. On the other hand, sexual propagation overcomes this limitation by reducing reliance on wild corals and enhancing genetic diversity. However, this method comes with high costs

Fig. 12.3 The presence of the crown-of-thorns starfish, *Acanthaster planci*, poses a threat to *Acropora* hard corals, hindering their survival

due to the mortality of coral recruits during the early settlement phase, often caused by algal competition (Leal et al. 2016; Neil et al. 2021).

To address these challenges, research efforts have focused on improving sexual restoration methods by understanding spawning timing, gamete collection, and successful larvae settlement (Chamberland et al. 2015; Schutter et al. 2023). Nonetheless, optimizing parameters for young coral growth in aquariums and ensuring their long-term survival remain challenging. Studies have indicated that specific feeding treatments, such as *Artemia nauplii* and bicarbonate supplementation, enhance the survival of sexual recruits in captive environments (Schutter et al. 2023). Besides, co-culturing with grazers, such as juvenile *Trochus niloticus* and *Thalotia gastropods*, which feed on algae increases coral recruit survivability and reduces costs with minimal aquarist intervention. For instance, the survival rate of 6-month-old recruits of *Acropora millepora* in the presence of *Thalotia gastropods* was 23.6 times greater than the survival rate in the control tanks (Neil et al. 2021).

Designing culture systems considers qualitative aspects, including shape, coloration, and resilience of coral species (Leal et al. 2016). However, various industries such as marine ornamental trade, pharmaceuticals, and reef restoration have specific criteria for defining coral quality. For instance, in the marine ornamental trade, chosen corals must not only be visually appealing with vibrant colors but also easily maintainable and resilient to various conditions such as water quality fluctuations and varying light intensities (Delbeek 2008, 2009). Coral species like *Turbinaria* and *Pocillopora*, known for their resilience and high growth rates, are often preferred choices in this context. In contrast, massive coral species such as *Dipsastraea* and *Porites* have slower growth rates, making them fewer appealing candidates for aquaculture (Leal et al. 2016).

In the pharmaceutical industry, the focus has shifted to corals with abundant secondary metabolites (e.g., sesquiterpenes, diterpenes, steroids, and norditerpenes), which are often found in soft coral species (Fig. 12.4) such as *Sinularia*, *Lithophyton*, and *Sarcophyton* genera (Koh et al. 2000; Khalesi et al. 2008). These metabolites exhibit a wide range of biological activities, including cytotoxic, anti-inflammatory, anti-tumor, and neuroprotective activities (Thao et al. 2013; Yang et al. 2020; Sun et al. 2021).

Conversely, when it comes to reef restoration, the selection of corals is guided by conservation status and ecological functions that contribute to enhancing reef complexity and creating habitats (Leal et al. 2016). Reef restoration involves the deliberate effort to rehabilitate coral reefs that have been damaged or degraded due to various factors such as climate change, pollution, or human activities. Notably, innovative approaches like the Mars Assisted Reef Restoration System (MARRS) utilize rapid-growing species like staghorn *Acropora* coral (Fig. 12.5), enhancing reef complexity and providing intricate habitats for marine organisms (International Coral Reef Initiative, ICRI 2021).

Fig. 12.4 Soft corals are favored by the pharmaceutical industry for their abundance of secondary bioactive metabolites, including (**a** and **b**) *Sinularia*, (**c**) *Lithophyton*, and (**d**) *Sarcophyton*

Fig. 12.5 Coral restoration of degraded reefs in Maliangin Island, Kudat using the Mars Assisted Reef Restoration System (MARRS) involves the use of staghorn *Acropora* coral

12.3 Coral Culture Conditions

Cultural conditions related to biotic and abiotic elements can be adjusted to enhance coral growth and reproduction. Light, water circulation, and nutrition are the three main variables that can be controlled and manipulated to maximize yield in coral aquaculture.

12.3.1 Light and Illumination

Light plays a fundamental role in promoting coral growth due to their symbiotic relationship with photosynthetic algae known as zooxanthellae. These unicellular dinoflagellates engage in symbiosis with coral hosts and use photosynthesis to convert light and carbon dioxide into oxygen and organic compounds. The excess photosynthetic products are transferred to the coral host for growth and other activities once the needs of zooxanthellae are met (Kuanui et al. 2020). Hence, maintaining an optimal light intensity and duration is vital in coral aquaculture as excessive light can reduce the efficiency of the photosystem and lead to bleaching (Richier et al. 2008; Chow et al. 2009; Kuanui et al. 2020). For instance, *Acropora cervicornis* exhibits its maximum growth rate when exposed to light intensities ranging from 38 to 44 μmol m^2s^{-1}. Nevertheless, its growth rate decreases when the light intensity exceeds 49 μmol m^2s^{-1}. Additionally, corals in the equatorial zone are exposed to 12 h of light and 12 h of darkness. Therefore, considering the photoperiod duration is crucial in coral growth. Extra hours of light could provide corals with a higher daily gain of photon irradiance (Kuanui et al. 2020).

In coral aquaculture, light manipulation involves qualitative elements (light spectrum) and technological components (types of light sources) (Leal et al. 2016). Light quantity is often measured as quantum irradiance within the range of photosynthetic active radiation (PAR). Light quality, typically determined by the spectrum of light, varies with depth and light source. Fluorescent lamps, metal halide lamps, and light-emitting diodes (LEDs) are commonly used as light sources in aquarium settings (Mass et al. 2010; Rocha et al. 2013; Leal et al. 2016). The use of artificial illumination is useful in determining the total operational costs. Therefore, utilizing energy-efficient light sources is essential to minimize production expenses. Research comparing light-emitting plasma (LEP) and LED in ex situ cultured corals found that blue LED light sources contribute to the growth of coral species like *Acropora formosa* and *Stylophora pistillata* (Rocha et al. 2013).

12.3.2 Water Circulation

Understanding the circulation of water flow is essential, as different coral species react differently to varying water currents. The relationship between coral health and water flow requires a careful balance. Increased water flow can enhance the delivery of particles to the coral surface but may reduce particle capture efficiency (Piniak 2002). Therefore, maintaining an intermediate level of water flow is considered ideal for ex situ facilities. This level ensures a high rate of particle delivery, allowing coral polyps to effectively capture and retain food particles. In addition, manipulating water flow significantly influences coral morphology. Branching corals, characterized by their arborescent forms, can adjust branch spacing to adapt to specific flow conditions (Todd 2008; Leal et al. 2016). In low water flow conditions, wider branch spacing leads to higher particle capture rates. However, in high water flow conditions, particle capture efficiency decreases with greater branch spacing.

Consequently, compact colony formations are typically observed in environments with high exposure to water motion, while open branching shapes are found in locations with low water movement (Leal et al. 2016). This relationship between water flow and coral morphology highlights the importance of a well-designed water flow system in ex situ coral aquaculture, ensuring both effective particle capture and the desired coral colony formations.

12.3.3 Organic Nutrients

Organic nutrients are crucial components within coral reef ecosystems that can be categorized into dissolved organic matter and particulate matter. Dissolved organic matter includes substances like sugars, free amino acids, and urea, while particulate matter exists as marine snow, sediment particles, and various sizes of planktonic organisms. However, not all coral species can ingest all particle sizes due to physical limitations or selective feeding behaviors, leading to varied food choices based on prey capture sizes (Leal et al. 2016).

Research has revealed that heterotrophic feeding significantly boosts calcification, protein, and lipid contents in cultured corals when compared to those deprived of nutrition. Certain coral species, such as *Stylophora pistillata*, *Turbinaria reniformis*, *Pocillopora damicornis*, *Seriatopora caliendrum*, and *Galaxea fascicularis*, exhibit greater skeletal growth when provided with adequate nutrition (Leal et al. 2016). Heterotrophic feeding is also important in enhancing coral resilience to stress, aiding in their recovery and adaptation to changing water quality and light variations (Leal et al. 2016). Additionally, it contributes essential nutrients that cannot be obtained through photosynthesis which are crucial for the energy budget of cultured corals (Tagliafico et al. 2018b).

In coral aquaculture, finding the optimal diet for achieving the fastest growth rate remains a significant challenge. Artemia, a widely accepted food source, offers advantages due to its ability to remain viable in cyst form for extended periods. However, its suitability for different cultured coral species should be explored, considering varying polyp sizes that affect the upper prey size limit (Tagliafico et al. 2018b). Moreover, the coral feeding rate is directly related to prey concentration, where exceeding the maximum feeding rate can lead to increasing nitrogen and phosphorus levels in the water which negatively impacts coral growth and survival (Hii et al. 2009; Tagliafico et al. 2018a). Research conducted in 2018 explored the optimal feeding rates for commercial corals (*Acropora millepora*, *Hydnophora rigida*, and *Duncanopsammia axifuga*), indicating an increase in feeding rates with higher food concentrations until a saturation point was reached. Interestingly, trained corals showed consistent feeding rates during both day and night, suggesting the feasibility of feeding corals during standard working hours without compromising their nutrition (Tagliafico et al. 2018a). Therefore, understanding the complexities of coral nutrition, including heterotrophic feeding patterns, prey selection, and feeding rates, is fundamental for successful coral cultivation in aquaculture.

12.3.4 Inorganic Nutrients

Nitrogen and phosphorus are essential nutrients for both corals and their symbiotic dinoflagellates, used in the synthesis of proteins and various biomass components. The symbiotic relationship between corals and zooxanthellae involves a complex nutrient exchange process. Zooxanthellae acquire nitrogen and phosphorus in inorganic forms, while corals obtain these essential elements through heterotrophic feeding and the organic substances produced by the zooxanthellae (Osinga et al. 2011). Additionally, corals obtain their nitrogen from various sources, both internally and externally. Urea and ammonia, by-products of coral host metabolism, are recycled and utilized as photosynthates by the symbiotic association (Leal et al. 2016). Externally, corals acquire nitrogen in the forms of nitrate and ammonia from seawater, which are fundamental components for coral growth. In the seawater, phosphates are present in minuscule amounts and are absorbed by corals. However, in aquarium settings, phosphates tend to accumulate, posing challenges for corals as they can hinder calcification processes (Leal et al. 2016). Hence, the balance of nitrogen and phosphorus intake is essential for the health and vitality of cultured corals.

12.4 Applications of Coral Aquaculture

Coral aquaculture has emerged as a promising solution, fostering drug discovery through the exploration of metabolites produced by cultured corals and symbiotic microbes, while simultaneously enhancing coral adaptation to bolster their overall health and reproductive potential. Hence, this shift toward coral aquaculture not only offers a discovery of new drugs but also plays a crucial role in preserving these invaluable marine organisms.

12.4.1 Drug Discovery

Traditionally, natural products from marine organisms were obtained primarily by collecting specimens from the wild. However, this conventional approach carries significant drawbacks, particularly concerning sustainability and reproducibility. Large-scale marine resource extraction for drug discovery has been demonstrated to negatively impact ecological balance, particularly when it comes to coral reefs, where the high biomass requirements can cause ecosystems to become unstable when harvested directly from the wild. Furthermore, the consistency of the supply of target metabolites is also hampered by differences in the chemical composition of specimens collected from various locations (Leal et al. 2012, 2013).

In response to these challenges, coral aquaculture emerges as a promising solution. This method allows for the continuous and controlled production of coral biomass in controlled environmental conditions (Yang et al. 2020; Sun et al. 2021). Cultured corals, particularly soft corals, belonging to specific genera such as

Sacrophyton, *Xenia*, *Sinularia*, *Lobophytum*, *Cespitularia*, and *Alcyonium* have proven to be a good source of secondary metabolites. These metabolites, which are diverse and include compounds like diterpenes, cembranoids, sterols, and alkaloids, play a crucial role in the biological functions of corals, such as defense mechanisms and interactions with symbiotic organisms, exhibiting a range of bioactivities, including anti-microbial, anti-inflammatory, and anti-cancer (Nguyen et al. 2022). Some of the metabolites are discussed below.

Diterpenes

The most metabolites found in soft coral cultures were diterpenes. There are six subclasses with diverse structure variations. The first subclass is eunicellin-based diterpene which includes klysimplexins A–X and klysimplexin sulfoxides A–C, isolated from genera *Klyxum*. Klysimplexin derivatives displayed a significant cytotoxic activity toward hepatocellular carcinoma (Hep G2, Hep 3B), human breast cancer cell line (MDA-MB-231, MCF-7), human salivary cancer cell line (Ca9–22), and human lung adenocarcinoma cell line (A549) (Chen et al. 2009, 2011a, b). The second subclass is membrane-type diterpene, which includes flexibilisin A, flexibilisolide A, and 11-epi-sinulariolide acetate, isolated from *Sinularia*, *Litophyton*, *Sarcophyton*, and *Briareum* genera (Su et al. 2009). The metabolites exhibited cytotoxicity against human cervical epitheloid (Hela), laryngeal (Hep 2), medulloblastoma, and breast (MCF-7) carcinoma cell lines (Su et al. 2009). The third subclass is spatane-type diterpene which includes leptoclalin A, isolated cultured soft coral *Sinularia leptoclados*. The metabolite displayed weak cytotoxicity against human tumor cell lines T-47 D and K-562 (Tsai et al. 2013). The fourth subclass is norcembranoidal diterpene which includes 5-epi-sinuleptolide, sinuleptolide, 5-epi-sinuleptolide, and 4α-hydroxy-5-epi-sinuleptolide, identified from cultured soft coral *Sinularia numerosa*. The compound exhibited moderate cytotoxicity toward the human T lymphoblast (CCRF-CEM) cell line (Tsai et al. 2013). The fifth subclass is briarane-type diterpene which includes briaexcavatins I–L, isolated from cultured scleraxonia *Briareum excavatum*, and exhibited inhibitory effects on superoxide anion production by human neutrophils (Sung et al. 2008), and the sixth subclass is aquariane-type diterpene which includes aquariolide A, isolated from the cultured *Erythropodium caribaeorum* (Taglialatela-Scafati et al. 2002).

Biscembranoid

Biscembranoids are a class of special tetraterpenoids which include glaucumolides A and B ximaolide A. They are isolated from soft corals belonging to the genera *Lobophytum*, *Sarcophyton*, and *Sinularia*. The compound exhibited anti-cancer and anti-inflammatory activities (Huang et al. 2015).

Sterols

Sterols such as (24*S*)-24-methylcholest-5-en-3*β*,4*α*-diol, 3*β*,4*α*-dihydroxyergosta-5,24(28)-diene, and gorgosterol have been isolated from *Sinularia* genus of soft coral. The sterols exhibited a slight inhibition effect on the expression of the inducible nitric oxide synthase (iNOS) protein in RAW 264.7 mouse macrophage cells that were stimulated with lipopolysaccharides (LPS) (Phan et al. 2021).

Metabolites from Corals and Symbiotic Microbes
Recent studies have shown the intricate relationship between corals and their symbiotic microbes. Coral tissues, skeletons, and mucus layers host a diverse array of microorganisms, including bacteria and fungi, thriving at significantly higher densities, reaching up to 100–1000 times more than those found in the surrounding seawater (Leal et al. 2013). These microorganisms provide many benefits to corals, including nutritional support and protection against harmful pathogens (Leal et al. 2013). They also provide important metabolites; for instance, Sun et al. (2014) described the isolation of isoechinulin-type alkaloids from a soft coral–derived fungus *Nigrospora oryzae*. The alkaloid exhibited bioactivates including an antifouling effect against the barnacle *Balanus amphitrite*'s larval settlement.

12.4.2 Coral Adaptation

Climate change disturbances have raised concerns about the potential decline of global coral reefs in the coming decades (Rinkevich 2015, 2021). Hence, coral aquaculture offers a mitigating strategy, not only by cultivating coral recruits for restoration but also by enhancing coral adaptation to changing environmental conditions. Adaptation refers to the alterations in physical and genetic characteristics of organisms in response to their surroundings to improve their survival. This process involves the modification of biological mechanisms and processes at various levels, whether individual, population, or community. Nonetheless, the lengthy process can be expedited through the application of the adaptation-based reef restoration (ABRR) method. ABRR comprises two primary phases: firstly, the cultivation of coral colonies in controlled conditions within coral nurseries and, secondly, the transplantation of matured corals from the nurseries to specific reef locations. This approach yields corals with enhanced health and reproductive capacity (Rinkevich 2021).

ABRR encompasses five distinct classes including assisted migration, host modification, coral chimerism, coral epigenetics, and phenotypic plasticity that can be implemented during the nursery phase. Variations in phenotypic plasticity can significantly enhance the fitness, functionality, and survivability of organisms under various pressures, such as climate change. Moreover, assisted migration, also known as assisted gene flow, is an active restoration strategy, aimed at addressing responses to environmental shifts by relocating populations of ecologically significant species (Rinkevich 2021). This approach serves as a means of genetically enhancing natural populations to increase their resistance to environmental challenges. Additionally, corals maintain a relationship with symbiotic dinoflagellates and the associated microbial consortium, forming a host that responds and adapts to stressors (Filbee-Dexter and Smajdor 2019). However, impacted biological properties of the coral host include changes in gene repertoires, biodiversity, taxa distributions, and gene expressions (Rinkevich 2021).

12.5 Challenges in Coral Aquaculture

Coral aquaculture faces significant challenges, notably the impact of sediments, the emergence of various coral diseases, and compromised health.

12.5.1 Water Clarity and Sedimentation

Water clarity and sedimentation can significantly affect both in situ and ex situ coral aquaculture. In the in situ coral aquaculture, water clarity is crucial for the health and growth of corals in their natural environment. Clear water allows for optimal penetration of sunlight, which is essential for photosynthesis, a process that provides energy for coral growth and calcification. In areas with poor water clarity, such as those affected by sedimentation or runoff from land, corals may experience reduced light availability, leading to decreased growth rates and increased susceptibility to stressors such as bleaching. Hence, maintaining good water clarity is vital for successful in situ coral aquaculture practices. Efforts to reduce sedimentation, pollution, and other factors contributing to poor water quality are essential for supporting healthy coral reef ecosystems (Elsaeed 2011; Jones et al. 2015, 2016).

In addition, dredging activities for land and coastal development have impacted the in situ coral aquaculture (Bahr et al. 2015). Sediments from land-based sources smother corals, hindering the photosynthesis process and impeding coral recruitment and growth (Stender et al. 2014; Tuttle et al. 2020). Dredged sediments typically consist of finer grains which take longer to settle out of the water column and disperse over larger distances (Duclos et al. 2013; Jones et al. 2015; Duckworth et al. 2017). The coral mortality rate to dredging significantly affects aquaculture production (Smith et al. 2016, 2021). It is estimated that over half a million corals have directly succumbed to burial by sediments associated with dredging activities (Spadafore et al. 2021).

For ex situ coral aquaculture, water clarity is equally important, particularly in coral nurseries or research laboratories where corals are cultivated in tanks. Clear water is necessary to ensure adequate light penetration for photosynthesis and optimal growth conditions for cultivated corals. High-quality artificial lighting systems are often used to replicate natural light regimes in aquarium settings. Hence, filtration systems are used to maintain water clarity by removing particulate matter, organic debris, and excess nutrients that can contribute to cloudiness and reduced water quality (Sheridan et al. 2013).

12.5.2 Coral Diseases and Compromised Health

Coral aquaculture faces significant challenges related to diseases and other compromised health (Sheridan et al. 2013). They are often characterized by changes in coloration, tissue bleaching, and necrosis (Raymundo et al. 2008). The impacts of disease outbreaks potentially result in the loss of coral cultures. In general, coral disease emerges from interactions between coral hosts, environmental stressors, and bacterial pathogens (Bruno 2015). It is defined as any disruption in coral health resulting in physiological impairment caused by biotic and abiotic stresses (Woodley et al. 2016). It can also be described as any abnormal signs compromising coral health (Raymundo et al. 2008). Approximately 40 disease signs have been identified globally, with more than 35 diseases recorded in the Caribbean region (Work and Aeby 2006; Bruckner 2015; Morais et al. 2022). Disease outbreaks have affected more than 200 hard coral species and are often attributed to human-induced factors like sedimentation, pollution, and intensive tourism (Woodley et al. 2016; El-Naggar 2020; Akmal and Shahbudin 2020).

In the context of coral aquaculture, diseases observed in cultivated corals are mostly similar to those found in natural habitats. Corals are susceptible to specific diseases such as brown jelly syndrome (Borneman 2001; Borneman and Lowrie 2001; Sweet et al. 2012), rapid tissue necrosis (Luna et al. 2007), skeleton eroding band (Riegl and Antonius 2003; Riegl et al. 2009), and white syndrome (Luna et al. 2009; Sussman et al. 2008; Sweet and Bythell 2012; Sweet et al. 2014). In addition to ex situ coral cultures, they are commonly exposed to signs like red slime algae, flatworm, and red bug infestations (Haapkyla et al. 2007; Rawlinson et al. 2011; Sweet et al. 2012). These signs of diseases increase the risk of transmission among corals, posing a significant threat to aquaculture production (Sheridan et al. 2013). Furthermore, the decline in coral health is often linked to dysbiosis, disrupting the natural microbial community and facilitating their prevalence (Egan and Gardiner 2016; Ainsworth and Gates 2016; Van der Loos et al. 2019; Moriarty et al. 2020). Some of the diseases and compromised health in coral aquaculture are discussed below.

Brown Jelly Syndrome

Brown jelly syndrome (BJS) is a common disease observed in aquariums and reef tanks (Sweet et al. 2012; Androschuk and Shaw 2015). This disease primarily affects *Euphyllia* coral (Cróquer et al. 2006). The signs of BJS can be identified by a distinctive brown jelly–like substance surrounding the affected corals and the rapid loss of coral tissue (Cróquer et al. 2006).

The main agent causing BJS is a single-celled microscopic protozoan known as a ciliate (Sweet et al. 2012; Randall et al. 2015). Microscopic observations have shown the presence of zooxanthellae within ciliates extracted from coral tissue (Ulstrup et al. 2007). However, it remains uncertain whether ciliates consume zooxanthellae for nutrition (Cheng et al. 2021). Nevertheless, it is known that ciliate infection can lead to bleaching and coral death (Cheng et al. 2021). When zooxanthellae are expelled from the gastrodermal cavity of the coral host, it results in discoloration, revealing calcium carbonate skeletal structure and leading to coral death (Curran and Barnard 2021).

In aquariums, a ciliate species associated with BJS is known as *Helicostoma nonatum* (Borneman and Lowrie 2001). This ciliate resembles those found in brown band disease (BrB) in the wild. Despite the different disease signs observed in BrB and BJS, it remains challenging to determine whether these are two separate diseases. Willis et al. (2004) initially proposed a similarity between the ciliate linked to BrB and *Helicostoma* species. However, research conducted by Bourne et al. (2008) revealed that the protozoan associated with BrB is related to different ciliate species within the class Oligohymenophora. Hence, further research should be conducted to understand its causes, mechanisms, and potential treatments.

Aquarists have suggested the use of antibiotics such as chloramphenicol, sulfonamides, synergists, fluoroquinolones, and macrolides to effectively treat BJS (Zhang et al. 2019). However, antibiotic treatments face limitations, including challenges in light sensitivity and short half-life, necessitating repeated treatments which may lead to water quality problems and the risk of other emerging diseases. Additionally, antibiotics can harm beneficial bacteria within the coral host and the surrounding water (Sweet et al. 2012). Therefore, monitoring water quality, proper husbandry practices, and quarantine procedures for the addition of new corals are essential to prevent the spread of BJS in aquariums.

Rapid Tissue Necrosis

Rapid tissue necrosis (RTN), also known as the shutdown reaction, is characterized by the rapid degradation of coral tissue (Borneman 2001; Anthony 2004). This condition contributes to coral mortality, both in natural environments and aquarium settings (Anthony 2004). Despite many reported cases, the causative agents of RTN remain unidentified (Luna et al. 2007). Although some field studies have noted the occurrences of coral tissue degradation and the presence of pathogenic bacteria, such as *Vibrio coralliilyticus* (Ben-Haim and Rosenberg 2002; Ben-Haim et al. 2003; Khodzori et al. 2021; Akmal et al. 2023), the exact cause and progression rate of the disease remains unclear.

In a study conducted by Gibbin et al. (2019), researchers investigated the effect of RTN on coral species *Pocillopora damicornis* under increased temperature conditions. This research utilized various techniques, including microfluidics, fluorescence microscopy, stable isotopes, electron microscopy, and NanoSIMS isotopic imaging to examine the progression rate of pathogenic *V. coralliilyticus* at both tissue and cellular levels during daytime and nighttime conditions. During the day, coral polyps increase mucus production and actively release pathogens. NanoSIMS analysis revealed an increase in carbon assimilation in the symbiotic algae but a decrease in the coral host. Nighttime incubations displayed different behaviors, with mucus formation on the coral surface and reduced carbon turnover in symbiotic algae upon *Vibrio* infection. Gibbin et al. (2019) concluded that rising temperatures can trigger the abundance of *Vibrio* spp. into a pathogenic state, leading to tissue lysis and expulsion of the symbiotic algae. Another study was done by Luna et al. (2009) using both culture-dependent and culture-independent techniques to examine healthy and RTN-affected coral tissues. The presence of *Vibrio harveyi*, a common bacterial pathogen in marine environments, was identified in diseased specimens (Austin and Zhang 2006; Chen et al. 2022). However, it is important to consider other factors such as coral responses to stressful conditions due to climate change disturbances (Borneman 2001).

Skeletal Eroding Band

Coral diseases associated with ciliates pose a significant threat, with the skeletal eroding band (SEB) being prevalent in the Indo-Pacific region (Page and Willis 2008; Page et al. 2015). SEB is characterized by skeletal eroding lesions marked by a distinctive speckled black band (Fig. 12.6), composed of the folliculinid ciliate, *Halofolliculina corallasia*, which can cause damage to coral polyps, coenosarc, and the surface of the coenosteum (Winkler et al. 2004).

A study conducted by Palmer and Gates (2010) used a macroscopic approach to identify SEB by observing the presence of a black line or band surrounding healthy coral tissue. This progression was characterized by the algae colonization on the exposed coral skeleton, resulting in a speckled black appearance due to the clustering of ciliates (Antonius and Lipscomb 2000). Additionally, SEB was found to be susceptible to *Montipora* and *Pocillopora* corals (Page and Willis 2008).

Another study by Lamb et al. (2014) suggested that injured corals are more susceptible to SEB, particularly in high-use sites for reef recreational activities. The increased vulnerability is attributed to additional stressors associated with intense human activity (Lamb et al. 2014). Page et al. (2015) also found that SEB is highly prevalent in various Indo-Pacific localities, especially in areas affected by high coastal development and human activities. These findings align with an earlier study by Winkler et al. (2004) in the northern Red Sea, where SEB prevalence reached up to 51% of colonies in sites exposed to physical damage due to tourism and nearby industrial developments in the Gulf of Aqaba.

Halofolliculina ciliates may also enter aquariums through untreated seawater, corals, and other substrates (Page et al. 2015). Once these ciliates are present in the aquarium, the risk of infections can be reduced by practicing measures such as

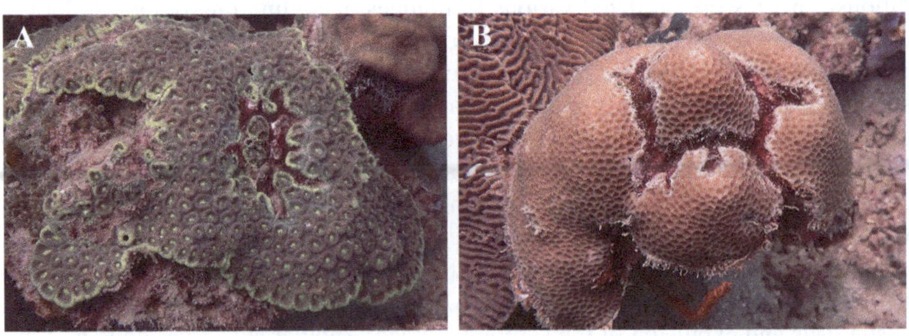

Fig. 12.6 Skeletal eroding band affected corals such as (**a**) *Dipsastraea* and (**b**) *Goniastrea*

filtering seawater and closely inspecting and removing ciliates from all items added to the aquarium (Page and Willis 2008; Rodríguez et al. 2009; Page et al. 2015).

White Syndrome

White syndrome (WS) is a term used in the Indo-Pacific region to describe white signs and syndromes such as white band or plague observed in the Caribbean region (Bythell et al. 2004; Ainsworth et al. 2007; Sussman et al. 2008). It can be identified by the presence of white bands, spots, or patches on corals (Fig. 12.7) (Luna et al. 2009; Bythell et al. 2004; Bourne et al. 2015). This disease involves rapid tissue loss, ranging from 1 to 100 cm^2 per day, with a clear boundary between healthy tissue and exposed skeleton (Luna et al. 2009). Previous studies have linked WS to the presence of *Vibrio* spp., including *V. harveyi*, *V. mediterranei*, *V. owensii*, and *V. coralliilyticus* (Luna et al. 2009; Piskorska et al. 2007; Ushijima et al. 2012; Sussman et al. 2008; Khodzori et al. 2021; Akmal et al. 2023).

To understand the role of *Vibrio* spp. in WS development, Luna et al. (2009) collected specimens displaying WS signs from natural reefs and aquariums. Using culture-based and culture-independent techniques, they found that 73% of WS cases were linked to a high bacterial presence, particularly *Vibrio* spp. with *Vibrio harveyi* as the most prevalent. Sweet and Bythell (2012) conducted further research, confirming an increased *V. harveyi* in diseased and healthy coral samples. Sussman et al. (2009) also demonstrated the increased zinc-metalloprotease produced by pathogenic *Vibrio* species, responsible for breaking down coral tissue and inactivating zooxanthellae. Additionally, Piskorska et al. (2007) observed an increase in *V. mediterranei* in the mucus of *Echinopora lamellosa* affected by WS.

Other factors like ciliates, helminths, and apoptosis have also been proposed as potential drivers of WS diseases (Pollock et al. 2017). Recent studies indicate that certain ciliates, such as *Philasterine scuticociliate*, may enter corals through injuries or lesions or indirect interactions with other microbes affecting coral health. Ciliates have been associated with various coral diseases globally, with diverse ciliate associations within WS lesions, particularly in diseased *Acropora* corals. However, it remains unclear whether ciliates play a primary or opportunistic secondary

Fig. 12.7 White syndromes affected corals such as (**a**) *Cyphastrea* and (**b**) *Acropora*

pathogenic role in coral diseases, necessitating further experimental validation of their causative agents (Ravindran et al. 2022).

Red Slime Algae

Red slime algae (RSA) poses a common challenge in aquariums afflicted the cultured corals. The algae are formed by a consortium of cyanobacteria, creating dense mats on coral surfaces. Similar to dinoflagellates, these mats can detach and suffocate corals (Lehmann 2022). Cyanobacteria can thrive in diverse environments, including fresh and brackish waters as well as sea and thermal waters (Greer et al. 2016). They can proliferate rapidly, leading to blooms and potentially producing harmful substances like cyanotoxins. Additionally, these cyanobacteria can rapidly multiply in environments rich in food sources, nitrate, and phosphate, forming a film that covers sand and corals. If left untreated, this film can deprive corals of light and oxygen (Buratti et al. 2017; Fiore et al. 2020).

RSA appears in various colors, ranging from red to black and blue-green, and the exact cause of this condition remains unidentified (Sweet et al. 2012). Nevertheless, significant similarities can be observed between the appearance of this disease and the cyanobacterial mats associated with black band disease (BBD) in natural coral populations. Cyanobacterial overgrowth is also frequently observed in the wild, especially in the Caribbean coral communities (Weil 2004; Weil et al. 2009). The increase in the prevalence of benthic cyanobacterial mats and the occurrence of coral diseases represent significant factors contributing to the degradation of coral reefs in this region (Weil 2004; Weil et al. 2009). Recent developments have discovered the potential interactions between the expanding presence of cyanobacterial mats and coral diseases, particularly BBD (Cissell et al. 2022). This intersection presents an urgent concern for conservation and a critical challenge for future research in this field.

Aquarists typically consider RSA as an indicator of poor water quality. Therefore, they recommend improving water quality by reducing nitrate and phosphate levels, monitoring light exposure, and enhancing water flow to effectively reduce red slime algae. Additionally, there is a phenomenon known as new tank syndrome, where the introduction of new silicone in the new tank can increase nutrient levels, promoting the growth of algae and cyanobacteria, including RSA (Sweet et al. 2012). Addressing these factors is essential to maintain healthy corals in aquarium environments.

Flatworm Infestation

Coral aquaculture frequently faces challenges, particularly from flatworm species like *Convolutriloba* sp. and *Waminoa* sp., causing brown rust disease (Shannon and Achatz 2007). *Waminoa* sp. commonly affects *Plerogyra* (Fig. 12.8) and *Euphyllia* corals and has been recorded in the Indo-Pacific coral communities (Sweet et al. 2012; Kunihiro et al. 2019). In contrast, *Convolutriloba* sp. primarily inhabits marine aquariums, displaying unique asexual reproduction methods (Zabotin and Evtugyn 2021). Only one species, *Convolutriloba hastifera*, has been identified in natural habitats, while the others were discovered in aquariums (Shannon and

Fig. 12.8 Flatworm infestation on *Plerogyra* corals

Achatz 2007). Although these flatworm species do not directly harm corals, extensive infestations can disrupt photosynthesis, affecting overall coral health (Sweet et al. 2012).

Another significant concern is the presence of *Acropora*-eating flatworms (AEFs), known as *Prosthiostomum acroporae* (previously identified as *Amakusaplana acroporae*), posing a common threat to *Acropora* corals in both aquariums and natural environments (Barneah et al. 2007; Haapkyla et al. 2007; Rawlinson et al. 2011). AEF infestations can rapidly deteriorate the health of *Acropora* colonies (Nosratpour 2008). The first documented case of *P. acroporae* infestation in the wild occurred off Lizard Island, Australia (Rawlinson et al. 2011). Additionally, this species has been widely observed, infecting *Acropora* colonies at multiple sites along the Great Barrier Reef (Barton et al. 2020). *P. acroporae* typically feed on coral tissue, leaving behind circular pale feeding scars approximately 1 mm in size (Nosratpour 2008; Rawlinson et al. 2011). The consumed tissue is digested within a branching gut that covers almost the entire lateral surface area of the flatworm, providing effective camouflage and making starved specimens appear opaque white (Rawlinson et al. 2011).

Red Bug Infestation

Red bug infestations are caused by a tiny parasitic crustacean species, *Tegastes acroporanus*, which can be identified by a distinct red dot sign. The infestations may include poor polyp extension and loss of coloration in affected corals. These pests primarily affect *Acropora* corals and pose significant threats in coral culture environments (Sweet et al. 2012; Christie and Raines 2016). Although the infestations are not typically lethal to the coral hosts, they can induce stress and secondary infections within coral colonies. For example, severely affected *Acropora* colonies may exhibit signs such as tissue loss, unusual polyp extension, loss of pigmentation, increased mucus production, and disrupted growth development (Carl 2008; Sweet et al. 2012; Christie and Raines 2016). Researchers have explored different methods to control *Tegastes* infestations in coral aquaculture. One approach involves the use of milbemycin oxime, which is effective against crustaceans without harming other invertebrates. This solution can be directly applied to entire coral tanks, making it a practical option for aquarists. Additionally, a biological control method, involving the introduction of the dragon-faced pipefish, *Corythoichthys haematopterus*, which feeds on crustaceans, including *T. acroporanus*. These treatments offer promising solutions for public aquariums, coral farmers, and clinicians, providing options for preventing undesirable parasitic crustaceans (Sweet et al. 2012; Christie and Raines 2016).

12.6 Recommendations

1. Emphasize future research in developing innovative techniques to optimize parameters for coral growth, enhance genetic diversity, and mitigate environmental stressors in both in situ and ex situ coral aquaculture.
2. Manage and optimize culture conditions for optimal results by exploring innovative techniques in light manipulation, water flow dynamics, and nutrient management to further advance the sustainability and efficiency of coral aquaculture.
3. Discover new drugs by exploring valuable secondary metabolites from cultured corals and symbiotic microbes.
4. Utilize advance methods, such as adaptation-based reef restoration (ABRR) to enhance the fitness, functionality, and survivability of corals under changing environmental conditions.
5. Implement effective disease management strategies for sustaining coral aquaculture. A holistic approach focusing on environmental impacts, conservation, and disease prevention is essential for the long-term success of coral aquaculture initiatives.

12.7 Conclusion

In conclusion, coral aquaculture has emerged as an innovative solution to the increasing threats faced by coral reefs globally. By employing both in situ and ex situ cultivation methods, coral aquaculture offers innovative approaches to rejuvenate declining coral cover and improve health conditions. This industry has transitioned from small-scale hobbyist pursuits to industrial-scale systems, driven by demands from marine ornamental trade, pharmaceutical industries, and reef restoration sectors. Coral aquaculture not only supports economic growth by generating employment opportunities but also contributes to the discovery of novel drugs. Despite challenges such as environmental fluctuations and disease occurrences, careful management strategies and ongoing research are crucial for ensuring the success of coral aquaculture initiatives.

Acknowledgement This study was funded by the Fundamental Research Grant Scheme (Grant Code: FRGS/1/2022/WAB05/UMS/02/4; FRG0574–1/2022) under the Ministry of Higher Education Malaysia.

References

Abaya LM, Wiegner TN, Beets JP, Colbert SL, Kaile'a MC, Kramer KL (2018) Spatial distribution of sewage pollution on a Hawaiian coral reef. Mar Pollut Bull 130:335–347

Ainsworth TD, Gates RD (2016) Corals' microbial sentinels. Science 352:1518–1519

Ainsworth TD, Kvennefors EC, Blackall LL, Fine M, Hoegh-Guldberg O (2007) Disease and cell death in white syndrome of Acroporid corals on the Great Barrier Reef. Mar Biol 151:19–29

Akmal KF, Shahbudin S (2020) Baseline assessment of coral health and disease in Tioman Island Marine Park, Malaysia. Community Ecol 21:285–301

Akmal KF, Shahbudin S, Abdul Muhaimin Z, Shah MD, Chong WS (2023) Application of biotechnology in white syndrome coral disease identification. In: Marine biotechnology: applications in food, drugs and energy. Springer, Singapore, pp 271–297

Androschuk T, Shaw R (2015) Brown jelly syndrome in captive coral (*Euphyllia* spp.). In: Proceedings of the undergraduate research in science conference of Alberta (URSCA), p 1

Anthony SL (2004) Coral disease in a large-scale reef mesocosm. In: Proceedings of the 10th international coral reef symposium, Okinawa, Japan, p 379

Antonius A, Lipscomb D (2000) First protozoan coral-killer identified in the Indo-Pacific. Atoll Res Bull 481:1–21

Austin B, Zhang XH (2006) *Vibrio harveyi*: a significant pathogen of marine vertebrates and invertebrates. Lett Appl Microbiol 43:119–124

Bahr KD, Jokiel PL, Toonen RJ (2015) The unnatural history of Kāne 'ohe Bay: coral reef resilience in the face of centuries of anthropogenic impacts. PeerJ 3:e950

Barneah O, Ben-Dov E, Kramarsky-Winter E, Kushmaro A (2007) Characterization of black band disease in Red Sea stony corals. Environ Microbiol 9:1995–2006

Barton JA, Willis BL, Hutson KS (2017) Coral propagation: a review of techniques for ornamental trade and reef restoration. Rev Aquac 9:238–256

Barton JA, Bourne DG, Humphrey C, Hutson KS (2020) Parasites and coral-associated invertebrates that impact coral health. Rev Aquac 12:2284–2230

Ben-Haim Y, Rosenberg E (2002) A novel *Vibrio* sp. pathogen of the coral *Pocillopora damicornis*. Mar Biol 141:47–55

Ben-Haim Y, Thompson FL, Thompson CC, Cnockaert MC, Hoste B, Swings J, Rosenberg E (2003) *Vibrio coralliilyticus* sp. nov, a temperature-dependent pathogen of the coral *Pocillopora damicornis*. Int J Syst Evol Microbiol 53:309–315

Borneman EH (2001) Aquarium corals: selection, husbandry and natural history. T.F.H. Publications, Inc., Neptune City

Borneman EH, Lowrie J (2001) Advances in captive husbandry and propagation: an easily utilized reef replenishment means from the private sector? Bull Mar Sci 69:897–913

Bourne DG, Boyett HV, Henderson ME, Muirhead A, Willis BL (2008) Identification of a ciliate (Oligohymenophorea: Scuticociliatia) associated with brown band disease on corals of the Great Barrier Reef. Appl Environ Microbiol 74:883–888

Bourne DG, Ainsworth TD, Pollock FJ, Willis BL (2015) Towards a better understanding of white syndromes and their causes on Indo-Pacific coral reefs. Coral Reefs 34:233–242

Bradley P, Jessup B, Pittman SJ, Jeffrey CF, Ault JS, Carrubba L, Lilyestrom C, Appeldoorn RS, Schärer MT, Walker BK, McField M (2020) Development of a reef fish biological condition gradient model with quantitative decision rules for the protection and restoration of coral reef ecosystems. Mar Pollut Bull 159:111387

Bruckner AW (2015) History of coral disease research. In: Woodley CM, Downs CA, Bruckner AW, Porter JW, Galloway SB (eds) Diseases of coral. Wiley, pp 52–84

Bruno JF (2015) The coral disease triangle. Nat Clim Chang 5:302–303

Buratti FM, Manganelli M, Vichi S, Stefanelli M, Scardala S, Testai E, Funari E (2017) Cyanotoxins: producing organisms, occurrence, toxicity, mechanism of action, and human health toxicological risk evaluation. Arch Toxicol 91:1049–1130

Bythell J, Pantos O, Richardson L (2004) White plague, white band, and other "white" diseases. In: Coral health and disease. Springer, Berlin, pp 351–365

Cabral RB, Geronimo RC (2018) How important are coral reefs to food security in the Philippines? Diving deeper than national aggregates and averages. Mar Policy 91:136–141

Carl M (2008) Predators and pests of captive corals. In: Leewis RJ, Janse M (eds) Advances in coral husbandry in public aquariums. Burger's Zoo, Arnhem, pp 31–36

Chamberland VF, Vermeij MJ, Brittsan M, Carl M, Schick M, Snowden S, Schrier A, Petersen D (2015) Restoration of critically endangered elkhorn coral (*Acropora palmata*) populations using larvae reared from wild-caught gametes. Glob Ecol Conserv 4:526–537

Chen BW, Wu YC, Chiang MY, Su JH, Wang WH, Fan TY, Sheu JH (2009) Eunicellin-based diterpenoids from the cultured soft coral *Klyxum simplex*. Tetrahedron 65:7016–7022

Chen BW, Chao CH, Su JH, Tsai CW, Wang WH, Wen ZH, Huang CY, Sung PJ, Wu YC, Sheu JH (2011a) Klysimplexins I–T, eunicellin-based diterpenoids from the cultured soft coral *Klyxum simplex*. Org Biomol Chem 9:834–844

Chen BW, Huang CY, Wen ZH, Su JH, Wang WH, Sung PJ, Wu YC, Sheu JH (2011b) Klysimplexins U–X, eunicellin-based diterpenoids from the cultured soft coral *Klyxum simplex*. Bull Chem Soc Jpn 84:1237–1242

Chen H, Siu SWI, Wong CTT, Qiu J, Cheung AKK, Lee SMY (2022) Anti-epileptic Kunitz-like peptides discovered in the branching coral *Acropora digitifera* through transcriptomic analysis. Arch Toxicol 96:2589–2608

Cheng CM, Cheng YR, Ding DS, Chen YT, Sun WT, Pan CH (2021) Effects of ciliate infection on the activities of two antioxidant enzymes (SOD and CAT) in captive coral (*Goniopora columna*) and evaluation of drug therapy. Biology 10:1216

Chow AM, Ferrier-Pages C, Khalouei S, Reynaud S, Brown IR (2009) Increased light intensity induces heat shock protein Hsp60 in coral species. Cell Stress Chaperones 14:469–476

Christie BL, Raines JA (2016) Effect of an otic milbemycin oxime formulation *on Tegastes acroporanus* infesting corals. J Aquat Anim Health 28:235–239

Cissell EC, Eckrich CE, McCoy SJ (2022) Cyanobacterial mats as benthic reservoirs and vectors for coral black band disease pathogens. Ecol Appl 32:e2692

Cooper EL, Hirabayashi K, Strychar KB, Sammarco PW (2014) Corals and their potential applications to integrative medicine. Evid Based Complement Alternat Med 2014:184959

Cróquer A, Bastidas C, Lipscomb D (2006) Folliculinid ciliates: a new threat to Caribbean corals? Dis Aquat Org 69:75–78

Curran A, Barnard S (2021) What is the role of zooxanthellae during coral bleaching? Review of zooxanthellae and their response to environmental stress. S Afr J Sci 117:1–7

De'ath G, Fabricius K (2010) Water quality as a regional driver of coral biodiversity and macroalgae on the Great Barrier Reef. Ecol Appl 20:840–850

Delbeek JC (2001) Coral farming: past, present and future trends. Aquar Sci Conserv 3:171–181

Delbeek JC (2008) Collecting and shipping live coral: techniques, tips and headaches. Public aquarium husbandry series. In: Leewis RJ, Janse M (eds) Advances in coral husbandry in public aquarium, vol 2, pp 363–373

Delbeek J (2009) The perfect coral: Euphyllia. Reef Life, EcoSystem Publications, Santa Ana, CA, pp 60–61

Duckworth A, Giofre N, Jones R (2017) Coral morphology and sedimentation. Mar Pollut Bull 125:289–300

Duclos PA, Lafite R, Le Bot S, Rivoalen E, Cuvilliez A (2013) Dynamics of turbid plumes generated by marine aggregate dredging: an example of a macrotidal environment (the Bay of Seine, France). J Coast Res 29:25–37

Duprey NN, Yasuhara M, Baker DM (2016) Reefs of tomorrow: eutrophication reduces coral biodiversity in an urbanized seascape. Glob Chang Biol 22:3550–3565

Egan S, Gardiner M (2016) Microbial dysbiosis: rethinking disease in marine ecosystems. Front Microbiol 7:991

El-Naggar HA (2020) Human impacts on coral reef ecosystem. In: Natural resources management and biological sciences. IntechOpen

Elsaeed GH (2011) The impact of dredging on coastal environments. Aust J Basic Appl Sci 5:74–81
Filbee-Dexter K, Smajdor A (2019) Ethics of assisted evolution in marine conservation. Front Mar Sci 6:20
Fiore M, Parisio R, Filippini T, Mantione V, Platania A, Odone A, Signorelli C, Pietrini V, Mandrioli J, Teggi S, Costanzini S (2020) Living near water bodies as a proxy of cyanobacteria exposure and risk of amyotrophic lateral sclerosis: a population-based case-control study. Environ Res 186:109530
Gibbin E, Gavish A, Krueger T, Kramarsky-Winter E, Shapiro O, Guiet R, Meibom A (2019) *Vibrio coralliilyticus* infection triggers a behavioral response and perturbs nutritional exchange and tissue integrity in a symbiotic coral. ISME J 13:989–1003
Gibbs MT, Newlands M (2022) Restoration heralds' new management challenges for coral reefs. Mar Policy 136:–104911
Greer B, McNamee SE, Boots B, Cimarelli L, Guillebault D, Helmi K, Campbell K (2016) A validated UPLC–MS/MS method for the surveillance of ten aquatic biotoxins in European brackish and freshwater systems. Harmful Algae 55:31–40
Haapkyla J, Seymour AS, Trebilco J, Smith D (2007) Coral disease prevalence and coral health in the Wakatobi Marine Park, south-east Sulawesi, Indonesia. J Mar Biol Assoc UK 87:403
Hii Y-S, Soo C, Liew H (2009) Feeding of scleractinian coral, *Galaxea fascicularis*, on Artemia salina nauplii in captivity. Aquac Int 17:363–376
Hoegh-Guldberg O, Poloczanska ES, Skirving W, Dove S (2017) Coral reef ecosystems under climate change and ocean acidification. Front Mar Sci 4:158
Hoegh-Guldberg O, Pendleton L, Kaup A (2019) People and the changing nature of coral reefs. Reg Stud Mar Sci 30:100699
Huang CY, Sung PJ, Uvarani C, Su JH, Lu MC, Hwang TL, Dai CF, Wu SL, Sheu JH (2015) Glaucumolides A and B, biscembranoids with new structural type from a cultured soft coral *Sarcophyton glaucum*. Sci Rep 5:15624
International Coral Reef Initiative, ICRI (2021) Mars assisted reef restoration system. https://icri-forum.org/mars-assisted-reef-restoration-system/. Accessed 19 Jan 2024
Jones R, Ricardo GF, Negri AP (2015) Effects of sediments on the reproductive cycle of corals. Mar Pollut Bull 100:13–33
Jones R, Bessell-Browne P, Fisher R, Klonowski W, Slivkoff M (2016) Assessing the impacts of sediments from dredging on corals. Mar Pollut Bull 102:9–29
Khalesi MK, Beeftink RH, Wijffels RH (2008) The soft coral *Sinularia flexibilis*: potential for drug development. In: Advances in coral husbandry in public aquariums, vol 2, pp 47–60
Khodzori FA, Saad S, Mansor NN, Nasir NANM, Khalid NNNA, Rawi FZ (2021) Pathogenic *Vibrio* spp. identified for white syndrome coral disease in Tioman Island Marine Park, Malaysia. Malays J Microbiol 17:69–79
Koh M, Iwanaga T, Hatanaka M, Nakano A, Morihara K, Takemura K (2000) Distribution of sarcophytol A in soft coral of the *Sarcophyton* genus. Biosci Biotechnol Biochem 64:858–861
Kuanui P, Chavanich S, Viyakarn V, Omori M, Fujita T, Lin C (2020) Effect of light intensity on survival and photosynthetic efficiency of cultured corals of different ages. Estuar Coast Shelf Sci 235:106515
Kunihiro S, Farenzena Z, Hoeksema BW, Groenenberg DS, Hermanto B, Reimer JD (2019) Morphological and phylogenetic diversity of Waminoa and similar flatworms (Acoelomorpha) in the western Pacific Ocean. Zoology 136:125692
Lamb JB, True JD, Piromvaragorn S, Willis BL (2014) Scuba diving damage and intensity of tourist activities increase coral disease prevalence. Biol Conserv 178:88–96
Leal MC, Puga J, Serodio J, Gomes NC, Calado R (2012) Trends in the discovery of new marine natural products from invertebrates over the last two decades—where and what are we bioprospecting? PLoS One 7:e30580
Leal MC, Calado R, Sheridan C, Alimonti A, Osinga R (2013) Coral aquaculture to support drug discovery. Trends Biotechnol 31:555–561
Leal MC, Ferrier-Pagès C, Petersen D, Osinga R (2016) Coral aquaculture: applying scientific knowledge to ex situ production. Rev Aquac 8:136–153

Lehmann W (2022) Coral reef aquarium husbandry and health. In: Invertebrate medicine, pp 107–127

Li B, Hong J, Yan S, Liu H, Ge L (2017) Generating optimal heat conduction paths based on bionic growth simulation. Int Commun Heat Mass Transf 83:55–63

Lin CH, Takahashi S, Mulla AJ, Nozawa Y (2021) Moonrise timing is key for synchronized spawning in coral *Dipsastraea speciosa*. Proc Natl Acad Sci U S A 118:e2101985118

Lin S, Chou N, Bao D, Zhang G, Xiong C, Fang J, Xie YM (2023) Design and fabrication of artificial brain coral: evolution principle, turbulent hydrodynamics and matter interchange. Comput Struct 276:106955

Luna GM, Biavasco F, Danovaro R (2007) Bacteria associated with the rapid tissue necrosis of stony corals. Environ Microbiol 9:1851–1857

Luna GM, Bongiorni L, Gili C, Biavasco F, Danovaro R (2009) *Vibrio harveyi* as a causative agent of the white syndrome in tropical stony corals. Environ Microbiol Rep 2(1):120–127

Mass T, Kline DI, Roopin M, Veal CJ, Cohen S, Iluz D, Levy O (2010) The spectral quality of light is a key driver of photosynthesis and photoadaptation in *Stylophora pistillata* colonies from different depths in the Red Sea. J Exp Biol 213:4084–4091

Morais J, Cardoso AP, Santos BA (2022) A global synthesis of the current knowledge on the taxonomic and geographic distribution of major coral diseases. Environ Adv 8:100231

Moriarty T, Leggat W, Huggett MJ, Ainsworth TD (2020) Coral disease causes, consequences, and risk within coral restoration. Trends Microbiol 28:793–807

Musco L, Prada F, D'Anna G, Galasso NM, Pipitone C, Fernández TV, Badalamenti F (2017) Turning casualty into opportunity: fragmenting dislodged colonies is effective for restoring reefs of a Mediterranean endemic coral. Ecol Eng 98:206–212

Neil RC, Humphrey C, Bourne DG, Heyward A (2021) Co-culture with grazers can improve survival and growth of multiple coral species. Aquaculture 544:737095

Nguyen NBA, Chen LY, El-Shazly M, Peng BR, Su JH, Wu HC, Lee IT, Lai KH (2022) Towards sustainable medicinal resources through marine soft coral aquaculture: insights into the chemical diversity and the biological potential. Mar Drugs 20:640

Nosratpour F (2008) Observations of a polyclad flatworm affecting acroporid corals in captivity. In: Leewis RJ, Janse M (eds) Advances in coral husbandry in public aquariums. Burgers' Zoo, Arnhem, pp 37–46

Osinga R, Schutter M, Griffioen B, Wijffels RH, Verreth JA, Shafir S, Henard S, Taruffi M, Gili C, Lavorano S (2011) The biology and economics of coral growth. Mar Biotechnol 13:658–671

Page CA, Willis BL (2008) Epidemiology of skeletal eroding band on the Great Barrier Reef and the role of injury in the initiation of this widespread coral disease. Coral Reefs 27:257–272

Page CA, Cróquer A, Bastidas C, Rodríguez S, Neale SJ, Weil E, Willis BL (2015) Halofolliculina ciliate infections on corals (skeletal eroding disease). In: Diseases of coral, pp 361–375

Palmer C, Gates R (2010) Skeletal eroding band in Hawaiian corals. Coral Reefs 29:469

Phan GH, Tsai YC, Liu YH, Fang LS, Wen ZH, Hwang TL, Chang YC, Sung PJ (2021) Sterol constituents from a cultured octocoral *Sinularia sandensis* (Verseveldt 1977). J Mol Struct 1246:131175

Piniak G (2002) Effects of symbiotic status, flow speed, and prey type on prey capture by the facultatively symbiotic temperate coral *Oculina arbuscula*. Mar Biol 141:449–455

Piskorska M, Smith G, Weil E (2007) Bacteria associated with the coral *Echinopora lamellosa* (Esper 1795) in the Indian Ocean—Zanzibar region. Afr J Environ Sci Technol 1:93–98

Pollock FJ, Wada N, Torda G, Willis BL, Bourne DG (2017) White syndrome-affected corals have a distinct microbiome at disease lesion fronts. Appl Environ Microbiol 83:e02799–e02816

Randall CJ, Jordán-Garza AG, van Woesik R (2015) Ciliates associated with signs of disease on two Caribbean corals. Coral Reefs 34:243–247

Ravindran C, Raveendran HP, Irudayarajan L (2022) Ciliated protozoan occurrence and association in the pathogenesis of coral disease. Microb Pathog 162:105211

Rawlinson KA, Gillis JA, Billings RE, Borneman EH (2011) Taxonomy and life history of the *Acropora*-eating flatworm *Amakusaplana acroporae* nov. sp. (Polycladida: Prosthiostomidae). Coral Reefs 30:693–705

Raymundo LJ, Couch CS, Harvell CD, Raymundo J, Bruckner AW, Work TM, Weil E, Woodley CM, Jordan-Dahlgren E, Willis BL, Sato Y (2008) Coral disease handbook: guidelines for assessment, monitoring & management. In: Coral reef targeted research & capacity building for management program. Currie Communications, Melbourne

Richier S, Cottalorda JM, Guillaume MM, Fernandez C, Allemand D, Furla P (2008) Depth-dependant response to light of the reef building coral, *Pocillopora verrucosa*: implication of oxidative stress. J Exp Mar Biol Ecol 357:48–56

Riegl B, Antonius A (2003) Halofolliculina skeleton eroding band (SEB): a coral disease with fossilization potential? Coral Reefs 22:48

Riegl B, Bruckner A, Coles SL, Renaud P, Dodge RE (2009) Coral reefs: threats and conservation in an era of global change. Ann N Y Acad Sci 1162:136–186

Rinkevich B (2008) Management of coral reefs: we have gone wrong when neglecting active reef restoration. Mar Pollut Bull 56:1821–1824

Rinkevich B (2014) Rebuilding coral reefs: does active reef restoration lead to sustainable reefs? Curr Opin Environ Sustain 7:28–36

Rinkevich B (2015) Climate change and active reef restoration—ways of constructing the "reefs of tomorrow". J Mar Sci Eng 3:111–127

Rinkevich B (2021) Augmenting coral adaptation to climate change via coral gardening (the nursery phase). J Environ Manag 291:112727

Rocha RJ, Pimentel T, Serôdio J, Rosa R, Calado R (2013) Comparative performance of light emitting plasma (LEP) and light emitting diode (LED) in ex situ aquaculture of scleractinian corals. Aquaculture 402:38–45

Rodríguez S, Croquer A, Guzmán HM, Bastidas C (2009) A mechanism of transmission and factors affecting coral susceptibility to Halofolliculina sp. infection. Coral Reefs 28:67–77

Schutter M, ter Hofstede R, Bloemberg J, Elzinga J, van Koningsveld M, Osinga R (2023) Enhancing survival of ex-situ reared sexual recruits of *Acropora palmata* for reef rehabilitation. Ecol Eng 191:106962

Shannon T, Achatz JG (2007) *Convolutriloba macropyga* sp nov., an uncommonly fecund acoel (Acoelomorpha) discovered in tropical aquaria. Zootaxa 1525:1

Sheridan C, Kramarsky-Winter E, Sweet M, Kushmaro A, Leal MC (2013) Diseases in coral aquaculture: causes, implications, and preventions. Aquaculture 396:124–135

Smith JE, Brainard R, Carter A, Grillo S, Edwards C, Harris J, Lewis L, Obura D, Rohwer F, Sala E, Vroom PS (2016) Re-evaluating the health of coral reef communities: baselines and evidence for human impacts across the central Pacific. Proc R Soc B Biol Sci 283:20151985

Smith A, Hancock B, Cook N, Vaughan D (2021) History of reef restoration. In: Vaughan DE (ed) Active coral restoration: techniques for a changing planet. Ross Publishing, Plantation FL

Souter D, Planes S, Wicquart J, Logan M, Obura D, Staub F (2020) Status of coral reefs of the world: 2020: summary for policymakers. Global Coral Reef Monitoring Network (GCRMN) and International Coral Reef Initiative (ICRI)

Spadafore R, Fura R, Precht WF, Vollmer SV (2021) Multi-variate analyses of coral mortality from the 2014–2015 stony coral tissue loss disease outbreak off Miami-Dade County, Florida. Front Mar Sci 8:723998

Stender Y, Jokiel PL, Rodgers KS (2014) Thirty years of coral reef change in relation to coastal construction and increased sedimentation at Pelekane Bay, Hawai'i. PeerJ 2:e300

Su JH, Lin YF, Lu Y, Yeh HC, Wang WH, Fan TY, Sheu JH (2009) Oxygenated cembranoids from the cultured and wild-type soft corals *Sinularia flexibilis*. Chem Pharm Bull 57:1189–1192

Sun XP, Xu Y, Cao F, Xu RF, Zhang XL, Wang CY (2014) Isoechinulin-type alkaloids from a soft coral-derived fungus *Nigrospora oryzae*. Chem Nat Compd 50:1153–1155

Sun LL, Li WS, Li J, Zhang HY, Yao LG, Luo H, Guo YW, Li XW (2021) Uncommon diterpenoids from the South China Sea soft coral *Sinularia humilis* and their stereochemistry. J Org Chem 86:3367–3376

Sung PJ, Lin MR, Su YD, Chiang MY, Hu WP, Su JH, Cheng MC, Hwang TL, Sheu JH (2008) New briaranes from the octocorals *Briareum excavatum* (Briareidae) and *Junceella fragilis* (Ellisellidae). Tetrahedron 64:2596–2604

Sussman M, Willis BL, Victor S, Bourne DG (2008) Coral pathogens identified for white syndrome (WS) epizootics in the Indo-Pacific. PLoS One 3:e2393

Sussman M, Mieog JC, Doyle J, Victor S, Willis BL, Bourne DG (2009) Vibrio zinc-metalloprotease causes photoinactivation of coral endosymbionts and coral tissue lesions. PLoS One 4:e4511

Sweet M, Bythell J (2012) Ciliate and bacterial communities associated with white syndrome and brown band disease in reef-building corals. Environ Microbiol 14:2184–2199

Sweet M, Jones R, Bythell J (2012) Coral diseases in aquaria and in nature. J Mar Biol Assoc U K 92:791–801

Sweet MJ, Croquer A, Bythell JC (2014) Experimental antibiotic treatment identifies potential pathogens of white band disease in the endangered Caribbean coral *Acropora cervicornis*. Proc R Soc B Biol Sci 281:20140094

Tagliafico A, Rangel S, Kelaher B, Christidis L (2018a) Optimizing heterotrophic feeding rates of three commercially important scleractinian corals. Aquaculture 483:96–101

Tagliafico A, Rangel S, Kelaher B, Scheffers S, Christidis L (2018b) A new technique to increase polyp production in stony coral aquaculture using waste fragments without polyps. Aquaculture 484:303–308

Taglialatela-Scafati O, Deo-Jangra U, Campbell M, Roberge M, Andersen RJ (2002) Diterpenoids from cultured *Erythropodium c aribaeorum*. Org Lett 4:4085–4088

Tan F, Yang H, Xu X, Fang Z, Xu H, Shi Q, Zhang X, Wang G, Lin L, Zhou S, Huang L (2020) Microplastic pollution around remote uninhabited coral reefs of Nansha Islands, South China Sea. Sci Total Environ 725:138383

Thao NP, Nam NH, Cuong NX, Quang TH, Tung PT, Chae D, Kim S, Koh YS, Van Kiem P, Van Minh C, Kim YH (2013) Anti-inflammatory norditerpenoids from the soft coral *Sinularia maxima*. Bioorg Med Chem Lett 23:228–231

Thatcher C, Høj L, Bourne DG (2022) Probiotics for coral aquaculture: challenges and considerations. Curr Opin Biotechnol 73:380–386

Todd PA (2008) Morphological plasticity in scleractinian corals. Biol Rev 83:315–337

Tsai TC, Wu YJ, Su JH, Lin WT, Lin YS (2013) A new spatane diterpenoid from the cultured soft coral *Sinularia leptoclados*. Mar Drugs 11:114–123

Tuttle LJ, Johnson C, Kolinski S, Minton D, Donahue MJ (2020) How does sediment exposure affect corals? A systematic review protocol. Environ Evid 9:1–7

Ulstrup KE, Kühl M, Bourne DG (2007) Zooxanthellae harvested by ciliates associated with brown band syndrome of corals remain photosynthetically competent. Appl Environ Microbiol 73:1968–1975

Ushijima B, Smith A, Aeby GS, Callahan SM (2012) *Vibrio owensii* induces the tissue loss disease *Montipora* white wyndrome in the Hawaiian reef coral *Montipora capitata*. PLoS One 7:e46717

Van der Loos LM, Eriksson BK, Salles JF (2019) The macroalgal holobiont in a changing sea. Trends Microbiol 27:635–650

Weil E (2004) Coral reef disease in the wider Caribbean: status and prognosis. In: Rosenberg E, Loya Y (eds) Coral health and disease. Springer, Berlin, pp 35–64

Weil E, Croquer A, Urreiztieta I (2009) Temporal variability and consequences of coral diseases and bleaching in La Parguera, Puerto Rico from 2003–2007. Caribb J Sci 45:221–246

Wilkinson C (2008) Status of coral reefs of the world: 2008. Global Coral Reef Monitoring Network and Reef and Rainforest Research Centre, Townsville

Willis BL, Page CA, Dindsdale EA (2004) Coral disease in the Great Barrier Reef. In: Rosenberg E, Loya Y (eds) Coral health and disease. Springer, Berlin, pp 69–104

Winkler R, Antonius A, Abigail Renegar D (2004) The skeleton eroding band disease on coral reefs of Aqaba, Red Sea. Mar Ecol 25:129–144

Woodley CM, Downs C, Bruckner AW, Porter JW, Galloway SB (2016) Diseases of coral. Wiley Blackwell, Hoboken

Work TM, Aeby GS (2006) Systematically describing gross lesions in corals. Dis Aquat Org 70:155–160

Yang M, Cui WX, Li H, Li SW, Yao LG, Tang W, Mudianta IW, Guo YW (2020) Sinulasterols A–C, three new bioactive oxygenated steroids from the South China Sea soft coral *Sinularia depressa*. Steroids 157:108598

Young CN, Schopmeyer SA, Lirman D (2012) A review of reef restoration and coral propagation using the threatened genus *Acropora* in the Caribbean and Western Atlantic. Bull Mar Sci 88:1075–1098

Zabotin YI, Evtugyn VG (2021) Ultrastructure of spermatozoa and female copulatory organs in preferably asexually reproducing acoel *Convolutriloba retrogemma* (Acoelomorpha). Zoomorphology 140:19–26

Zaidi PN, Mustaffa FAA, Hong TC (2023) Influence of different feeding regimes on the survival and growth of *Acropora digitifera* in husbandry. J Sustain Sci Manag 18:118–132

Zhang R, Yu K, Li A, Wang Y, Huang X (2019) Antibiotics in corals of the South China Sea: occurrence, distribution, bioaccumulation, and considerable role of coral mucus. Environ Pollut 250:503–510

13 Conversion Mangrove Area into Fishponds: A Case Study of Ambong Bay, Kota Belud, Sabah, Malaysia

Ejria Saleh, Annita Seok Kian Yong, Julian Ransangan, Sitti Raehanah Muhamad Shaleh, Norfazreena Mohd Faudzi, Ching Fui Fui, Rafidah Othman, Mohammad Tamrin Mohamad Lal, and Wei Sheng Chong

Abstract

The mangrove forest is one of the most productive marine ecosystems supporting coastal communities. It is one of the richest storehouses of biological diversity and primary productivity in the tropical region. This chapter provides a case study on small-scale farming from former mangrove areas and its challenges. In Ambong Bay, Kota Belud, Sabah, Malaysia, mangrove areas are owned by the coastal community and have been developed for small-scale fish farming. The objectives were to identify mangrove areas to be converted to fish ponds, to measure water quality and fish health in the pond, and to estimate the profit gained from the fish pond. At the project's end, the profit earned from the fish farm activities was calculated. The case study was carried out between October 2021 and June 2023. The Borneo Marine Research Institute, Universiti Malaysia Sabah distributed about 330 juvenile fish of Asian sea bass to the farmers. The conversion of mangrove areas to fish ponds involved more than ten families. The measured water parameters and fish health fall within the range of common pond water quality. The growth rate of Asian sea bass was approximately 3.7 cm per month (128 g per month), with a fish survival rate of approximately 67%.

E. Saleh
Higher Institution Centre of Excellence (HICoE), Borneo Marine Research Institute, Universiti Malaysia Sabah, Jalan UMS, Kota Kinabalu, Sabah, Malaysia

Natural Disaster Research Centre, Faculty of Science and Natural Resources, Universiti Malaysia Sabah, Jalan UMS, Kota Kinabalu, Sabah, Malaysia

A. S. K. Yong (✉) · J. Ransangan · S. R. Muhamad Shaleh · N. Mohd Faudzi · C. Fui Fui · R. Othman · M. T. Mohamad Lal · W. S. Chong
Higher Institution Centre of Excellence (HICoE), Borneo Marine Research Institute, Universiti Malaysia Sabah, Jalan UMS, Kota Kinabalu, Sabah, Malaysia
e-mail: annitay@ums.edu.my

N. M. Faudzi et al. (eds.), *Essentials of Aquaculture Practices*, https://doi.org/10.1007/978-981-97-6699-4_13

Predators and the lack of pond management pose challenges for farmers to optimize their harvest. The baseline information gathered can be utilized for awareness and education programs highlighting the importance of sustainable farming and mangrove conservation. The program is part of the activities in implementing the ten essentials to achieve the National Goals of ensuring food security for the nation (Goal 5), Sustainable Development Goal (SDG) 11 (sustainable cities and communities), and SDG 14 (life below water).

Keywords

Mangrove · Sustainable farming · Coastal community · Food security

13.1 Mangrove in Malaysia

The mangrove ecosystem plays an important role in stabilizing Earth's temperature. Mangrove forests are critical for carbon sequestration and storage, rivaling other forest types' ability to store high amounts of carbon (Nyanga 2020). Malaysia encompasses an estimated 629,038 ha of mangrove areas, with Sabah holding the majority at nearly 57%, while Sarawak and Peninsular account for 26% and 17%, respectively (Omar and Misman 2020; Abdulaali et al. 2022). In 1975, Malaysia's mangrove coverage was estimated at 700,000 ha and declined to 572,000 ha in 2000 (Omar and Misman 2020). Agriculture, aquaculture, settlements, and poaching contribute to the reduction of mangrove areas.

Mangroves are one of the vital ecosystem services in coastal areas and should be protected by the government. The mangroves in Malaysia are mostly found in Sabah (60%) (Hamdan et al. 2020), located in the northern part of Borneo Island. This area is also situated within the Coral Triangle, which boasts the highest marine diversity in the world. The mangrove forest in Sabah is estimated at 341,000 ha, with 95.7% of the area gazetted as a forest reserve under the forest enactment administered by the Sabah Forestry Department (Nilus et al. 2010). However, mangrove areas do not fall under the protected mangrove forest, such as the fringing mangrove in Ambong Bay, Kota Belud, Sabah. Nearly all mangrove areas with land titles, except those within the green belt reserve area, are reclaimed for development or converted to fishponds by the communities in Ambong Bay.

As Saleh et al. (2011) reported, the mangrove in Ambong Bay is impacted by road construction and other human activities such as human settlement and agriculture. The catchment areas are also developed for small rubber plantations, which contribute to high sedimentation at its shoreline area. This chapter discusses the transformation of unprotected mangrove areas into fish ponds and its contribution to the livelihood of Ambong Bay, Kota Belud, Sabah coastal communities. The objectives are to identify mangrove areas converted to small-scale fish farming, to measure water quality and fish health in the pond, and to estimate the profit gained from the fish pond. Conversion of mangrove areas into fish ponds without proper planning can have significant environmental impacts, including habitat loss, disruption of coastal ecosystems, and loss of biodiversity.

13.2 Small-Scale Fish Farming

The mangrove areas in Ambong are situated in the inner part of the bay (Fig. 13.1). They are designated with land titles, granting owners the right to reclaim and develop the land for housing, agriculture, and aquaculture activities. Additionally, the remaining mangroves designated for the green belt have been harvested for fishing materials. Communities surrounding Ambong Bay, such as those in Kg. Ambong and Kg. Tolus, have utilized parts of the mangrove area, converting them into fish ponds. This chapter aims to identify mangrove areas that have been converted into fish ponds in Kg. Tolus, examine the management of these fish ponds, and assess their contribution to the social economy of the coastal community. Fish farming is a common activity carried out in mangrove areas due to the demand for seafood, which contributes to additional income. Accessibility to water resources facilitates the establishment and maintenance of fish ponds.

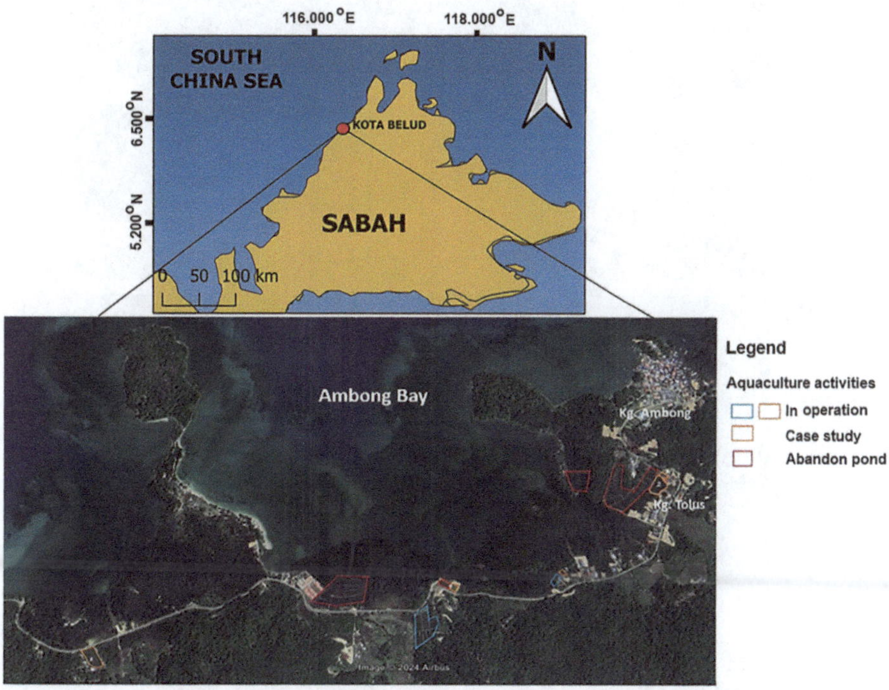

Fig. 13.1 Location of the small-scale brackish water ponds at the inner part of Ambong Bay

13.2.1 Fish Pond Criteria

The community involved in this case study resides in Kg. Tolus, Kota Belud (Fig. 13.1), Sabah, Malaysia. The criteria for selecting ponds include being operational, formerly a mangrove area with land title, and converted for aquaculture activities. Ponds 1 and 2 were constructed within approximately 6 months before the commencing of the project (Fig. 13.2). Pond 1 (Fig. 13.2a) has an estimated 18 m × 20 m in size. The primary fish species in this pond were Tilapia (*Oreochromis niloticus*) and climbing perch (*Anabas testudineus* or "ikan puyu"). The water in Pond 1 was stagnant, relying on rainfall as its water source.

Pond 2 was built around the same time as Pond 1 (Fig. 13.2b). However, it housed various brackish water species, such as tilapia and milkfish (*Chanos chanos*). The water depth in Pond 2 ranges between 1.5 and 2.0 meters, depending on the tidal water level. According to Ong and Ransangan (2018), water quality in Ambong Bay meets the standards set by the Malaysia Marine Water Quality Standard (Class 2)

(a)
(b)

Fig. 13.2 Condition of (**a**) Pond 1 and (**b**) Pond 2 during the site visit

for pollution and suitability for seafood farming for human consumption. It is noteworthy that only Pond 2 utilizes seawater as its water source for the pond.

Pond 3 (Fig. 13.3a) was constructed over 20 years ago. It was a shallow pond with a water depth of less than 1 m. The dominant fish species in this pond was Tilapia. In the 2000s, high tides flooded the pond area, releasing fish into the sea. Pond 4, depicted in Fig. 13.3b, was newly built as of 2022. Its water supply primarily came from rainfall. Seawater was pumped into the pond during high tides to increase the water level.

Fig. 13.3 Condition of (**a**) Pond 3, (**b**) Pond 4, and (**c**) Pond 5 during the site visit. Mangroves are seen behind the Ponds 3 and 4

(a)

(b)

(c)

Table 13.1 The criteria of the ponds selected as a case study

No. of pond	Location (N/E)	Pond size (m^2)	Previous fish species status	Depth of the pond (m)	Number of fish introduced
1	6° 18′ 09.35″/116° 19′ 41.88″	18 m × 20 m = 360 m^2	Tilapia, climbing perch (ikan puyu)	<0.9	50
2	6° 18′ 07.93″/116° 19′ 41.60″	30 m × 60 m = 1800 m^2	Indo-Pacific tarpon (ikan bulan), crescent perch (kerong jalur), milkfish (baulu)	<2.0	130
3	6° 17′ 55.16″/116° 19′ 28.64″	22 m × 27 m = 594 m^2	Tilapia	<1	65
4	6° 17′ 53.14″/116° 19′ 14.91″	13 m × 40 m = 520 m^2	Tilapia	<1	55
5	6° 17′ 43.96″/116° 18′ 30.53″	75 m × 36 m = 2700 m^2	Indo-Pacific tarpon, crescent perch, milkfish	<1.2	30

Pond 5 (Fig. 13.3c) was originally a paddy field built in the 1980s. Over time, the rising water levels inundated the area, eventually transforming it into a fish pond. Seawater entered the pond during high tides, carrying fry or fish eggs into the area. Common fish species identified in this pond included the Tilapia, Indo-Pacific tarpon (*Megalops cyprinoides* or "ikan bulan"), Crescent perch (*Terapon jarbua* or "kerong jalur"), and milkfish (*Chanos chanos* or "baulu"). These marine fishes were also commonly found in the coastal area.

All ponds selected for this case study were operational, with black tilapia as the dominant stocked fish. This species has a high tolerance range for water parameters such as water temperature and salinity. The optimum temperature range for normal development, reproduction, and growth of tilapia is about 25–32 °C (El-Sayed and Kawanna 2008). The field survey was conducted on October 8, 2021, to identify suitable ponds, engage with the community, and measure water quality. Pond owners had agreed to receive Asian seabass juvenile fish from the Borneo Marine Research Institute for the case study, but no budget was allocated for fish food. To address this, the current Tilapia fry population served as a food source for the new juvenile fish released into the ponds. The locations and criteria of the ponds are outlined in Table 13.1.

13.2.2 Water Quality Measurement

Water quality measurements inside the pond were conducted using a YSI Meter on October 8, 2021; September 23, 2022; and November 24, 2022. Fish body length and weight size were measured to assess fish growth and to identify any issues and challenges in managing the small-scale fish farm. Universiti Malaysia Sabah (UMS) researchers investigated reports of fish diseases and changes in water quality within the pond upon receiving reports from the pond's owner.

13.2.3 Supply of Juvenile Fish

Approximately 330 juvenile Asian sea bass (*Lates calcarifer*), each measuring about 10 cm in size and 50 g in weight, were distributed among the pond owners based on Table 13.1. The market price of fish of this size was MYR 3.50 (USD 0.74) per fish. In this case study, the juvenile fish used for stocking the ponds were provided free of charge by the Borneo Marine Research Institute, UMS (Fig. 13.4). This program was part of the institute's social community responsibility project benefiting the local community. The number of juvenile fish distributed to the communities was based on the pond size and stocking density (Table 13.1 and Fig. 13.5). The juvenile fish were also able to survive in water salinity of 0 ppt, suiting the water condition of the pond.

The highest number of juvenile fish was allocated to Pond 2 (130 fish), following the pond size ratio and the existing fish population (Table 13.1). The depth of the pond was generally less than 1 m, depending on rainfall. These pond criteria were

Fig. 13.4 Team of Borneo Marine Research Institute with the pond owners during the handover of the fish to the community on September 23, 2022

Fig. 13.5 Release of juvenile Asian sea bass (*Lates calcarifer*) in Pond 2

taken into account when determining the number of new juvenile fish introduced into the pond. However, a smaller quantity of juvenile fish (30 fish) was provided to Pond 5 as requested by the pond owner.

13.3 Mangrove Area for Fish Pond

Malaysia has embarked on significant initiatives to conserve and restore its mangrove ecosystem, with the government demonstrating a commitment to bolstering protection efforts, particularly following the tsunami on December 26, 2004 (Omar and Misman 2020). However, the communities in Kg. Tolus are converting the mangrove area into a fish pond or reclaim for housing. There were more than 20 fish ponds visible in the satellite image dated February 8, 2023 (Fig. 13.1). The number of fish ponds had increased during the COVID-19 pandemic. During the field trip in 2021, a large area of mangroves was cut due to a lack of awareness regarding the role of mangroves as natural habitats, and there was little understanding of the impact of global environmental changes that can lead to natural disasters. The communities perceive mangroves as not contributing to their social economy. Therefore, the mangrove area has been converted into fish ponds, serving as an alternative food source and a sign of land ownership.

Table 13.2 Range of the water parameters inside the pond

Pond	Temperature (°C)	Salinity (ppt)	DO (mg/L)	pH
1	31.1–31.5	0.37–1.11	3.28–4.29	3.16–3.17
2	31.4–32.7	23.24–27.25	3.01–4.35	5.71–7.20
3	30.7–31.1	0.46–5.20	2.87–3.23	5.19–7.69
4	31.7–32.4	1.21–6.21	4.19–5.40	7.62–8.44
5	30.2–30.9	0.02–9.04	4.48–5.24	8.24–8.55

13.3.1 Water Quality of the Fish Pond

The water quality inside the pond significantly affects the fish's growth, health, and survival rate (Boyd 2017). The surface water temperature in the pond ranged from 30.7 to 32.4 °C (Table 13.2). Salinity levels indicate that the water in the ponds was primarily brackish to freshwater, except for Pond 2. Rainfall served as the main source of water for the ponds. The efficient water intake system in Pond 2 allowed seawater to enter the pond, increasing the salinity of the water. Overall, the measured water quality parameters fell within the range typical for fish farms. However, dissolved oxygen (DO) levels were relatively low and may affect fish growth in the pond. DO levels (<4.30 mg/L) and pH (<3.17) in Pond 1 were lower compared to the other ponds. There were no complaints from pond owners regarding water quality issues such as eutrophication or mass fish deaths due to disease or other causes. The only complaints were related to fish predators and decreased water levels in the pond.

13.3.2 Fish Harvesting

The pond owners harvested fish from the ponds, but they found it difficult to detect the presence of Asian sea bass fish. Despite the high number of tilapia fish in Ponds 1 and 3, there were no reports of Asian sea bass fish in these ponds. This was likely due to their fishing method, which used small hooks and lines. Most of the fish caught were tilapia, known for their rapid breeding in the pond. Attacks by sea otters and monitor lizards may have also contributed to the disappearance of the Asian sea bass released into these ponds. No fish harvesting took place in Pond 2 during the case study period. However, a single Asian sea bass was caught using hook and line in Pond 4, estimated to weigh 1 kg and measured 40 cm long. Between June 2 and June 7, 2023, fish in Pond 5 were harvested using gill nets (Fig. 13.6a). A total of 20 Asian sea bass were retrieved from the pond, with an average weight of 1.2 kg (Table 13.3) and a length of 43 cm (Fig. 13.6b).

According to information from the pond owners, the food for the Asian sea bass juvenile released into the ponds was primarily sourced from the fries available in the pond specifically tilapia fries. Ponds 1, 3, and 4 were far from the owners' houses, providing an environment where predators such as sea otters, monitor lizards, and birds could freely access the ponds. Ponds 2 and 5, on the other hand, were situated in front of the houses and can be easily monitored.

Fig. 13.6 (**a**) Fish harvesting in the pond and (**b**) measuring size of the fish

Table 13.3 Estimate income return based on the survival rate from Pond 5

Items	Price (MYR/USD)	Total (MYR/USD)
Market price of the juvenile fish	3.50/0.74 × 30 fish	105.00/22.14
Market price of live Asian seabass/kg	20.00/4.22 × 20 fish × 1.4 kg	560.00/118.08
	Gross income return	**455.00/95.94**

Based on Pond 5 fish data harvested, after 9 months in the pond, the survival rate of Asian sea bass was approximately 67%. Based on this survival rate, a simple calculation of the income estimated a return of MYR 455.00/USD 95.94 (Table 13.3). The income return would be higher if the number of fish released matched the optimum capacity of the pond and the stocking density of the fish. The growth rate of the fish was approximately 3.7 cm per month or 128 g per month. According to Sen et al. (2019), Asian seabass is fast growing, with the potential to gain more than 1 kg per year and attain market size within 8 months.

13.3.3 Issues and Challenges

Generally, the water source for the ponds was seawater during high tide (Ponds 2 and 3). Other ponds primarily rely on rainwater or water pumped during high tide. During the rainy season, erosion of the surface area divider by surface runoff contributed to high levels of suspended solids in the pond, thereby affecting the water quality. However, during the dry season, the water level in the pond drastically reduces, leaving the fish exposed to predators. Additionally, the water temperature increased as all ponds were directly exposed to the sun. The pond's low dissolved oxygen (DO) levels may be due to a lack of water circulation and possibly high fish stock. Additionally, the pond's water level decreased due to high sedimentation during the rainy season and high evaporation during the dry season (Fig. 13.7). Improvements to water quality can be made by implementing water circulation to increase dissolved oxygen levels in the pond and maintain fish health.

Fish predators, such as sea otters (*Enhydra lutris*) and monitor lizards (*Varanus salvator*) were commonly found in all ponds. However, the community has yet to take action to prevent these predators, except in Pond 1 (Fig. 13.8). In Pond 1, metal roofing has been used as a fence to deter sea otters and monitor lizards from accessing the pond (Fig. 13.8a). During the dry season, various bird species, including the kingfisher (*Alcedo atthis*), little egret (*Egretta garzetta*), and fish owl (*Ketupa blakistoni*), are frequently observed around the ponds. To prevent birds from entering the ponds, black shading nets were installed on the top of the pond (Fig. 13.8b). Additionally, these nets provided shade for the fish.

Fig. 13.7 The water level in Pond 4 decreased during the dry season

(a)

(b)

Fig. 13.8 (**a**) Metal roofing was utilized as a barrier to prevent fish predators from accessing the pond, while (**b**) birds were deterred from reaching the pond through the installation of black shading nets on its surface

Preserving the remaining mangrove forests in Ambong Bay is imperative, as they contribute to reducing global greenhouse gas emissions and supporting climate change mitigation efforts. Mangrove ecosystem restoration and conservation minimize vulnerability and are powerful allies in the battle against climate change (Ramli et al. 2023). Reduction of mangrove cover along its shoreline may expose coastal communities to sea-level rise and storms. The case study identified that the mangrove area converted to fish ponds was not fully utilized. Pond maintenance is not a priority; some ponds are abandoned, while the mangroves, serving as marine habitat and natural shoreline protection, have already disappeared. Conversion of mangrove areas into aquaculture requires awareness and education on the importance of mangroves as habitat, shoreline protection, and climate mitigation measures. The impacts of mangrove loss should be carefully considered and mitigated through sustainable management practices to ensure the long-term health and resilience of coastal environments.

13.4 Conclusion

Based on field observations in September 2022, more than ten families with fish ponds originate from the mangrove area in the inner part of Ambong Bay. However, older ponds have been abandoned due to a lack of maintenance and high operational costs. Recently, during the COVID-19 pandemic, mangrove areas were cleared, and the land was converted into ponds for aquacultural activities. These ponds were also constructed on land as proof of ownership, with no serious pond management implemented. Generally, the fish receive little attention as pond owners release fish into the ponds without providing feeding or water circulation. Sometimes, leftover food, such as rice, is put in the ponds as fish feed. The communities perceive the prominent role of the ponds as a source of food rather than a livelihood.

The role of ponds as a commercial means for earning money and meeting protein needs should be optimized. Based on the experience in Pond 5, fish ponds can be a potential source of financial stability, food security, and social stability for the community. Introducing pond-based fish farming with proper management, such as suitable fish fry and modern technology to control fish predators, would contribute to greater economic and food security at the household level of the communities. It is suggested that basic training related to good pond management practices would help ensure sustainable aquaculture development. Pond management practices, including daily feed input, stocking density, and water exchange, significantly impact water quality. An adequate buffer zone is needed to reduce the effects of stressors (storm surges, extreme high tide) from the sea.

The baseline information gathered from this case study is valuable for awareness and education programs on the importance of sustainable farming and mangrove ecosystem conservation for the coastal community. This program is part of the activities in implementing the ten essentials to achieve National Goals, particularly Goal 5 (ensuring food security for the nation), and two Sustainable Development Goals (SDGs): SDG 11 (sustainable cities and communities) and SDG 14 (life below water).

Acknowledgement This chapter is supported by the project entitled "*Rakan Akuakultur dan Rakan Laut sebagai Platform Penghubung Universiti Malaysia Sabah dan Komuniti Akuakultur dan Pesisir Pantai*" (Grant Code: SDK0002-2017). We appreciate the community in Kg. Tolus for participating in this case study.

References

Abdulaali HS, Usman IMS, Basung JE, Abdulhasan MJ, Hameed MS (2022) Conceptual design of ecotourism products proposal at mangrove forest in Ketam Island, Malaysia. Int J Glob Optim Appl 1(2):111–119. https://doi.org/10.56225/ijgoia.vli2.20

Boyd CE (2017) General relationship between water quality and aquaculture performance in ponds. In: Jeney G (ed) Fish diseases. Academic Press, Massachusetts, pp 147–166

El-Sayed AF, Kawanna M (2008) Optimum water temperature boosts the growth performance of Nile tilapia (*Oreochromis niloticus*) fry reared in a recycling system. Aquac Res 39:670–672

Hamdan O, Mubarak TH, Ismail P (2020) Status of mangrove in Malaysia. Forest Research Institute Malaysia, Kuala Lumpur, p 288

Nilus R, Chung AYC, Pereira JT, Sugau JB, Tangah J, Sabran S, Chong RFY (2010) Mangrove of Sabah: an introduction to flora and fauna. Sabah Forestry Department, Sandakan

Nyanga C (2020) The role of mangrove forests in decarbonizing the atmosphere. In: Bartoli M, Frediani M, Rosi L (eds) Carbon-based material for environmental protection and remediation. IntechOpen. https://doi.org/10.5772/intechopen.92249

Omar H, Misman MA (2020) Extents and distribution of mangroves in Malaysia. In: Omar H, Husin TM, Parlan I (eds) Status of mangroves in Malaysia. Forest Research Institute, Kuala Lumpur

Ong F, Ransangan J (2018) Assessment of spatial and temporal variations of water quality for future mariculture operation in Ambong Bay, Sabah, Malaysia. Open J Mar Sci 8:1–19. https://doi.org/10.4236/ojms.2017.81001

Ramli NA, Krishna G, Aiman N (2023) Mangrove conservation behaviour from environmental psychological perspectives. Malays J Sustain Environ 10:193–215. https://doi.org/10.24191/myse.v0iJune.22680

Sen S, Terai A, Sreyrum P, Lundh T, Barnes AC, Da CT, Kiessling A (2019) Growth performance of fry and fingerling Asian Seabass (*Lates calcarifer*) from Cambodian broodstock reared at different salinities. Livest Res Rural Dev 31:39. http://www.lrrd.org/lrrd31/3/sorph31039.html. Accessed 18 Mar 2024

Unraveling Sustainable Aquaculture Through Pillars of Practices

14

Imelus Nius, Muhammad Dawood Shah, Fikri Akmal Khodzori, Rossita Shapawi, Nurzafirah Mazlan, and Wei Sheng Chong

Abstract

Aquaculture is one of the most rapidly expanding industries in the world. It plays a central role in global food production and serves as the primary source of protein for human consumption. To navigate and excel in this dynamic field, a deep understanding of the fundamental principles of aquaculture is essential to ensure long-term success and sustainability. This book chapter provides an overview that presents the essence of aquaculture through ten key pillars. From strategic site selection to careful feeding strategies, each pillar contributes to a holistic understanding and provides the foundation for success in this ever-evolving industry. The global perspective emphasizes aquaculture's adaptability in different landscapes, from small pond systems to extensive and intensive methods. Despite its inherent challenges, aquaculture is emerging as a cornerstone in the global food security edifice, receiving support and cooperation from national and international organizations. In addition to its primary role in food production, aquaculture actively contributes to economic growth, job creation, and resource management. Adherence to ethical practices and implementing responsible policies are essential to avoid social conflict, with the "planet-first" approach at the forefront of decision-making. The harmonious interplay of these ten pillars forms the basis for successful and sustainable aquaculture that aligns with the overarching global goals of environmental sustainability, economic growth, and food security. This chapter serves as a guide for practitioners and emphasizes aquaculture's pivotal role in achieving broader societal goals.

I. Nius · M. D. Shah · F. A. Khodzori · R. Shapawi · N. Mazlan · W. S. Chong (✉)
Higher Institution Centre of Excellence (HICoE), Borneo Marine Research Institute, Universiti Malaysia Sabah, Kota Kinabalu, Sabah, Malaysia
e-mail: chong@ums.edu.my

N. M. Faudzi et al. (eds.), *Essentials of Aquaculture Practices*,
https://doi.org/10.1007/978-981-97-6699-4_14

Keywords

Aquaculture · Sustainability · Food security · Global perspective · Responsible practices

14.1 Introduction

In aquaculture, the choice of location and the pond's design are crucial to any project's success. The complex interplay of factors influencing site selection, such as water quality, temperature, salinity, substrate type, and availability of natural resources, requires a sophisticated understanding. Soil quality, in particular, is a critical factor that controls both pond productivity and water quality while also determining the feasibility of construction (Rahman and Yakupitiyage 2006; Guong and Hoa 2012; Shafi et al. 2021). Equally crucial is the adequacy and quality of the water supply, which plays a decisive role in the site selection (Folke and Kautsky 1992; Yoo and Boyd 2012). The topography of the terrain also plays an important role, as it determines the contours of the pond construction, whether it is a quarry pond, a dyke, or a gully (Hossain et al. 2009; Yokoyama 2003; Oh et al. 2017).

In the ever-evolving landscape of aquaculture practices, integrating geographic information systems (GISs) takes center stage (Hossain et al. 2009; Falconer et al. 2020). GIS tools facilitate spatial analysis, mapping, and informed decision-making by analyzing topography, land use, market proximity, and potential risks (Silva et al. 2011; Gimpel et al. 2018). This technological synergy is becoming an indispensable ally in the search for optimal site suitability (Nath et al. 2000; Longdill et al. 2008). Meanwhile, the principles of hydrodynamics, a cornerstone of pond design, influence water exchange rates, oxygenation, flow patterns, and sediment management (Peterson 2000). A well-thought-out pond design based on hydrodynamic principles not only improves water circulation but also lays the foundation for a robust aquatic life base (Masaló Llorà 2008; Klebert et al. 2013; Campbell and Hall 2019; Li et al. 2019; Cheng et al. 2020; Zhang et al. 2022). Ultimately, combining ecological acumen with technological and engineering considerations forms the basis for the sustainability and longevity of aquaculture systems and pond designs (Bosma and Verdegem 2011).

In the species selection phase, careful evaluation of the criteria is essential to ensure the sustainability and profitability of the operation. Market demand is proving to be the linchpin that directly impacts the economic viability of aquaculture projects. At the same time, growth rate and feed conversion efficiency are becoming increasingly important and determine the time and resources required to bring products to market (Henryon et al. 2002; Besson et al. 2016). Disease resistance is essential to avoid economic losses in potential disease outbreaks, while ecological compatibility ensures harmonious integration without disturbing existing ecosystems (Leung and Bates 2013; Lafferty et al. 2015; Naylor et al. 2021).

Diversification and polyculture further unfold their impact on aquaculture systems by heralding benefits such as higher productivity, better nutrient utilization, and greater ecosystem resilience (Thomas et al. 2021; Nissar et al. 2023). The strategic fusion of multiple species capitalizes on different ecological niches and mitigates the risks associated with monoculture. The symbiotic relationships between the other species promote an efficient and sustainable system, emphasizing the importance of appropriate management practices in monocultures. Factors such as stocking density, feeding regime, and environmental conditions become the compass for optimal growth and healthy outcomes (Beveridge and Brummett 2015; Naylor et al. 2021).

The narrative threads of informed decision-making and strategic planning run seamlessly through the fabric of aquaculture practices. On closer inspection, the efficiency of aquaculture in converting feed into edible protein becomes apparent, with nutrition and feeding becoming the fundamental cornerstones. Adapting feed formulations to the specific nutritional needs of different aquaculture species, considering life stages and growth rates becomes the cornerstone of sustainable development (Glencross et al. 2007; Kandathil Radhakrishnan et al. 2020). Selecting feed ingredients, protein sources, binders, and additives is a careful ballet where availability, cost, and nutritional quality must be considered (Olmos Soto et al. 2015). Feeding strategies ranging from demand feeding to automatic feeders are used to optimize consumption, growth, and feed conversion. The link between nutrition and health emphasizes the role of a balanced diet in reducing reliance on antibiotics and other treatments (Rombenso et al. 2022).

In the quest for sustainability, feeding practices are taking center stage. Strategies to minimize feed waste and nutrient pollution and use wild fish for feed production are becoming essential components. The "FIFO" concept takes center stage and governs the management of feed freshness (Kok et al. 2020). As the narrative expands to the global landscape, the dominance of carp and algae in semi-intensive, small-pond aquaculture in developing countries paints a vivid picture. Cultivated fisheries in various water bodies, from inland lakes to rivers and reservoirs, are an example of the adaptability of aquaculture (Naylor et al. 2021; Abisha et al. 2022; Zhang et al. 2023a, b). The dichotomy between extensive and intensive systems underlines the importance of optimal management practices to minimize environmental risks in all areas (Naylor et al. 2021). Zhang et al. (2023)

This book chapter, which has been carefully compiled and is about to unfold, summarizes the essentials of aquaculture practices into ten pillars. From strategic site selection to precise feeding strategies, each pillar paves the way for a comprehensive understanding and lays the foundation for success in this dynamic and evolving field.

14.2 Pillars of Practices in Aquaculture

14.2.1 Pillar 1: Strategic Site Selection for Optimal Aquaculture Success

The cornerstone of successful aquaculture is careful site selection, which is the first pillar of aquaculture. Selecting a suitable site forms the basis for the entire project and influences its design, organization, and management. This is particularly important for fish ponds, especially those intended for coastal or brackish water aquaculture and focusing on valuable species such as shrimp and finfish (Benetti et al. 2010; Falconer et al. 2016; Aguilar-Manjarrez et al. 2017).

When choosing an ideal location for a fish farm, several factors must be considered, including the availability and quality of water, soil quality, and accessibility. The most crucial component, water supply, must be sufficient throughout the year for ponds, tanks, and other applications on the farm. Optimal water quality, characterized by being well oxygenated, free of impurities, at a suitable temperature, and with an optimal pH, is crucial for promoting the best conditions for fish growth (Assefa and Abebe 2018; Zhang et al. 2023a, b). Accessibility, including the convenience of transport and the timely availability of labor, electricity, construction materials, and proximity to markets, also contributes to the success of aquaculture.

Before planning an aquaculture farm, a thorough site investigation is essential to ensure the best possible water supply and productivity. Applying a multi-criteria evaluation (MCE) method based on a geographic information system (GIS) allows stakeholders to assess bathymetry, soil condition, water quality, and accessibility. The results of this study are visualized on a map and serve as a starting point for the pond's design. Studies, such as the one on an oyster farm in Terengganu, Malaysia, show the positive results of using GIS and MCE methods in site selection (Mamat et al. 2014). Although GIS-MCEM is a systematic and cost-effective method for site selection, it is vital to recognize its limitations, especially when social infrastructure factors are not considered (Ibrahim et al. 2021).

Through careful site investigation, aquaculture operations can strategically select an appropriate pond design while incorporating optimal water management practices. This increases production rates and lays the foundation for a successful and sustainable aquaculture operation.

14.2.2 Pillar 2: Water Quality Management in Aquaculture

Water quality is the second pillar, as it profoundly impacts aquatic organisms' health and productivity. Recognizing its critical role, this pillar emphasizes the need for careful management of water quality to ensure an ideal environment for fish growth in ponds, reduce stress levels, and prevent adverse effects on farmed organisms (de los Santos et al. 2020; Tumwesigye et al. 2022).

Monitoring various water quality parameters is essential for the success of aquacultures. Dissolved oxygen (DO) levels, ideally maintained between 5 and 7 ppm, play a crucial role in the growth and mortality of aquatic species. At the same time, factors such as temperature, pH, and salinity have a significant impact on water quality and require careful assessment to prevent stress-induced mortality, especially in soft-shell crabs (Roy et al. 2010; Hernández et al. 2011; Carbajal-Hernández et al. 2013).

Advanced monitoring and management techniques, such as recirculating aquaculture systems (RASs) and integrated multitrophic aquaculture, are becoming solutions to reduce the accumulation of waste and contaminants and ensure sustainable and environmentally friendly practices (de los Santos et al. 2020; Wang et al. 2022). Educating fish farmers on maintaining optimal water quality is critical to increasing fish farm productivity and meeting the global demand for high-quality seafood (Tumwesigye et al. 2022).

The interaction between water quality and soil selection or pond age makes this pillar even more complex. Soil quality, which is influenced by pond age, affects water quality parameters and pond productivity. Traditional ponds, which act as biological filters, require careful sediment management through regular removal or modification to deepen ponds, increase volume, and optimize harvesting processes (Boyd 1995; Yuvanatemiya and Boyd 2006). Understanding the suitability of different soil types for aquaculture substrates is crucial, highlighting the importance of natural pH, low water permeability, and rapid mineralization of organic matter for constructing optimal aquaculture ponds (Hasibuan et al. 2023).

14.2.3 Pillar 3: Infrastructure-Optimized Site Selection with GIS Integration

Ensuring the accessibility of aquaculture ponds is a crucial aspect that significantly impacts overall efficiency, income generation, and environmental sustainability. Optimized site selection, facilitated by the geographic information system (GIS), is critical for improving aquaculture practices. GIS is a powerful planning tool that can comprehensively cover large areas and visualize data effectively (McLeod et al. 2002).

Integrating GIS with various software applications can create a detailed map layout that includes key features such as rivers, streams, lakes, road networks, and market facilities. This advanced system surpasses traditional site selection methods by significantly reducing the time it takes to identify a suitable location. Recent studies using GIS and other software tools have demonstrated improved site selection methods. These advances have provided insight into infrastructural considerations, including proximity to water bodies, roads, markets, and hatcheries (Nayak et al. 2014).

14.2.4 Pillar 4: Innovative Pond Designs for Sustainable Aquaculture

Pond design is a critical aspect of aquaculture and requires a harmonious blend of scientific expertise and practical experience to ensure optimal water management and production. The selection of an appropriate pond design not only takes into account the specific needs of aquatic species, but also considers issues such as the escape of animals from marine cages and the discharge of waste into the environment (Martins et al. 2010; Badiola et al. 2012).

Various aquaculture systems, including cage-based systems, flow-through systems (FTSs), recirculating aquaculture systems (RASs), and partitioned aquaculture systems (PASs), have been used (Martins et al. 2010; Badiola et al. 2012; Kamali et al. 2022). While traditional systems face challenges such as untreated wastewater and high water demand, innovative approaches such as the partitioned aquaculture system (PAS) offer a promising solution. PAS divides a pond into sections for waste treatment and fish growth, allowing water recirculation to improve quality while minimizing water consumption (Brown et al. 2016; Jescovitch and Boyd 2017).

However, challenges arise with PAS due to the more extensive land requirements. Competing land uses, limited freshwater availability, and insufficient space hinder suitability for commercial aquaculture (Badiola et al. 2012; Brown et al. 2016; Kamali et al. 2022). The regulating aquaculture system (RAS) has emerged as a commercially viable alternative. RAS optimizes the use of water resources, energy, and land use while reducing water exchange in indoor ponds (Badiola et al. 2012; Kamali et al. 2022; Wang et al. 2022; Taufik et al. 2023). Although RAS is less common than pond aquaculture, it is recognized for its efficiency, environmental friendliness, and sustainability (Boyd and McNevin 2015). The controlled indoor RAS environment minimizes water consumption by 90–99%, occupies less land area, and supports high-density aquaculture activities (Kamali et al. 2022; Li et al. 2023).

The design of ponds for efficient water management and productivity in aquaculture requires an integrated approach. As the regulating aquaculture system (RAS) demonstrates, innovative pond designs offer the potential for thriving ecosystems that enable sustainable aquaculture practices while protecting the surrounding environment through effective water exchange, filtration technologies, and selective stocking strategies.

14.2.5 Pillar 5: Strategic Stocking and Species Management

The profitability of aquaculture projects depends on smart choices in species selection and stocking strategies. The constant search for new species to meet evolving market demand emphasizes the dynamic nature of aquaculture growth (Quéméner et al. 2002; Alvarez-Lajonchère and Ibarra-Castro 2013). Different aquatic species' different physiologies and metabolic characteristics require considering cultural or site conditions carefully. For example, due to the solitary nature of lobsters, it is necessary to breed them separately from crayfish to avoid cannibalism (Henriksson et al. 2021).

Species selection is closely linked to site conditions, emphasizing the need for a close link between species selection and site characteristics. The type of aquaculture system used at a location also influences the selection of suitable species for the farm (Ahmed et al. 2007). Analyzing market demand is central to matching species selection to environmental and socio-economic conditions and affects the profitability of aquaculture and the ecological balance of local ecosystems (Tisdell 2007; Alvarez-Lajonchère and Ibarra-Castro 2013).

Once species selection is finalized, careful stocking techniques come into play, with stocking density being a critical element for fish production, farm profitability, and water quality. Monitoring water quality, fish health, and growth trends is essential for effective stocking strategies. High stocking density, if unplanned, can lead to detrimental effects such as stress, reduced water quality, and adverse effects on the fish final weight and specific growth rate (Ellis et al. 2002; Ullah et al. 2018; Hossain et al. 2022). A recent study on orange mud crabs in Malaysia shows the importance of considering the sex ratio in population density to avoid higher mortality during mating (Waiho et al. 2015).

In essence, strategic stocking and species management are necessary for achieving economic profitability and environmental sustainability in aquaculture operations. The synergy between well-informed species selection and carefully planned stocking measures forms the cornerstone of this fifth pillar of aquaculture practice.

14.2.6 Pillar 6: Strategic Species Selection for Sustainable Aquaculture

In the field of aquaculture, the selection of suitable species and systems is a complex decision that is influenced by several key factors. Development goals, target groups, marketability, means of production, support facilities, and environmental aspects are crucial in determining the ideal aquaculture species and systems for a given scenario (Baluyut 1989). A widely used economic strategy for species selection involves prioritizing market demand, comprehensive economic feasibility assessments, and thorough evaluations of socio-cultural and environmental variables (Leung et al. 2008).

Environmental conditions, especially site conditions, have a profound impact on the economic success of aquaculture. They influence the monetary value of farming a particular species, the returns from genetic improvement, and the economic importance of selecting certain genetic traits. The predominant production conditions in Asian countries include marine, freshwater, and brackish water systems, with inland freshwater cultures having the highest value, followed by brackish water and marine aquaculture (Ahmed et al. 2007).

In freshwater aquaculture, different production techniques are used to select species, which are influenced by farm parameters such as size, water depth, seeding and feeding rate management, and the overall pond system. India and Bangladesh operate extensive systems that rely on natural nutrients and lower seeding rates, while countries such as Malaysia, mainland China, and Taiwan use higher seeding rates and supplemental feeding to increase production and yields (Ahmed et al. 2007). With Asia dominating over 94% of global freshwater aquaculture production, strategic species selection is becoming increasingly essential to ensure sustainable aquaculture practices and address the increasing global demand for food fish.

14.2.7 Pillar 7: Optimal Stocking Density Management for Enhanced Aquaculture Productivity

Efficient stocking density management is a cornerstone of aquaculture, influencing physiological response, growth performance, and the overall success of fish farming. The dynamic nature of stocking density, affected by environmental conditions, market fluctuations, and the fish species, requires a differentiated approach to achieve optimal results (Seginer 2009; Refaey et al. 2018).

A balanced stocking density is central to achieving productivity and profitability in aquaculture. The interplay of factors such as water quality, labor requirements, and equipment needs directly impacts industry yields and costs. The challenge is to find a delicate balance that maximizes the profit margin for the harvestable crop while avoiding undue stress on the fish, which ultimately affects the final weight, specific growth rate, and overall sustainability of the farm (Villanueva et al. 2013; Haque et al. 2015; Hossain et al. 2022).

Environmental factors, particularly water temperature, are critical determinants affecting farmed organisms' metabolism, growth, survival, and feeding rates. The intricate relationship between stocking density and water temperature reveals its profound effects on shrimp mortality rates and fish growth and immunity. Thyroid hormone levels, critical to fish growth and metabolic activity, are sensitive to changes in stocking density. Understanding these dynamics is vital to making informed decisions that maintain optimal thyroid hormone levels, gut mucosal cell counts, and metabolic enzyme activity to ensure the overall health of farmed fish (Malham et al. 2009; Villanueva et al. 2013; Refaey et al. 2018; Fatima et al. 2021).

Furthermore, the intricate balance extends to feeding practices, with higher stocking densities requiring higher feeding rates. This link requires careful attention to avoid water quality deterioration and feed waste, which can lead to parasitic diseases and microbial growth. Innovative solutions, such as pond flume systems, show the potential for improved growth and survival rates and emphasize the need for a symbiotic relationship between stocking density and feeding strategies to achieve improved access to food and overall growth (Abdel-Aziz et al. 2021; Fatima et al. 2021; Hossain et al. 2022).

14.2.8 Pillar 8: Advancing Nutritional Efficiency and Feed Sustainability

Fish is an important food source for humans and animals, providing high-quality proteins, essential fatty acids, vitamins, and minerals. The global increase in aquaculture production, contributing an additional 6.7 million tonnes to aquatic animal production from 2018 to 2020, has increased the importance of feed management (Agriculture Organization of the United Nations. Fisheries Department 2000). Despite the undeniable increase in aquaculture productivity, this growth has significantly driven up feed costs, with the total cost of feeding in intensive aquaculture increasing by over 50% of total production costs (Li et al. 2022).

This escalation in feed consumption while increasing productivity requires an accurate understanding of feeding efficiency and the development of cost-effective feeding strategies throughout the production process (Li et al. 2022). Recent studies investigating changes in amino acid patterns to achieve optimal protein-to-energy ratio have shown improved protein consumption efficiency and increased production yield (Khan et al. 2020). Considering the effects of nutrient sources on the physical characteristics and muscle quality of farmed fish, adopting appropriate feeding and nutrient delivery techniques in aquaculture systems is imperative (Ma et al. 2023).

The paradigm of "precision feeding" has emerged and focuses on species-specific, developmental stage-orientated nutritional management tailored to the farming environment. This approach aims to provide fish with the necessary nutrients, minimize environmental impact, reduce food waste, and increase yields (Yuru et al. 2020). Using sensors, cameras, and water quality monitors in aquaculture facilities to collect data on fish behavior enables the development of feeding techniques based on image analysis. Understanding fish behavior, nutritional needs, and the factors that influence these needs becomes pivotal in implementing precision management measures, which are essential for promoting a healthy and sustainable fisheries sector.

Aquaculture companies are exploring alternative plant and animal-based ingredients when faced with challenges such as high-quality protein shortages, rising fish oil and fishmeal costs, and feed supply chain issues. Marine microalgae considered an effective substitute provide valuable nutritional components, including proteins, carbohydrates, amino acids, polyunsaturated fatty acids (PUFAs), carotenoids, and vitamins. The use of mixed microalgae in the diet is proving to be more beneficial for the growth of aquatic animals and emphasizes the importance of nutritional value for optimal growth performance. Beyond the healthy aspect, solving problems with feed traders proposing direct links between feed producers and farmers is crucial to ensure feed availability with high nutritional value and realize the full potential of aquaculture production.

14.2.9 Pillar 9: Precision Nutrition and Feeding Strategies in Aquaculture

In the complex aquaculture landscape, understanding the different nutritional needs of other fish and aquatic organisms is of central importance. The specific needs, which include macronutrients (proteins, fats, carbohydrates) and micronutrients (vitamins and minerals), vary from species to species and are critical for optimal growth and good health. Carnivorous fish, for example, require higher amounts of proteins and fats than their omnivorous counterparts (Wilson and Halver 1986).

Supplements are crucial in aquaculture feed as they promote growth and overall health. Enzymes, vitamins, minerals, probiotics, and Chinese medicinal herbs are strategically incorporated into fish feed to improve protein utilization and performance (Yuru et al. 2020). Enzymes effectively increase feed conversion and utilization, while minerals such as copper, iron, and zinc increase amylase activity (Li et al. 2007; Yuru et al. 2020). Specific additives such as salt to promote growth in carp and shrimp and thiamine to improve growth and enzyme activity in young carp emphasize the nuanced approach to meeting nutrient needs (Tocher et al. 2008; Yuru et al. 2020).

Given carnivorous species' higher amino acid requirements and the cost constraints associated with fishmeal in commercial feeds, continued research to explore alternatives and substitutes for a sustainable diet is essential (Glencross et al. 2023). This imperative points to the future of aquaculture nutrition and emphasizes the need for innovative approaches to balance the nutritional needs of different species with economic considerations.

In addition, feeding strategies are evolving through the integration of sensor technologies. Cameras and water quality sensors are widely used, but susceptibility to environmental influences requires further development. The emergence of machine vision, particularly near-infrared computer vision, enables the automatic determination of feeding requirements without affecting fish growth or causing damage (Li et al. 2017; Zhou et al. 2018). As aquaculture companies become involved in precision nutrition and sophisticated feeding strategies, the interplay between nutritional science, technological innovation, and sustainability is becoming increasingly important.

14.2.10 Pillar 10: Precision Feeding Strategies for Enhanced Aquaculture Productivity

Effective feeding techniques in aquaculture play a critical role in optimizing growth, survival, and feed conversion ratio (FCR), thereby minimizing feed wastage and ensuring consistent production (Kubitza and Lovshin 1999; Dwyer et al. 2002;). Realizing these goals requires a thorough understanding of the intricate interplay between genetic background, habits, metabolic traits, and the aquatic ecology relevant to the fish species (Costello et al. 2015).

Each phase of a fish's life cycle comprises four interrelated components: breeding units, feeding management, stock management, and water quality management. The interaction of these factors requires careful consideration of water temperature, dissolved oxygen concentration, waste content, stocking density, and fish health (Hecht 2013). Feed and feed management systems in aquaculture need to consider three main elements: ingredients, producers, and on-farm feeding practices.

Adapting feeding plans based on the ontogenetic characteristics of the digestive system of fish species is crucial. Understanding the morphological characteristics of the digestive systems found in stomachless fish, such as cyprinids, cichlids, and silurids, helps formulate suitable diets in the early stages of ontogenetic development (Dabrowski and Portella 2005).

The amount and effectiveness of food intake significantly affect growth, with malnourished fish growing more slowly and exhibiting potential problems such as dominance hierarchies and size differences. Optimizing feeding frequency is crucial, with feeding slightly below satiation being a common approach. Managing feed intake according to optimal feeding rates helps mitigate problems such as fluctuating growth and increased FCR (Hecht 2013; Patel and Yakupitiyage 2003).

The importance of changing feeding frequency is exceptionally high for slow-feeding organisms such as freshwater and marine shrimp. Studies show that fish fed twice daily have higher specific growth rates and lower feed conversion ratios, emphasizing the need for further research into the effects of feeding frequency on size variance (Dwyer et al. 2002). In summary, precision feeding strategies are a cornerstone for increasing productivity in aquaculture.

14.3 Challenges and Opportunities in Aquaculture: Navigating Toward Sustainability

Although aquaculture struggles with environmental impacts, disease management, and social issues, it is a hub for realizing sustainable global food security goals. The 122% increase in fish consumption between 1990 and 2018, outpacing other protein sources, makes aquaculture a key player in meeting the world's growing demand for seafood (Henchion et al. 2017; Boyd et al. 2022). For this reason, national governments and international organizations are actively supporting the sustainable development and scaling of aquaculture to improve supply and ensure food security on a global scale (Hua et al. 2019).

Ensuring environmental sustainability in aquaculture requires continuous cooperation between all stakeholders. Governments, farmers, ecologists, and consumers must come together to manage aquaculture practices with a risk assessment approach to increase the potential for sustainability. Addressing critical issues such as eutrophication, habitat destruction, and the responsible use of antibiotics and chemicals is essential to achieving the environmental aspects of sustainability (Carballeira Braña et al. 2021). The expansion of sustainable aquaculture is critical to improving food production in a dynamically changing climate. Organizations such as NOAA are committed to the sustainable growth of domestic aquaculture (Craig 2002; Salin and Arome Ataguba 2018; Boyd et al. 2020; Froehlich et al. 2022; Jiang et al. 2022).

In addition to its central role in food production, aquaculture drives economic growth, job creation, and sustainable resource management, particularly in industrialized countries (Subasinghe et al. 2009; Gephart et al. 2020; Cottrell et al. 2021). Partnerships in marine aquaculture play a key role in sharing costs, products, equipment, and technologies and fostering a broad international scientific community that continuously explores new methods and management approaches (Slater 2017). Given that wild-caught fisheries are no longer sustainable, aquaculture is essential to meet the increasing global demand for fish (Borromeo Jr et al. 2022).

The diverse landscape of aquaculture methods requires careful consideration depending on the target species and geographical factors. Robust government oversight is essential to ensure ethical practices, as a lack of regulation increases the risk of human rights violations and unfair labor practices in the aquaculture industry (Chamberlain and Rosenthal 1995; Liao 2000; Asche et al. 2008; Blanchard et al.

2017). Cases of social conflict over water supply between land users and aquaculture farmers emphasize the urgent need for a responsible policy framework. Adopting a "planet-first" approach to production is essential to ensure the continued supply of healthy food from aquaculture while maintaining sustainability (Boyd et al. 2020; Couture et al. 2021; Schøning et al. 2023).

14.4 Conclusions

In aquaculture, prosperous and sustainable practices depend on a carefully orchestrated interplay of ten essential pillars. The first pillar emphasizes the central role of strategic site selection in shaping the foundation of an aquaculture project. A thorough site survey, utilizing advanced methods such as GIS-MCE, enables optimal water supply and productivity. This strategic site selection guides the subsequent pillar of water quality management, where maintaining optimal conditions for aquatic organisms plays a crucial role. This includes monitoring various parameters, introducing advanced techniques such as aquaculture recirculation systems, and training fish farmers to increase productivity and meet the global demand for seafood. The intricate link between water quality and soil selection and innovative pond designs emphasizes the complexity of the second and fourth pillars. When aquaculture operations strategically choose pond design, they lay the foundation for thriving ecosystems and sustainable practices.

The third pillar integrates GIS technologies to optimize infrastructure and improve overall efficiency, income generation, and environmental sustainability. The next pillar introduces the importance of innovative pond designs, showcasing different aquaculture systems and highlighting the efficiency of the regulatory aquaculture system (RAS). Strategic stocking and species management, the fifth pillar, is essential for economic profitability and environmental sustainability. The cornerstone of this pillar is the synergy between well-informed species selection and carefully planned stocking measures.

The following pillars deal with the finer points of aquaculture: strategic species selection, optimizing stocking density management, increasing nutritional efficiency, precision nutrition, and feeding strategies. The eighth pillar looks at escalating feed costs in aquaculture and argues in favor of precision feeding techniques to balance nutritional requirements and economic considerations. The final pillar, precision nutrition and feeding strategies, looks at the intricate details of meeting the dietary needs of different species through strategic supplementation and evolving feeding approaches. Overall, the culmination of these pillars emphasizes the need for precision in every aspect of aquaculture, from site selection and pond design to stocking strategies and nutritional management, paving the way for greater productivity and sustainability in the aquaculture sector.

Acknowledgement This study was funded by the Higher Institution Centre of Excellence (HICoE) Grant (HIC2404), Malaysian Ministry of Higher Education's Fundamental Research

Grant Scheme (FRGS/1/2022/WAB05/UMS/02/1; FRG0568–1/2022), and the Universiti Malaysia Sabah's Skim Penyelidikan Lantikan Baharu (SLB2229) to Wei Sheng Chong.

References

Abdel-Aziz MFA, Hassan HU, Yones A-M, Abdel-Tawwab YA, Metwalli AAA-T (2021) Assessing the effect of different feeding frequencies combined with stocking density, initial weight, and dietary protein ratio on the growth performance of tilapia, catfish and carp. Sci Afr 12:e00806. https://doi.org/10.1016/j.sciaf.2021.e00806

Abisha R, Krishnani KK, Sukhdhane K, Verma AK, Brahmane M, Chadha NK (2022) Sustainable development of climate-resilient aquaculture and culture-based fisheries through adaptation of abiotic stresses: a review. J Water Climate Chang 13(7):2671–2689. https://doi.org/10.2166/wcc.2022.045

Agriculture Organization of the United Nations. Fisheries Department (2000) The state of world fisheries and aquaculture, 2000, vol 3. Food and Agriculture Organization

Aguilar-Manjarrez J, Soto D, Brummett R (2017) Aquaculture zoning, site selection and area management under the ecosystem approach to aquaculture. A handbook. Food and Agriculture Organization of the United Nations (FAO)

Ahmed M, Dey MM, Garcia YT (2007) The role of species and systems in the development and growth of aquaculture in Asia: needs and prospects. In: Species and system selection for sustainable aquaculture. Wiley, pp 85–102. https://doi.org/10.1002/9780470277867.ch7

Alvarez-Lajonchère L, Ibarra-Castro L (2013) Aquaculture species selection method applied to marine fish in the Caribbean. Aquaculture 408–409:20–29. https://doi.org/10.1016/j.aquaculture.2013.05.020

Asche F, Guttormsen AG, Tveteras R (2008) Aquaculture—opportunities and challenges special issue introduction. Mar Resour Econ 23(4):395–400. https://doi.org/10.1086/mre.23.4.42629670

Assefa WW, Abebe WB (2018) GIS modeling of potentially suitable sites for aquaculture development in the Lake Tana basin, Northwest Ethiopia. Agric Food Secur 7(1):72. https://doi.org/10.1186/s40066-018-0222-0

Badiola M, Mendiola D, Bostock J (2012) Recirculating aquaculture systems (RAS) analysis: main issues on management and future challenges. Aquac Eng 51:26–35. https://doi.org/10.1016/j.aquaeng.2012.07.004

Baluyut, E. A., United Nations Development Programme, & Food and Agriculture Organization of the United Nations. (1989). Aquaculture systems and practices : a selected review. United Nations Development Programme : Food and Agriculture Organization of the United Nations

Benetti DD, Benetti GI, Rivera JA, Sardenberg B, O'Hanlon B (2010) Site selection criteria for open ocean aquaculture. Mar Technol Soc J 44(3):22–35. https://doi.org/10.4031/MTSJ.44.3.11

Besson M, Aubin J, Komen H, Poelman M, Quillet E, Vandeputte M, van Arendonk JAM, de Boer IJM (2016) Environmental impacts of genetic improvement of growth rate and feed conversion ratio in fish farming under rearing density and nitrogen output limitations. J Clean Prod 116:100–109. https://doi.org/10.1016/j.jclepro.2015.12.084

Beveridge MCM, Brummett RE (2015) Aquaculture and the environment. In: Freshwater fisheries ecology. Wiley, pp 794–803. https://doi.org/10.1002/9781118394380.ch55

Blanchard JL, Watson RA, Fulton EA, Cottrell RS, Nash KL, Bryndum-Buchholz A, Büchner M, Carozza DA, Cheung WW, Elliott J (2017) Linked sustainability challenges and trade-offs among fisheries, aquaculture and agriculture. Nat Ecol Evol 1(9):1240–1249. https://doi.org/10.1038/s41559-017-0258-8

Borromeo RC Jr, Urbano JM, Rivera AT (2022) Eco-friendly fish farming: responsibly-sustainable aquaculture among ASEAN countries. Soc Polit 23(1):42–59

Bosma RH, Verdegem MCJ (2011) Sustainable aquaculture in ponds: principles, practices and limits. Livest Sci 139(1):58–68. https://doi.org/10.1016/j.livsci.2011.03.017

Boyd CE (1995) Pond bottom soil analyses. In: Bottom soils, sediment, and pond aquaculture. Springer, New York, pp 304–340. https://doi.org/10.1007/978-1-4615-1785-6_10

Boyd C, McNevin A (2015) Aquaculture, resource use, and the environment. Wiley

Boyd CE, D'Abramo LR, Glencross BD, Huyben DC, Juarez LM, Lockwood GS, McNevin AA, Tacon AG, Teletchea F, Tomasso JR Jr (2020) Achieving sustainable aquaculture: historical and current perspectives and future needs and challenges. J World Aquacult Soc 51(3):578–633. https://doi.org/10.1111/jwas.12714

Boyd CE, McNevin AA, Davis RP (2022) The contribution of fisheries and aquaculture to the global protein supply. Food Secur 14(3):805–827. https://doi.org/10.1007/s12571-021-01246-9

Brown TW, Tucker CS, Rutland BL (2016) Performance evaluation of four different methods for circulating water in commercial-scale, split-pond aquaculture systems. Aquac Eng 70:33–41. https://doi.org/10.1016/j.aquaeng.2015.12.002

Campbell MD, Hall SG (2019) Hydrodynamic effects on oyster aquaculture systems: a review. Rev Aquac 11(3):896–906. https://doi.org/10.1111/raq.12271

Carbajal-Hernández JJ, Sánchez-Fernández LP, Villa-Vargas LA, Carrasco-Ochoa JA, Martínez-Trinidad JF (2013) Water quality assessment in shrimp culture using an analytical hierarchical process. Ecol Indic 29:148–158. https://doi.org/10.1016/j.ecolind.2012.12.017

Carballeira Braña CB, Cerbule K, Senff P, Stolz IK (2021) Towards environmental sustainability in marine finfish aquaculture. Front Mar Sci 8. https://doi.org/10.3389/fmars.2021.666662

Chamberlain G, Rosenthal H (1995) Aquaculture in the next century: opportunities for growth, challenges of sustainability. World Aquac 26(1):21–25

Cheng H, Li L, Aarsæther KG, Ong MC (2020) Typical hydrodynamic models for aquaculture nets: a comparative study under pure current conditions. Aquac Eng 90:102070. https://doi.org/10.1016/j.aquaeng.2020.102070

Costello MJ, Claus S, Dekeyzer S, Vandepitte L, Tuama ÉÓ, Lear D, Tyler-Walters H (2015) Biological and ecological traits of marine species. PeerJ 3:e1201. https://doi.org/10.7717/peerj.1201

Cottrell RS, Ferraro DM, Blasco GD, Halpern BS, Froehlich HE (2021) The search for blue transitions in aquaculture-dominant countries. Fish Fish 22(5):1006–1023. https://doi.org/10.1111/faf.12566

Couture JL, Froehlich HE, Buck BH, Jeffery KR, Krause G, Morris JA Jr, Pérez M, Stentiford GD, Vehviläinen H, Halpern BS (2021) Scenario analysis can guide aquaculture planning to meet sustainable future production goals. ICES J Mar Sci 78(3):821–831. https://doi.org/10.1093/icesjms/fsab012

Craig RK (2002) The other side of sustainable aquaculture: mariculture and nonpoint source pollution. Wash UJL Pol'y 9:163

Dabrowski K, Portella MC (2005) Feeding plasticity and nutritional physiology in tropical fishes. In: Fish physiology, vol 21. Elsevier, pp 155–224. https://doi.org/10.1016/S1546-5098(05)21005-1

de los Santos CB, Olivé I, Moreira M, Silva A, Freitas C, Araújo Luna R, Quental-Ferreira H, Martins M, Costa MM, Silva J, Cunha ME, Soares F, Pousão-Ferreira P, Santos R (2020) Seagrass meadows improve inflowing water quality in aquaculture ponds. Aquaculture 528:735502. https://doi.org/10.1016/j.aquaculture.2020.735502

Dwyer KS, Brown JA, Parrish C, Lall SP (2002) Feeding frequency affects food consumption, feeding pattern and growth of juvenile yellowtail flounder (Limanda ferruginea). Aquaculture 213(1–4):279–292. https://doi.org/10.1016/S0044-8486(02)00224-7

Ellis T, North B, Scott AP, Bromage N, Porter M, Gadd D (2002) The relationships between stocking density and welfare in farmed rainbow trout. J Fish Biol 61(3):493–531. https://doi.org/10.1111/j.1095-8649.2002.tb00893.x

Falconer L, Telfer TC, Ross LG (2016) Investigation of a novel approach for aquaculture site selection. J Environ Manag 181:791–804. https://doi.org/10.1016/j.jenvman.2016.07.018

Falconer L, Middelboe AL, Kaas H, Ross LG, Telfer TC (2020) Use of geographic information systems for aquaculture and recommendations for development of spatial tools. Rev Aquac 12(2):664–677. https://doi.org/10.1111/raq.12345

Fatima S, Komal W, Manzoor F, Latif AA, Liaqat R, Ameen S, Janjua RS (2021) Analysis of the growth performance, stress, profile of fatty acids and amino acids and cortisol in Tilapia (Oreochromis niloticus), cultured at high stocking density using in-pond raceway system. Saudi J Biol Sci 28(12):7422–7431. https://doi.org/10.1016/j.sjbs.2021.08.048

Folke C, Kautsky N (1992) Aquaculture with its environment: prospects for sustainability. Ocean Coast Manag 17(1):5–24. https://doi.org/10.1016/0964-5691(92)90059-T

Froehlich HE, Gentry RR, Lester SE, Rennick M, Lemoine HR, Tapia-Lewin S, Gardner L (2022) Piecing together the data of the US marine aquaculture puzzle. J Environ Manag 308:114623. https://doi.org/10.1016/j.jenvman.2022.114623

Gephart JA, Golden CD, Asche F, Belton B, Brugere C, Froehlich HE, Fry JP, Halpern BS, Hicks CC, Jones RC (2020) Scenarios for global aquaculture and its role in human nutrition. Rev Fish Sci Aquac 29(1):122–138. https://doi.org/10.1080/23308249.2020.1782342

Gimpel A, Stelzenmüller V, Töpsch S, Galparsoro I, Gubbins M, Miller D, Murillas A, Murray AG, Pınarbaşı K, Roca G, Watret R (2018) A GIS-based tool for an integrated assessment of spatial planning trade-offs with aquaculture. Sci Total Environ 627:1644–1655. https://doi.org/10.1016/j.scitotenv.2018.01.133

Glencross B, Booth M, Allan GL (2007) A feed is only as good as its ingredients—a review of ingredient evaluation strategies for aquaculture feeds. Aquac Nutr 13(1):17–34. https://doi.org/10.1111/j.1365-2095.2007.00450.x

Glencross B, Fracalossi DM, Hua K, Izquierdo M, Mai K, Øverland M, Robb D, Roubach R, Schrama J, Small B, Tacon A, Valente LMP, Viana M, Xie S, Yakupityage A (2023) Harvesting the benefits of nutritional research to address global challenges in the 21st century. J World Aquacult Soc 54(2):343–363. https://doi.org/10.1111/jwas.12948

Guong VT, Hoa NM (2012) Aquaculture and agricultural production in the Mekong Delta and its effects on nutrient pollution of soil and water. In: Renaud FG, Kuenzer C (eds) The Mekong Delta system: interdisciplinary analyses of a river delta. Springer, Dordrecht, pp 363–393. https://doi.org/10.1007/978-94-007-3962-8_14

Haque M, Rahman M, Islam M, Ali M, Islam M, Shahabuddin A (2015) Production and growth performances of carps in different stocking densities of polyculture. World J Zool 10(4):247–251

Hasibuan S, Syafriadiman S, Aryani N, Fadhli M, Hasibuan M (2023) The age and quality of pond bottom soil affect water quality and production of Pangasius hypophthalmus in the tropical environment. Aquac Fish 8(3):296–304. https://doi.org/10.1016/j.aaf.2021.11.006

Hecht T (2013) A review of on-farm feed management practices for North African catfish (Clarias gariepinus) in sub-Saharan Africa. In: On-farm feeding and feed management in aquaculture, pp 463–479

Henchion M, Hayes M, Mullen AM, Fenelon M, Tiwari B (2017) Future protein supply and demand: strategies and factors influencing a sustainable equilibrium. Food Secur 6(7):53. https://doi.org/10.3390/foods6070053

Henriksson PJG, Troell M, Banks LK, Belton B, Beveridge MCM, Klinger DH, Pelletier N, Phillips MJ, Tran N (2021) Interventions for improving the productivity and environmental performance of global aquaculture for future food security. One Earth 4(9):1220–1232. https://doi.org/10.1016/j.oneear.2021.08.009

Henryon M, Jokumsen A, Berg P, Lund I, Pedersen PB, Olesen NJ, Slierendrecht WJ (2002) Genetic variation for growth rate, feed conversion efficiency, and disease resistance exists within a farmed population of rainbow trout. Aquaculture 209(1):59–76. https://doi.org/10.1016/S0044-8486(01)00729-3

Hernández JJC, Fernández LPS, Pogrebnyak O (2011) Assessment and prediction of water quality in shrimp culture using signal processing techniques. Aquac Int 19(6):1083–1104. https://doi.org/10.1007/s10499-011-9426-z

Hossain MS, Chowdhury SR, Das NG, Sharifuzzaman SM, Sultana A (2009) Integration of GIS and multicriteria decision analysis for urban aquaculture development in Bangladesh. Landsc Urban Plan 90(3):119–133. https://doi.org/10.1016/j.landurbplan.2008.10.020

Hossain MA, Hossain MA, Haque MA, Mondol MMR, Harun-Ur-Rashid M, Das SK (2022) Determination of suitable stocking density for good aquaculture practice-based carp fattening in ponds under drought-prone areas of Bangladesh. Aquaculture 547:737485. https://doi.org/10.1016/j.aquaculture.2021.737485

Hua K, Cobcroft JM, Cole A, Condon K, Jerry DR, Mangott A, Praeger C, Vucko MJ, Zeng C, Zenger K (2019) The future of aquatic protein: implications for protein sources in aquaculture diets. One Earth 1(3):316–329. https://doi.org/10.1016/j.oneear.2019.10.018

Ibrahim GRF, Hamid AA, Darwesh UM, Rasul A (2021) A GIS-based Boolean logic-analytical hierarchy process for solar power plant (case study: Erbil Governorate—Iraq). Environ Dev Sustain 23(4):6066–6083. https://doi.org/10.1007/s10668-020-00862-3

Jescovitch LN, Boyd CE (2017) A case study: impacts of deviating from model research design to the commercial industry for split-pond aquaculture. Aquac Eng 79:35–41. https://doi.org/10.1016/j.aquaeng.2017.09.001

Jiang Q, Bhattarai N, Pahlow M, Xu Z (2022) Environmental sustainability and footprints of global aquaculture. Resour Conserv Recycl 180:106183. https://doi.org/10.1016/j.resconrec.2022.106183

Kamali S, Ward VCA, Ricardez-Sandoval L (2022) Dynamic modeling of recirculating aquaculture systems: effect of management strategies and water quality parameters on fish performance. Aquac Eng 99:102294. https://doi.org/10.1016/j.aquaeng.2022.102294

Kandathil Radhakrishnan D, AkbarAli I, Schmidt BV, John EM, Sivanpillai S, Thazhakot Vasunambesan S (2020) Improvement of nutritional quality of live feed for aquaculture: an overview. Aquac Res 51(1):1–17. https://doi.org/10.1111/are.14357

Khan KU, Gous RM, Sakomura NK, Azevedo JM, Nascimento TMT, Mansano CFM, de Souza Romaneli R, da Silva Oliveira T, Boaratti AZ, Fernandes JBK (2020) Modeling optimum levels of balanced digestible protein for adult pacu (Piaractus mesopotamicus) and the development of an economic optimization procedure for different marketing strategies of the final yield. Aquaculture 526:735368. https://doi.org/10.1016/j.aquaculture.2020.735368

Klebert P, Lader P, Gansel L, Oppedal F (2013) Hydrodynamic interactions on net panel and aquaculture fish cages: a review. Ocean Eng 58:260–274. https://doi.org/10.1016/j.oceaneng.2012.11.006

Kok B, Malcorps W, Tlusty MF, Eltholth MM, Auchterlonie NA, Little DC, Harmsen R, Newton RW, Davies SJ (2020) Fish as feed: using economic allocation to quantify the fish in: fish out ratio of major fed aquaculture species. Aquaculture 528:735474. https://doi.org/10.1016/j.aquaculture.2020.735474

Kubitza F, Lovshin LL (1999) Formulated diets, feeding strategies, and cannibalism control during intensive culture of juvenile carnivorous fishes. Rev Fish Sci 7(1):1–22. https://doi.org/10.1080/10641269991319171

Lafferty KD, Harvell CD, Conrad JM, Friedman CS, Kent ML, Kuris AM, Powell EN, Rondeau D, Saksida SM (2015) Infectious diseases affect marine fisheries and aquaculture economics. Annu Rev Mar Sci 7(1):471–496. https://doi.org/10.1146/annurev-marine-010814-015646

Leung TLF, Bates AE (2013) More rapid and severe disease outbreaks for aquaculture at the tropics: implications for food security. J Appl Ecol 50(1):215–222. https://doi.org/10.1111/1365-2644.12017

Leung P, Lee C-S, O'Bryen PJ (2008) Species and system selection for sustainable aquaculture. Wiley

Li J-S, Li J-L, Wu T-T (2007) The effects of copper, iron and zinc on digestive enzyme activity in the hybrid tilapia Oreochromis niloticus (L.) ×Oreochromis aureus (Steindachner). J Fish Biol 71(6):1788–1798. https://doi.org/10.1111/j.1095-8649.2007.01643.x

Li D, Xu L, Liu H (2017) Detection of uneaten fish food pellets in underwater images for aquaculture. Aquac Eng 78:85–94. https://doi.org/10.1016/j.aquaeng.2017.05.001

Li W, Cheng X, Xie J, Wang Z, Yu D (2019) Hydrodynamics of an in-pond raceway system with an aeration plug-flow device for application in aquaculture: an experimental study. R Soc Open Sci 6(7):182061. https://doi.org/10.1098/rsos.182061

Li H, Chatzifotis S, Lian G, Duan Y, Li D, Chen T (2022) Mechanistic model based optimization of feeding practices in aquaculture. Aquac Eng 97:102245. https://doi.org/10.1016/j.aquaeng.2022.102245

Li H, Cui Z, Cui H, Bai Y, Yin Z, Qu K (2023) Hazardous substances and their removal in recirculating aquaculture systems: a review. Aquaculture 569:739399. https://doi.org/10.1016/j.aquaculture.2023.739399

Liao IC (2000) Aquaculture development: challenges for the 21st century. Aquac Sci 48(2):273–278

Longdill PC, Healy TR, Black KP (2008) An integrated GIS approach for sustainable aquaculture management area site selection. Ocean Coast Manag 51(8):612–624. https://doi.org/10.1016/j.ocecoaman.2008.06.010

Ma F, Wang L, Huang J, Chen Y, Zhang L, Zhang M, Yu M, Jiang H, Qiao Z (2023) Comparative study on nutritional quality and serum biochemical indices of common carp (Cyprinus carpio) aged 11 to 13 months aged cultured in traditional ponds and land-based container aquaculture systems. Food Res Int 169:112869. https://doi.org/10.1016/j.foodres.2023.112869

Malham SK, Cotter E, O'Keeffe S, Lynch S, Culloty SC, King JW, Latchford JW, Beaumont AR (2009) Summer mortality of the Pacific oyster, Crassostrea gigas, in the Irish Sea: the influence of temperature and nutrients on health and survival. Aquaculture 287(1):128–138. https://doi.org/10.1016/j.aquaculture.2008.10.006

Mamat N, Rasam ARA, Adnan NA, Abdullah IC (2014) GIS-based multi-criteria decision making system for determining potential site of oyster aquaculture in Terengganu. In: 2014 IEEE 10th international colloquium on signal processing and its applications, pp 71–76. https://doi.org/10.1109/CSPA.2014.6805723

Martins CIM, Eding EH, Verdegem MCJ, Heinsbroek LTN, Schneider O, Blancheton JP, d'Orbcastel ER, Verreth JAJ (2010) New developments in recirculating aquaculture systems in Europe: a perspective on environmental sustainability. Aquac Eng 43(3):83–93. https://doi.org/10.1016/j.aquaeng.2010.09.002

Masaló Llorà I (2008) Hydrodynamic characterisation of aquaculture tanks and design criteria for improving self-cleaning properties [doctoral thesis, Universitat Politècnica de Catalunya]. TDX (Tesis Doctorals en Xarxa). https://doi.org/10.5821/dissertation-2117-94374

McLeod I, Pantus F, Preston N (2002) The use of a geographical information system for land-based aquaculture planning. Aquac Res 33(4):241–250. https://doi.org/10.1046/j.1355-557x.2001.00667.x

Nath SS, Bolte JP, Ross LG, Aguilar-Manjarrez J (2000) Applications of geographical information systems (GIS) for spatial decision support in aquaculture. Aquac Eng 23(1):233–278. https://doi.org/10.1016/S0144-8609(00)00051-0

Nayak AK, Pant D, Kumar P, Mahanta P, Pandey N (2014) GIS-based aquaculture site suitability study using multi-criteria evaluation approach. Indian J Fish 61(1):108–112

Naylor RL, Hardy RW, Buschmann AH, Bush SR, Cao L, Klinger DH, Little DC, Lubchenco J, Shumway SE, Troell M (2021) A 20-year retrospective review of global aquaculture. Nature 591(7851):551. https://doi.org/10.1038/s41586-021-03308-6

Nissar S, Bakhtiyar Y, Arafat MY, Andrabi S, Mir ZA, Khan NA, Langer S (2023) The evolution of integrated multi-trophic aquaculture in context of its design and components paving way to valorization via optimization and diversification. Aquaculture 565:739074. https://doi.org/10.1016/j.aquaculture.2022.739074

Oh RRY, Friess DA, Brown BM (2017) The role of surface elevation in the rehabilitation of abandoned aquaculture ponds to mangrove forests, Sulawesi, Indonesia. Ecol Eng 100:325–334. https://doi.org/10.1016/j.ecoleng.2016.12.021

Olmos Soto J, Paniagua-Michel JJ, Lopez L, Ochoa L (2015) Functional feeds in aquaculture. In: Kim S-K (ed) Springer handbook of marine biotechnology. Springer, pp 1303–1319. https://doi.org/10.1007/978-3-642-53971-8_59

Patel AB, Yakupitiyage A (2003) Mixed feeding schedules in semi-intensive pond culture of Nile tilapia, Oreochromis niloticus, L.: is it necessary to have two diets of differing protein contents? Aquac Res 34(14):1343–1352. https://doi.org/10.1046/j.1365-2109.2003.00957.x

Peterson EL (2000) Observations of pond hydrodynamics. Aquac Eng 21(4):247–269. https://doi.org/10.1016/S0144-8609(99)00034-5

Quéméner L, Suquet M, Mero D, Gaignon J-L (2002) Selection method of new candidates for finfish aquaculture: the case of the French Atlantic, the channel and the North Sea coasts. Aquat Living Resour 15(5):293–302. https://doi.org/10.1016/S0990-7440(02)01187-7

Rahman MM, Yakupitiyage A (2006) Use of fishpond sediment for sustainable aquaculture—agriculture farming. Int J Sustain Dev Plan 1(2):192–202. https://doi.org/10.2495/SDP-V1-N2-192-202

Refaey MM, Li D, Tian X, Zhang Z, Zhang X, Li L, Tang R (2018) High stocking density alters growth performance, blood biochemistry, intestinal histology, and muscle quality of channel catfish Ictalurus punctatus. Aquaculture 492:73–81. https://doi.org/10.1016/j.aquaculture.2018.04.003

Rombenso A, Araujo B, Li E (2022) Recent advances in fish nutrition: insights on the nutritional implications of modern formulations. Animals 12(13):1705. https://doi.org/10.3390/ani12131705

Roy LA, Davis DA, Saoud IP, Boyd CA, Pine HJ, Boyd CE (2010) Shrimp culture in inland low salinity waters. Rev Aquac 2(4):191–208. https://doi.org/10.1111/j.1753-5131.2010.01036.x

Salin KR, Arome Ataguba G (2018) Aquaculture and the environment: towards sustainability. In: Sustainable aquaculture, pp 1–62. https://doi.org/10.1007/978-3-319-73257-2_1

Schøning L, Hausner VH, Morel M (2023) Law and sustainable transitions: an analysis of aquaculture regulation. Environ Innov Soc Trans 48:100753. https://doi.org/10.1016/j.eist.2023.100753

Seginer I (2009) Are restricted periods of over-stocking of recirculating aquaculture systems advisable? A simulation study. Aquac Eng. 41(3):194–206. https://doi.org/10.1016/j.aquaeng.2009.07.007

Shafi J, Waheed KN, Mirza ZS, Zafarullah M (2021) Variation in bottom soil quality with increasing pond age in freshwater aquaculture. Turk J Fish Aquat Sci 22(2). https://doi.org/10.4194/TRJFAS18305

Silva C, Ferreira JG, Bricker SB, DelValls TA, Martín-Díaz ML, Yáñez E (2011) Site selection for shellfish aquaculture by means of GIS and farm-scale models, with an emphasis on data-poor environments. Aquaculture 318(3):444–457. https://doi.org/10.1016/j.aquaculture.2011.05.033

Slater MJ (2017) Societal and economic impacts of aquaculture. J World Aquacult Soc 48(4):539–541. https://doi.org/10.1111/jwas.12445

Subasinghe R, Soto D, Jia J (2009) Global aquaculture and its role in sustainable development. Rev Aquac 1(1):2–9. https://doi.org/10.1111/j.1753-5131.2008.01002.x

Taufik M, Tuan Ismail TI, Manan H, Ikhwanuddin M, Abdul Salam AI, Abdul Rahim AI, Ishak AN, Kamaruzzan AS, Draman AS, Kasan NA (2023) Synergistic effects of recirculating aquaculture system (RAS) with combination of clear water, probiotic and biofloc technology: a review. Aquac Fish. https://doi.org/10.1016/j.aaf.2023.07.006

Thomas M, Pasquet A, Aubin J, Nahon S, Lecocq T (2021) When more is more: taking advantage of species diversity to move towards sustainable aquaculture. Biol Rev 96(2):767–784. https://doi.org/10.1111/brv.12677

Tisdell C (2007) The environment and the selection of aquaculture species and systems: an economic analysis. In: Species and system selection for sustainable aquaculture. Wiley, pp 57–66. https://doi.org/10.1002/9780470277867.ch5

Tocher DR, Bendiksen EÅ, Campbell PJ, Bell JG (2008) The role of phospholipids in nutrition and metabolism of teleost fish. Aquaculture 280(1–4):21–34. https://doi.org/10.1016/j.aquaculture.2008.04.034

Tumwesigye Z, Tumwesigye W, Opio F, Kemigabo C, Mujuni B (2022) The effect of water quality on aquaculture productivity in Ibanda District. Uganda 2:23. https://doi.org/10.3390/aquacj2010003

Ullah K, Emmanuel A, Anjum MZ (2018) Effect of stocking density on growth performance of Indus mahseer (tor macrotepis). Int J Fish Aquat Stud 6(3):49–52

Villanueva RR, Araneda ME, Vela M, Seijo JC (2013) Selecting stocking density in different climatic seasons: a decision theory approach to intensive aquaculture. Aquaculture 384–387:25–34. https://doi.org/10.1016/j.aquaculture.2012.12.014

Waiho K, Mustaqim M, Fazhan H, Wan Norfaizza WI, Megat FH, Ikhwanuddin M (2015) Mating behaviour of the orange mud crab, Scylla olivacea: the effect of sex ratio and stocking density on mating success. Aquac Rep 2:50–57. https://doi.org/10.1016/j.aqrep.2015.08.004

Wang C, Jiang C, Gao T, Peng X, Ma S, Sun Q, Xia B, Xie X, Bai Z, Xu S, Zhuang X (2022) Improvement of fish production and water quality in a recirculating aquaculture pond enhanced with bacteria-microalgae association. Aquaculture 547:737420. https://doi.org/10.1016/j.aquaculture.2021.737420

Wilson RP, Halver JE (1986) Protein and amino acid requirements of fishes. Annu Rev Nutr 6(1):225–244. https://doi.org/10.1146/annurev.nu.06.070186.001301

Yokoyama H (2003) Environmental quality criteria for fish farms in Japan. Aquaculture 226(1):45–56. https://doi.org/10.1016/S0044-8486(03)00466-6

Yoo KH, Boyd CE (2012) Hydrology and water supply for pond aquaculture. Springer, New York

Yuru Z, Ronghua L, Chaobin Q, Guoxing N (2020) Precision nutritional regulation and aquaculture. Aquac Rep 18:100496

Yuvanatemiya V, Boyd CE (2006) Physical and chemical changes in aquaculture pond bottom soil resulting from sediment removal. Aquac Eng 35(2):199–205. https://doi.org/10.1016/j.aquaeng.2006.02.001

Zhang J, Jia G, Wang M, Cao S, Mkumbuzi SG (2022) Hydrodynamics of recirculating aquaculture tanks with different spatial utilization. Aquac Eng 96:102217. https://doi.org/10.1016/j.aquaeng.2021.102217

Zhang J-X, Ran Z-S, Xie H-X, Kong F, Zhang M-Q, Zhou Y, Li Y-R, Liao K, Yan X-J, Xu J-L (2023a) A systematic analysis and evaluation of nutritional composition of 23 strains of marine microalgae commonly used in aquaculture. Algal Res 72:103122. https://doi.org/10.1016/j.algal.2023.103122

Zhang Z, Lin W, He D, Wu Q, Cai C, Chen H, Shang Y, Zhang X (2023b) Aquaculture environment changes fish behavioral adaptability directly or indirectly through personality traits: a case study. Rev Fish Biol Fish 33(4):1423–1441. https://doi.org/10.1007/s11160-023-09779-2

Zhou C, Lin K, Xu D, Chen L, Guo Q, Sun C, Yang X (2018) Near infrared computer vision and neuro-fuzzy model-based feeding decision system for fish in aquaculture. Comput Electron Agric 146:114–124. https://doi.org/10.1016/j.compag.2018.02.006